Rhetorical Animals

Ecocritical Theory and Practice

Series Editor: Douglas A. Vakoch, METI

Ecocritical Theory and Practice highlights innovative scholarship at the interface of literary/cultural studies and the environment, seeking to foster an ongoing dialogue between academics and environmental activists.

Recent Titles

Rhetorical Animals: Boundaries of the Human in the Study of Persuasion edited by Kristian Bjørkdahl and Alex C. Parrish
Ecofeminism in Dialogue edited by Douglas A. Vakoch and Sam Mickey
The Image of the River in Latin/o American Literature edited by Jeanie Murphy and Elizabeth G. Rivero
Seeing Animals after Derrida edited by Sarah Bezan and James Tink
Eco Culture: Disaster, Narrative, Discourse edited by Robert Bell and Robert Ficociello
Southeast Asian Ecocriticism: Theories, Practices, Prospects edited by John Charles Ryan
Victorian Ecocriticism: The Politics of Place and Early Environmental Justice edited by Dewey W. Hall
Bhopal's Ecological Gothic: Disaster, Precarity and the Biopolitical Uncanny by Pramod K. Nayar
Fallen Animals: Art, Religion, Literature edited by Zohar Hadromi-Allouche
Ecological Thought in German Literature and Culture edited by Gabriele Dürbeck, Urte Stobbe, Hubert Zapf, and Evi Zemanek
The Horse in Literature and Film: Uncovering a Transcultural Paradigm by Francisco LaRubia-Prado
Water in Medieval Literature: An Ecocritical Reading by Albrecht Classen
Sustainability and the City: Urban Poetics and Politics edited by Lauren Curtright and Doris Bremm

Rhetorical Animals

Boundaries of the Human in the Study of Persuasion

Edited by
Kristian Bjørkdahl and Alex C. Parrish

LEXINGTON BOOKS
Lanham • Boulder • New York • London

Published by Lexington Books
An imprint of The Rowman & Littlefield Publishing Group, Inc.
4501 Forbes Boulevard, Suite 200, Lanham, Maryland 20706
www.rowman.com

Unit A, Whitacre Mews, 26-34 Stannary Street, London SE11 4AB

Photos in chapter 10 taken by author.

British Library Cataloguing in Publication Information Available

Library of Congress Cataloging-in-Publication Data Available

ISBN 978-1-4985-5845-7 (cloth : alk. paper)
ISBN 978-1-4985-5846-4 (electronic)

∞™ The paper used in this publication meets the minimum requirements of American National Standard for Information Sciences—Permanence of Paper for Printed Library Materials, ANSI/NISO Z39.48-1992.

Printed in the United States of America

Contents

Introduction

Contextualizing Human Rhetorical Practice as Animal Behavior: An Invitation to Animal Rhetorics

Kristian Bjørkdahl and Alex C. Parrish

Picture, for a moment, one of the paradigmatic sites of ancient rhetorical performance: the *Pnyx* of ancient Athens, the assembly area where public meetings took place beginning in the early fifth century BCE. We can easily imagine the *rhetor*, perched on the platform at the base of the hill. And we can probably call forth an image of his—mostly landed male—audience, an assembly of anywhere from 6000 to 13,000 of the speaker's citizen peers. To make the picture more concrete, let us say that the speaker, in this instance, is no lesser celebrity than Pericles, who is about to give his last speech to the city before the outbreak of the Peloponnesian war. "Among them came forward Pericles, son of Xanthippus," wrote Thucydides about the event, "the first man of his time at Athens, ablest alike in counsel and in action."[1] Pericles comes forward to advise his fellow Athenians to prepare for war with the Lacedaemonians. The Spartans have been making unfair demands on Athens, argues the *strategos*, and these demands "have only one meaning, and that is slavery." The proud city of Athens should revoke any further claims that its counterpart might put forth, he says, and concludes: "If they march against our country, we will sail against theirs." As it must be, his oration is a speech in the deliberative genre, and as it should, it is infused with what we—anachronistically—might call "we-intentions."[2]

If we can create a mental image of this scene without much difficulty, that is because scenes like these are part of the mythology of rhetoric. As such they reveal what we intuitively tend to include in the rhetorical tradition, but they also illustrate what is *not* there, what we do *not* necessarily see when we imagine, to our mind's eye, a paradigmatic "rhetorical situation." Among the many groups—women, slaves, foreigners—who are famously absent from this scene, another conspicuous absence is that of *animals*. There are no nonhumans in this scene, nor does it acknowledge much animality in the human actors. In fact, the origin myths of rhetoric tend to imagine rhetorical performances as the pinnacle of our species' capacity, humans at our *most* human. From Isocrates to Habermas, public discourse has been seen as a paragon of human civilization, and the ability that engenders this activity—language, speech—is the mark of the human. No animals speak, therefore no animals persuade, therefore rhetoric is exclusively human.

What does the "we" of rhetoric look like—and why does it take this shape? And is there any way to imagine that nonhuman animals could form part of that community? These are the central questions of this book. The issue at stake, then, is how animals and their various forms of communication might somehow be of interest, or even inspiration, to rhetorical scholars and practitioners. Because of the relative novelty of this idea, we cannot start from preconceptions—and what follows is first and foremost an *exploration*. We know this from the start, however: If the prospect of a nonhuman conception of rhetoric should materialize, that would undoubtedly require a departure from the traditional image we have both of ourselves and of rhetoric. But what reasons do we have for not making—or at least testing—such a departure?

The democratic *polis* of ancient Athens was notoriously exclusive: No more than 10 to 20 percent of the population took active part in public politics. The principle of *isēgoría* (*ἰσηγορία*), the equal right to address the assembly, was held in high regard—indeed it was seen as one of the basic principles of democratic practice. Today we are perhaps more inclined to think that the ancient Greek conception of *isēgoría* was actually something of a smokescreen for highly *un*equal access to the public sphere. In this book, we direct the same sort of doubts at the established notion that humans are the superior, if not to say exclusive, rhetorical animal. Is this idea simply a smokescreen for other, nonhuman, forms of communication? What would happen if we cleared the air, as it were, and began to consider all forms of communication as equally interesting? What happens when we start to imagine that rhetoric is *more than human*?

In recent years, a similar urge has sparked increasing interest, among humanists and social scientists, in human-animal relations—to the point where many now speak of an "animal turn" in the humanities and social sciences. Across established fields like history, psychology, anthropology, literature, sociology, philosophy, and law, an interdisciplinary field of human-animal studies has been forming.[3] Certain common themes run through this diverse field. Animal studies scholars from most all disciplines are concerned to understand—and often critique—the conditions under which we reproduce human-animal difference, and they shed light on the moral-political as well as the academic implications of the boundary work that separates humans from other animals.

Despite the long history of language use as a marker of this boundary, the academic quest to investigate the reproduction of human-animal difference has, somewhat surprisingly, not taken root within rhetorical studies—at least not until very recently. The authors in this book investigate the place of nonhuman animals in the purview of rhetorical theory; they look at what it would mean to communicate beyond the human community; they consider how rhetoric reveals our "brute" roots. As such, this book aims to explore the implications of the animal turn within rhetorical studies. Would such a turn imply, for instance, that rhetoric needs a nonanthropocentric reconfiguration?

The broad backdrop for these questions is a process of rapprochement between humans and animals in our academic, as well as in popular, culture. This process has been driven forth by scientific advances into the social, emotional, and cognitive lives of animals[4] and also, conversely, by proposals to study human beings in terms of evolutionary psychology, sociobiology, or a "generalized symmetry" whereby humans are seen to interact in networks with all kinds of other things and concepts.[5] It comes further from moral-political appeals to alternatively *level* or *muddle* the human-animal distinction.[6] While one notable development in recent decades has been the "decentering of the (human) subject," another has been an increasing willingness to consider animals *as* subjects.[7] These strands of inquiry are in many ways unconnected and even opposed to each other, but when seen from afar, the result of them both is nevertheless a convergence in our ideas about human and nonhuman communicative being in the world. This "convergence on communication" connects human rhetoric to animal signaling from several sides at once: Increasingly, animal signaling is seen to resemble paradigmatic—instrumental—human rhetoric; its goal-oriented nature, its richness, and the diversity of purpose it serves have all opened up the idea that animals, too, *persuade*. At the same time, human rhetoric is now increasingly seen to be more like paradigmatic—"instinctual," uncontrolled—animal signaling than we have previously assumed, as presumptions of human agency and rhetorical control are put into question.

Recently, this convergence has taken place also within rhetorical studies, in a strand of research that cannot deny its debts to George Kennedy's classic article "A Hoot in the Dark."[8] As if in a symbolic ritual, Kennedy dissolved rhetoric into its basic constituents, revealing in the process that the elements of rhetoric did not all amass on the same side of a human-animal split—in fact, that these elements did not harmonize with the idea of a split at all, but had to be conceived as the results of various species adapting solutions to similar evolutionary pressures. Contemporary scholars of animal rhetorics are certainly no uniform group of "Kennedyans," though, but represent a movement with many, and largely separate, origins. In fact, the broad cultural rapprochement between humans and animals, the "animal turn" in the *academie*, and the even more specific turn to animal rhetorics are all diverse movements, which flow from several sources, voice various concerns, and further a multitude of practical agendas. *Convergence* might after all be the wrong word, since it tends to suggest a smooth coming together of different perspectives. *Confrontation* might be more apt, since what we are seeing in many cases is something more like a battlefield where different sides are fighting it out over how the human-animal distinction is to be reconfigured in our time. Regardless of the metaphor, a space has now opened up, in popular as well as in academic culture, where the role of language—and therefore rhetoric—as marker of human-animal difference is deconstructed and debated. Even those who defend the idea of human exceptionalism are now beginning to see the need to build that distinction anew.

In what follows, we will suggest that the idea of rhetoric as exclusively human forms one of rhetoric's origin myths, but that this idea has recently been challenged, from many sides. Although these challenges have not been uniform, there have been two main approaches to addressing this myth within our discipline. The first suggests that animals have always been objects of interest in rhetorical theory, even if they have not been recognized as rhetors in their own right. The other approach shows that we need to theorize the rhetorical practices of all animal agents, not just those human ones who persuade by various means. Whereas the first treats animals mostly as literary symbols or themes that crop up regularly in rhetorical texts, the second aims to recognize the rhetorical practices of nonhuman animals, who, research shows, are sophisticated communicators across many media.

RHETORIC AS THE MARK OF HUMANITY

If it is true that one can learn a great deal about a society or community—about a culture—by considering their myths of origin, the tales of the rhetorical community reveal that it has in many cases seen itself, from the very start, as an exclusively human endeavor. While recent efforts in the history of rhetoric have shown counterexamples that complicate this discussion, there is yet much evidence for human exceptionalism as the norm.[9] One of our discipline's founding fathers, Isocrates, wrote for instance:

> The art of discourse … is the source of most of our blessings … Because there has been implanted in us the power to persuade each other and to make clear to each other whatever we desire, not only have we escaped the life of wild beasts, but we have come together and founded cities and made laws and invented arts; and, generally speaking, there is no institution devised by man which the power of speech has not helped us to establish.[10]

In Isocrates, there is a dual significance to the coupling of humans to rhetoric: Human beings are the only "rhetorical animal," but this is not just an incidental fact about us, but rather our distinguishing feature, precisely that which sets us apart from other animals. In this way, "the art of discourse" acquires the status of an essential characteristic; so important is rhetoric for the development of all things human—there is "no institution devised by man which the power of speech has not helped us to establish"—that it takes on the status as the *source* of our humanity. Isocrates drives the point home when he says that, thanks to rhetoric, we have "escaped the life of wild beasts." Our capacity for rhetoric is quite simply what makes us human.

We can consider this a version of the "origin myth" of rhetoric, not just because it offers a striking, rhetoric-aggrandizing conception of the role of rhetoric, but because it resonates with a long line of rhetorical theorists, including other notables in the rhetorical tradition. For instance, Aristotle argued that "it is absurd to hold that a man ought to be ashamed of being unable to defend himself with his limbs, but not of being unable to defend himself with speech and reason, when the use of rational speech is more distinctive of a human being than the use of his limbs."[11] The interesting thing about this statement is not just that Aristotle thinks rational speech is "distinctive of a human being," but that he believes shame to be the obviously correct emotional response for a human being unable to "defend himself with speech and reason." For Aristotle, to be deficient at rhetoric is equivalent to being deficient as a human being—it is in a sense to be subhuman.

Cicero's variation on this theme was no less remarkable, portraying rhetoric, again, as the basic distinguishing mark of humanity. He referred to the faculty of eloquence as "that particular respect wherein chiefly men are superior to animals," and asked, in an echo of Aristotle: "Who therefore would not rightly admire this faculty, and deem it his duty to exert himself to the utmost in this field?" Cicero, however, presents us with a narrative that, in its grandiosity, even transcends that of his predecessors:

> To come, however, at length to the highest achievements of eloquence, what other power could have been strong enough either to gather scattered humanity into one place, or to lead it out of its brutish existence in the wilderness up to our present condition of civilization as men and as citizens, or, after the establishment of social communities, to give shape to laws, tribunals, and civic rights? And not to pursue any further instances—wellnigh countless as they are—I will conclude the whole matter in a few words, for my assertion is this: that the wise control of the complete orator is that which chiefly upholds not only his own dignity, but the safety of countless individuals and of the entire State.[12]

One might think that such programmatic statements are utterly at odds with the current *Zeitgeist*, and that similarly self-serving justifications for the supposed elevation of humans would not fly today. However, in his recent book *Kingdom of Speech*, the famous journalist and author Tom Wolfe is engaged in an endeavor which is remarkably similar to that of the great forefathers of rhetoric. Complaining that scientists "one hundred and fifty years since the Theory of Evolution was announced [...] had learned ... *nothing*" about the origins and evolution of human language, Wolfe volunteers to enlighten his audience about what it is precisely "that has left endless generations of academics, certified geniuses, utterly baffled when it comes to speech."[13] In short, he suggests, it is this: "Speech is not one of man's several unique attributes—speech is the attribute of all attributes! Speech is 95 percent plus of what lifts man above animal!"

> Physically, man is a sad case. His teeth, including his incisors, which he calls eyeteeth, are baby-size and can barely penetrate the skin of a too-green apple. His claws can't do anything but scratch him where he itches. His stringy-ligament body makes him a weakling compared to all the animals his size. Animals his size? In hand-to-paw, hand-to-claw, or hand-to-incisor combat, any animal his size would have him for lunch. Yet man owns or controls them all, every animal that exists, thanks to his superpower: speech.[14]

As Wolfe sees it, to assume that we must make sense of language, speech, as an evolutionary phenomenon is to fall into "the trap of cosmogonism" (that is, "world-birthing"), a *compulsion*—supposedly inherited from Darwin—to "find the ever-elusive Theory of Everything, an idea of narrative that reveals everything in the world to be part of a single and suddenly clear pattern."[15] Wolfe has no time for Darwin's (or for Chomsky's) "obsession" with fitting speech into an evolutionary history. Instead, he wants us to rejoice in the idea that speech is our "superpower" and our "attribute of all attributes." Here, then, all the classical elements of the idea of humanity's rhetorical exceptionalism are still very much alive: The mysterious, almost *unnatural*, origins of the capacity for speech, the place of this capacity as a prime distinguishing mark that makes all other attributes insignificant, and the move from this descriptive "insight" to the normative idea that *Man* (sic) may rightfully rule over the other animals.

As the reception of Wolfe's book indicates, however, there are no longer any shortcuts available for those who want to base human exceptionalism in language or rhetoric: "His white suit unsullied by research," ecology professor Jerry Coyne wrote, reviewing Wolfe's book in the *Washington Post*, "Tom Wolfe tries to take down Charles Darwin and Noam Chomsky." The key word for Coyne was obviously *tries*, as he concludes: "Somewhere on his mission to tear down the famous, elevate the neglected outsider and hit the exclamation-point key as often as possible, Wolfe has forgotten how to think."[16] Meanwhile, *The Guardian*'s Steven Poole characterized Wolfe's book as "an irresponsibly partial account, riddled with falsehoods."[17]

This instance is perhaps somewhat peculiar; Wolfe is a public figure, a celebrity even, and has no formal credentials as a scientist. He also sports a persona that clearly infuriates many. Wolfe's book might nevertheless illustrate two interesting features: first, that there are still people today who are invested in the uniqueness of rhetoric as an exclusively human activity. In other words, the myth propounded by Isocrates, Aristotle, Cicero and many more—the origin myth of rhetoric—which says that "the use of rational speech is more distinctive of a human being than the use of his limbs" and hence how we "escaped the life of wild beasts" is still very much alive. Second, this myth today meets an unprecedented wave of protest and challenge. As Coyne and Poole both exemplify, there is now the science of human-animal coevolution to consider, as well as the principle of good will in trying to understand the scope of the capacities of nonhumans. "Anthropomorphism," which for a long time was a dreaded word in scholarly circles, a scientific taboo, has slowly been losing its hold on the imaginations of scholars and scientists.[18] Increasingly, we are coming to understand that exceptionalism, or anthropocentric thought, is the Scylla to anthropomorphism's Charybdis. If one assumes in either direction—that animals are too human, or that humans are disconnected from or set above other animals—one risks giving in to assumptions that do not fit the mounting body of evidence that animals are thinking, feeling creatures trying to navigate their own environments in interesting ways.

HOOTS IN THE DARK

Although intellectual strands and academic traditions can never be neatly sorted, it seems safe to say that one particularly salient moment in our discipline's complication of its own origin myth—the narrative most recently represented by Wolfe—came with the publication of George Kennedy's article "A Hoot in the Dark: The Evolution of General Rhetoric" in *Philosophy & Rhetoric* in 1992.[19] Suggesting, famously, that rhetoric was not to be understood as a sort of "ornament" on top a "semantic base" but *vice versa*, as a basic "form of energy [that is] not found only in language [but which] is present also in physical actions, facial expressions, gestures, and signs generally." Kennedy arrived at a view where rhetoric was "prior to speech," and where, hence, humans could no longer be considered the only *rhetorical animal*. In Kennedy's piece, we find both of the two strands that have constituted rhetoric's recent challenge to the human-animal distinction. On the one hand, he argued that the behaviors of basically all animals—and perhaps even plants—could be thought of as “rhetorical”; by way of rhetorical energy, all animals worked on each other by means of signals and codes. On the other hand, he placed the interest in the communicative behaviors of animals into the rhetorical tradition, as it were, by describing various aspects of animal rhetoric as “deliberative,” “judicial,” and “epideictic,” and by insisting that their proto-rhetoric involved, among other things, “delivery” (i.e. *actio*).

For various reasons, however, Kennedy’s article did not immediately set off a wave of interest in animal rhetoric.[20] When such a field has indeed begun to materialize in recent years, its inspirations have been diverse. While some efforts start from Kennedy’s combination of an evolutionary outlook with classical rhetoric, others take off from other inspirations, including continental philosophy—not least Jacques Derrida’s much talked-about intervention.[21] Other approaches bring to bear ecological/ambient rhetoric, with one pivotal contribution being Thomas Rickert’s *Ambient Rhetoric: The Attunements of Rhetorical Being* (2013),[22] which, by its distribution of rhetorical agency onto a wider—extrahuman—field of influences, is seen by many to open up the prospect of animal rhetorics. A similar thrust is provided by materialist understandings of rhetoric. These are not strictly speaking a novelty within the discipline, having been part of the conversation at least since McGee’s “A Materialist’s Conception of Rhetoric,”[23] but this agenda has recently been spurred on by the work of Bruno Latour, which is fast becoming a commonplace resource for rhetorical scholars, exemplified by titles like *Thinking with Bruno Latour in Rhetoric and Composition* (2015)[24] and *Rhetoric, Through Everyday Things* (2016)[25] as central volumes. Finally, the new-found interest in the body and the “rhetoric of affect,” represented by books like Debra Hawhee’s *Moving Bodies: Kenneth Burke at the Edges of Language* (2009)[26] and Brian Massumi’s *Parables for the Virtual: Movement, Affect, Sensation* (2002),[27] have also provided academic incentives to consider rhetoric as a more-than-human phenomenon.

As a result of these and other developments, what was until recently a niche interest—animal rhetorics—is becoming a driving force in new research into persuasion. In 2011, the *Journal of Advanced Composition* published a special issue on the topic, and the journal *Philosophy & Rhetoric* also published a forum dedicated to “Creaturely Rhetorics” edited by Debra Hawhee and Diane Davis. *Philosophy & Rhetoric* followed up in 2014 with an additional special issue on extrahuman rhetorics. Several articles, workshops, and conference panels began to expand interest in this topic once again, leading to the publication of monographs and edited collections that have since attempted to treat rhetoric in terms of animal behavior. Alex Parrish’s *Adaptive Rhetoric: Evolution, Culture, and the Art of Persuasion* (2013) was the first to place human persuasion in the context of animal signaling, thus turning on its head the common assumption that human rhetorical practice is the primary case to which we compare the practices of "lesser" rhetorical beasts. Emily Plec (who is a contributor to this book) released an edited collection in the same year, treating human-animal cross-species communication.[28]

Additional contributions come from researchers like Louise Westling, whose *Logos of the Living World* (2014) treated issues of animal communication in terms of the philosophy of Maurice Merleau-Ponty. Greg Goodale's book *The Rhetorical Invention of Man: A History of Distinguishing Humans from Other Animals* (2016) has recently been issued by Lexington Press. In it he touches on some important topics close to the heart of this collection, such as the history of the philosophy of distinction between human and nonhuman animals. Ultimately, however, Goodale seems to subscribe to the specialness of human communication rather than viewing humans as one persuasive species among many. Debra Hawhee's *Rhetoric in Tooth and Claw* (2016) expands her previous treatments of animals in classical rhetorical theory, complicating our claim above that many early rhetoricians excluded the nonhuman in their origin myths. While this list is meant to be representative, it is not exhaustive. The animal turn in rhetorical theory is quickening at an exciting rate, with many new collections and monographs slated for publication over the next two years.

TRACKING ANIMALS: THE PRESENT AND FUTURE OF CROSS-SPECIES RHETORICS

For millennia, humans have worked to separate themselves rhetorically and philosophically from their fellow animals. The irony is that there is a growing sentiment among animal studies scholars that we might not even be smart enough to know how smart other animals truly are, so all of our efforts to elevate our species to this point—mostly based on perceptions of cognitive superiority—are based purely on our own desires to feel special or entitled.[29]

As the twenty-first century commences, scholars have increasingly begun to recognize the ubiquity of animals in the history of rhetoric, and many of us have started to realize how we have misjudged other animals. Among those who choose to address this slight, we see the beginnings of a new interdisciplinary field of animal rhetorics forming in two major research areas. These branches reflect both preexisting divisions in the study of rhetorical theory and also some interdisciplinary assumptions that closely related efforts in animal studies bring with them. One of the most important differences to note is the subject-agent divide.

Animal studies scholars in the humanities often adopt one of two major approaches to analyzing the place of the animal in human endeavors. These approaches are commonly known as human-animal studies (HAS) and critical animal studies (CAS). Critical animal studies is often associated with scholar-activism and the push for social justice of "critical theory" generally, but the more fundamental difference, we would argue, is one of approach. Do we, as many human-animal scholars tend to do, view animals as a subject of study—something existing for human observance, classification, consumption, and control—or are animals thinking, feeling agents who require liberation from the oppressive hierarchical thinking of humankind? Are we concerned about animals only when attended to by human beings, or are animals agents whose rights and desires should be accounted for? CAS scholars tend to be (but aren't always) scholar-activists who take up the argument for agency and antihierarchy, while many HAS scholars treat animals like one would treat any literary theme one found interesting.

Both approaches are able to lend valuable insight into the general project of raising awareness of animals' places in the human-dominated Anthropocene Epoch, and both are reflected in this book. For scholars in this community, it is easy to perpetuate existing divisions between a rhetoric of science and a science of rhetoric. That is, rhetoricians have already for some time subjected scientific communication to rhetorical analysis, and (more rarely) we have looked scientifically at the ways rhetoric functions. Either approach can illuminate our topic, but they act as obverse and reverse of the same coin. What we are able to learn from each approach will help us continue the important work of recontextualizing a field that has forgotten its animal origins.

Studying animals as subjects characterizes the traditional Western scientific approach. This also makes sense at first blush for the humanist, because humanists are concerned with those things that are particular to, or interesting for many, humans. Recent research concerns itself with the ways animals have been used in the history of rhetoric, especially as regards the classical tradition.[30] These efforts suggest questions about "animality" in a field that is historically preoccupied by the study of logos, which is presumed to exclude the animal (as well as the humans we seek to dehumanize by treating them as animal-like in their perceived inability to overcome their emotions).

Questions of animal agency are cropping up across several disciplines, and even the humanistic studies are heavily influenced by current scholarship in the life and social sciences. It is common for rhetoricians to discuss the agency of all sorts of people and things, following thinkers like Latour or Miller. Likewise, George Kennedy attempted to displace rationality as the guiding force of rhetorical practice, allowing that animals and plants (but not objects) had an instinctual, precommunicative rhetorical "energy" that was the font of all persuasion.[31] Recent studies have expanded this work, narrowing in on specific animal populations in ostensibly human-dominated settings.[32] Moreover, researchers have begun investigating overlooked life forms, including microscopic organisms such as gut bacteria and brewer's yeast, in hopes of identifying behaviors we consider rhetorical.[33] Finally, as mentioned above, one of the editors of this book has argued that human persuasiveness should be treated as one type among many forms of animal communication, instead of a special instance of something exceptional that no other creature can produce.[34]

If we begin from the premises that the communication strategies of many animals, including humans, may have evolved to overcome similar social or environmental challenges, and therefore should hold some important traits in common, and that human forms of communication are more instinctual, less overwhelmingly rational, than rhetoric scholars have heretofore believed, we can arrive at several conclusions that these considerations suggest. Two are of especial importance to this book. First is that *animal rhetorics contextualize human rhetoric*. There are many benefits to isolating particular features of larger phenomena for close study, which is what rhetorical theory has done in the past. What has not happened very often in the history of that discipline, however, is taking what we have learned and recontextualizing it in reference to the whole world of persuasion that goes on in the animal kingdom (and beyond). For instance, the idea of plant persuasion as being interesting to rhetoricians was suggested briefly in Kennedy's *Comparative Rhetoric*, among other places, but has yet to receive an extended treatment among rhetoricians. There is much scientific evidence suggesting some plants can communicate and persuade, so this could be a fruitful area of future study. Following broad thinkers like Latour, Thomas Rickert affords us a sort of New Materialist resituation by suggesting that every aspect of our environment can add to a whole ecology of persuasion.[35]

The second conclusion these arguments suggest is that *human exceptionalism has thoroughly penetrated post/modern rhetorical theory*, to the point that we must reexamine the inherited doctrine of the nineteenth and twentieth centuries. This edited collection has many fine essays that interrogate the history of rhetorical theory on behalf of animal studies and its many layers of influence. These essays follow several commenters on continental philosophy, through whose ideas we see the influence of animal minds in Derrida. Posthumanism and transhumanism are represented by several authors working in the traditions of Donna Haraway and Karen Barad. Evolutionary or continuist approaches to rhetoric follow in the footsteps of George Kennedy and Alex Parrish, who see human communication on a spectrum of persuasive abilities witnessed in nature. There are attempts to decenter language in the study of human communication, which draw inspiration from Kenneth Burke and Debra Hawhee on gesture-speech and body language. While some of the claims about entire species take on a normative approach (as a matter of convenience for making generalized comparisons, not a lack of nuance or awareness of non-"normal" otherness), several of the authors in this collection speak directly to the norms of abled communication. Drawing inspiration from such varied figures as Temple Grandin and Helen Keller, our views of persuasion and perception are challenged in ways that hopefully draw us ever closer to understanding the great variety of persuasive modes in nature, and the challenges inherent in constructing and enacting various modes of communication across lines of ability, culture, social status, sex and gender, and even species. A more detailed description of each chapter will follow.

FURTHERING ANIMAL RHETORICS

While we have introduced several of the thematic divisions and rearticulations represented by the essays in this collection, we have consciously divided the essays in a manner more fitting the embodied study of human and nonhuman animal persuasion. The first group of essays treats some of the smallest objects (and, perhaps the greatest subjects) we are able to identify as rhetorical, most of which are found internally—either internal to our bodies or internal to our conception of rhetoricity.

David Gruber's "Multiple Rhetorical Animals: Motivation and Fairness in a Paradigm of Rhetoric as Emotive Consciousness" pays heed to Jordynn Jack's (2010) call for "a neuroscience of rhetoric" by incorporating recent findings in neuroscientific research on the topics of motivation and fairness. These terms, Gruber argues, are important to understand an expanded role of pathos in animal rhetorics, a *pistis* that has played a subaltern role in many mainstream theories of contemporary rhetoric.

Hayley Zertuche continues the new materialist investigations into the internal processes involved in the practice of persuasion in her chapter, "A Humanimal Rhetorics of Biological Materiality." Her concern, however, is with the claims of "preoriginary rhetoricity." Tracing the developments following Kennedy's argument for a general rhetorical "energy," she follows the thread through Jo Liska's manipulation and the extrahuman rhetorics of Diane Davis. By thinking of animal minds in conversation with other minds, following Alex Parrish, Zertuche applies these very similar arguments from rhetoricians to a diffractory reading of animal body/minds, with the help of Haraway and Barad.

While much of this section deals with rhetorics originating inside the body, Kelin Loe's "Let's Listen With Our Feet: Animals, Neurodivergence, Vulnerability, and Haptic Rhetoricity" concerns itself primarily with the actual surface of the body and other loci of haptic sensation. Inspired by neurodiverse writers like Temple Grandin and Helen Keller, Loe suggests the additional constraints "disabled" rhetors operate under actually enable dynamic new sensory languages that could be useful to those studying both human and nonhuman animals whose sense perception can differ vastly from our own.

Jennifer Saltmarsh's chapter, "Human Boundary Seepage and Bacterial Rhetorics," further decenters normative views of sensation, cognition, and persuasion by traveling deep into the body to introduce the idea of quorum sensing, a collective form of bacterial communication sent via chemical signals, which allows the bacteria to make context-bound, decentralized decisions as if they were a single organism. Groups of bacteria in the human gut, especially, can be persuasive in some very surprising ways, influencing a person's mood, digestion, cognition, and sociality. The way this affects the common human claims of agency, individuality, and self-determination further complicates our views of cognition and persuasive behavior.

The second part of this collection ventures beyond the human body to explore the abstract theoretical and disciplinary boundaries that exist outside the self but weigh heavily on how the self and the rhetorical other are defined. In Ellen Gorsevski's "The Biotic Turn in Rhetoric: Ethical Internatural Communication as Suasory Peacebuilding," the author explores the strong, already existing social justice and peacebuilding sentiments among rhetorical scholars, and what those would look like in the context of human-animal studies. Specifically, if we consider the notion of humanimal hybridity, rhetoric's long history of pressing for the remediation of human suffering could lend itself to the ongoing fight for animal respect and rights.

Dustin Greenwalt follows Gorsevski's argument in his "Toward an Ethological Rhetoric" by further exploring the relationship between animal communication and animal bodies. Because sensory perception is vital in shaping the forms of communication each animal body produces, there is a need for an ethological view of rhetoric, which takes into account both form and function. Greenwalt concludes his chapter by suggesting some typologies of affects produced by nonhuman animals, allowing us to more fully understand their persuasive abilities.

In Iklim Goksel's "Beyond a Patriarchal Rhetorical Economy: Nonhuman Animals as Agents in Turkic Legends and Political Culture," the idea of a logos-dominated rhetoric is interrogated, which excludes both nonhuman and human female animals. On the other hand, Turkic legends and myths suggest that animals are in fact capable of reason, and that they may indeed speak to us in a very human way. This lends the animals agency in a way that Turkic women have not always enjoyed, providing a new model to broaden our view of rhetoric and how it can work to expand beyond a more patriarchal western tradition.

Alex Parrish, in his chapter "Humans, Dolphins, and Other People" suggests that many of the questions explored thus far in this collection are made possible because of a strong human drive, across various cultures, to set humans apart from at least some animals, and often all. But what happens when intelligent animals are granted the status of nonhuman persons in several countries around the world? For one, we need to rethink the patchwork of ways we attempt to exclude people (of various species) from the rights and responsibilities of personhood. The implications of this drive to "define away" personhood can affect us all, so it is important to make a strong case for dolphin personhood because of dolphins' intelligence, their individual identities, and their feats of superhuman perception and communication.

The final group of essays brings us full circle in an attempt to demonstrate the value of studying animal communication across the boundaries of culture and species. That we can learn so much about what is inside each of us by forming a better understanding of the relatedness of others allows us to see rhetorical theory as valuable on a broader scale. We contextualize ourselves by studying others, expanding our knowledge of both.

Emily Plec and Susan Hafen, in their article, "Learning to Howl: An Exercise in Internatural Abduction," note that the field of communication studies is gradually moving away from its anthropocentric traditions in order to include more voices in their research. They argue that human-animal communication is particularly suited to enrich both rhetorical theory and animal studies because of the already persuading relationships various human and nonhuman animals share when working together in various settings. In particular, Plec and Hafen explore the give and take between a sled dog team and their musher, as both sides must rely on the special skills of the other, often in order to survive in some very dangerous situations.

Jake Dionne takes an even more personal approach to human-animal communication in his "Touring the Sixth Persona: Dodos and the Rhetorical Effects of Missed Communication." While the rhetoric in this piece is often, sadly, one-sided, his visits to several museums of natural history in order to view the few extant remains of the extinct dodo bird (*Raphus cucullatus*) serve as an important exploration of the rhetoric being mobilized in order to direct public memory.

Suggesting that Karen Barad's performative understanding of discursive practices makes animals more interesting as subjects to rhetoricians, Marilyn Cooper explains "How Dogs (and Other Nonhuman Animals) Become Interesting." Leaning on her own experience of training her dog to herd sheep, Cooper also discusses Donna Haraway's essay concerning her dog, Cayenne, and the way their shared experiences open up an intra-action between partners rather than a rigid master-dog relationship.

Andrea Guiterrez focuses on avian voices in the Indian literary tradition in "How to Understand a Parrot's Words and What You Can Learn from Him: Early Indian Writers on Animal Speech." For at least 1600 years, Indian philosophers have debated the articulateness of animals, most often through the lens of bird voices. Many philosophers in that tradition concluded that while humans and parrots are not the same, they both possess a voice and therefore agency. How this connects with recent work in new materialism, then, offers us a new model for reading relationships between human and nonhuman animals in literary and philosophical traditions.

In his chapter, "The Rhetoric of Nonanthropocentric Rhetoric," Kristian Bjørkdahl directs our attention toward the rhetoric of recent calls for a nonanthropocentric turn in rhetoric. How—that is, with what rhetorical means and effects—have appeals for such a turn been made, he asks, and he begins to answer by way of a reception history of George Kennedy's article "A Hoot in the Dark." Bjørkdahl argues that the reception of this famous contribution was structured by two opposed incentives of the academic field, *conservation* and *innovation*. This has relegated Kennedy's program to a position between two stools, from which it should be rescued.

NOTES

1. Thucydides, I.139–146.
2. Wilfrid Sellars, *Science and Metaphysics: Variations on Kantian Themes* (New York: The Humanities Press, 1968).
3. One indication that animal studies is coming of age as an academic field is the appearance of textbooks and readers, see Margo DeMello, *Animals and Society: An Introduction to Human-Animal Studies* (New York: Columbia University Press, 2012); Paul Waldau, *Animal Studies: An Introduction* (New York: Oxford University Press, 2013); and Clifton P. Flynn, *Social Creatures: A Human and Animal Studies Reader* (New York: Lantern Books, 2008).
4. See, for instance, Marc Bekoff, *Minding Animals: Awareness, Emotions, And Heart* (New York: Oxford University Press, 2002); and Marc Bekoff, Colin Allen, and Gordon M. Burghardt, *The Cognitive Animal: Empirical and Theoretical Perspectives on Animal Cognition* (Cambridge, The MIT Press, 2002).
5. Bruno Latour, *Science in Action: How to Follow Scientists and Engineers through Society* (Cambridge: Harvard University Press, 1987), and Stuart Shapiro, "Caught in a Web: The Implications of Ecology for Radical Symmetry in STS," *Social Epistemology* 11, no. 1 (1997): 97–110.
6. See, respectively, Peter Singer, *Animal Liberation* (New York: Random House, 1995), and Donna Jeanne Haraway, *The Companion Species Manifesto: Dogs, People, and Significant Otherness*, vol. 1 (Chicago: The University of Chicago Press/Prickly Paradigm, 2003), and ———, *When Species Meet* (Minnesota: University of Minnesota Press, 2008).
7. Jodey Castricano (ed.), *Animal Subjects: An Ethical Reader in a Posthuman World* (Waterloo: Wilfrid Laurier University Press, 2008), and Jodey Castricano and

Lauren Corman (eds), *Animal Subjects 2.0* (Waterloo: Wilfrid Laurier University Press, 2016).

8. George A. Kennedy, "A Hoot in the Dark: The Evolution of General Rhetoric," *Philosophy & Rhetoric* 25, no. 1 (1992): 1–21.

9. For one person making a case for animal subjects being constitutive of early rhetorical theories, see Debra Hawhee, *Rhetoric in Tooth and Claw: Animals, Language, Sensation* (Chicago: University of Chicago Press, 2016).

10. Isocrates, quoted in James Jasinski, *Sourcebook on Rhetoric* (Thousand Oaks: Sage, 2001), xiv.

11. Aristotle, *The Art of Rhetoric*, The Loeb Classical Library ([S.l.]: Heinemann, 1925). I.1.1355b.

12. Cicero, *De Oratore*, I. viii. 34— ix. 37.

13. Aristotle, *The Art of Rhetoric*, The Loeb Classical Library ([S.l.]: Heinemann, 1925). I.1.1355b.

14. Ibid.

15. Ibid., 20.

16. https://www.washingtonpost.com/opinions/his-white-suit-unsullied-by-research-tom-wolfe-tries-to-take-down-charles-darwin/2016/08/31/8ee6d4ee-4936-11e6-90a8-fb84201e0645_story.html?utm_term=.d34ef24bc60d

17. https://www.theguardian.com/books/2016/sep/08/the-kingdom-of-speech-by-tom-wolfe-review

18. Lorraine Daston and Gregg Mitman (eds), *Thinking with Animals: New Perspectives on Anthropomorphism* (New York: Columbia University Press, 2005).

19. Kennedy, "A Hoot in the Dark."

20. Debra Hawhee, "Toward a Bestial Rhetoric," *Philosophy & Rhetoric* 44, no. 1 (2011): 81–87.

21. See Jacques Derrida and Marie-Louise Mallet, *The Animal That Therefore I Am* (New York: Fordham University Press, 2008).

22. Thomas J. Rickert, *Ambient Rhetoric: The Attunements of Rhetorical Being* (Pittsburgh: University of Pittsburgh Press, 2013).

23. Michael Calvin McGee, "A Materialist's Conception of Rhetoric," in Barbara A. Biesecker and John Louis Lucaites (eds), *Rhetoric, Materiality, and Politics* (New York: Peter Lang Publishing, 2009), 17–42.

24. Paul Lynch and Nathanial Rivers (eds), *Thinking with Bruno Latour in Rhetoric and Composition* (Carbondale: Southern Illinois University Press, 2015).

25. Scot Barnett and Casey Boyle (eds), *Rhetoric, Through Everyday Things* (Tuscaloosa: University of Alabama Press, 2016).

26. Debra Hawhee, *Moving Bodies: Kenneth Burke at the Edges of Language* (Columbia: University of South Carolina Press, 2009).

27. Brian Massumi, *Parables for the Virtual: Movement, Affect, Sensation* (Durham: Duke University Press, 2002).

28. Emily Plec (ed.), *Perspectives on Human-Animal Communication: Internatural Communication* (New York: Routledge, 2013).

29. Frans de Waal, *Are We Smart Enough to Know How Smart Animals Are?* (New York: W.W. Norton & Company, 2016).

30. Hawhee, *Rhetoric in Tooth and Claw*.

31. George A. Kennedy, *Comparative Rhetoric: An Historical and Cross-Cultural Introduction* (New York: Oxford University Press, 1998).

32. Natasha Seegert, "Play of Sniffication: Coyotes Sing in the Margins," *Philosophy and Rhetoric* 47, no. 2 (2014): 158-178.

33. Ehren Helmut Pflugfelder, "Rhetoric's New Materialism: From Micro-Rhetoric to Microbrew," *Rhetoric Society Quarterly* 45, no. 5 (2015): 441-461.

34. Alex C. Parrish, *Adaptive Rhetoric: Evolution, Culture, and the Art of Persuasion*, Routledge Studies in Rhetoric and Communication (New York: Routledge, 2013).

35. Thomas J. Rickert, *Ambient Rhetoric: The Attunements of Rhetorical Being* (Pittsburgh: University of Pittsburgh Press, 2013).

BIBLIOGRAPHY

Aristotle. *The Art of Rhetoric*. The Loeb Classical Library. Cambridge: Harvard University Press, 1926.

Barnett, Scot, and Casey Boyle, eds. *Rhetoric, through Everyday Things*. Tuscaloosa: University of Alabama Press, 2016.

Bekoff, Marc. *Minding Animals: Awareness, Emotions, and Heart*. New York: Oxford University Press, 2002.

Bekoff, Marc, Colin Allen, and Gordon M. Burghardt. *The Cognitive Animal: Empirical and Theoretical Perspectives on Animal Cognition*. Oxford, MIT Press, 2002.

Castricano, Jodey (ed.). *Animal Subjects: An Ethical Reader in a Posthuman World*. Waterloo: Wilfrid Laurier University Press, 2008.

Castricano, Jodey and Lauren Corman (eds). *Animal Subjects 2.0*. Wilfrid Laurier University Press, 2016.

Daston, Lorraine and Gregg Mitman (eds). *Thinking with Animals: New Perspectives on Anthropomorphism*. New York: Columbia University Press, 2005.

de Waal, Frans. *Are We Smart Enough to Know How Smart Animals Are?* New York: W.W. Norton & Company, 2016.

DeMello, Margo. *Animals and Society: An Introduction to Human-Animal Studies*. New York: Columbia University Press, 2012.

Derrida, Jacques and Marie-Louise Mallet. *The Animal That Therefore I Am*. New York: Fordham University Press, 2008.

Flynn, Clifton P. *Social Creatures: A Human and Animal Studies Reader*. New York: Lantern Books, 2008.

Haraway, Donna Jeanne. *The Companion Species Manifesto: Dogs, People, and Significant Otherness*. Chicago: The University of Chicago Press/Prickly Paradigm, 2003.

———. *When Species Meet*. Minneapolis: University of Minnesota Press, 2008.

Hawhee, Debra. *Moving Bodies: Kenneth Burke at the Edges of Language*. Columbia: University of South Carolina Press, 2009.

———. *Rhetoric in Tooth and Claw: Animals, Language, Sensation*. Chicago: University of Chicago Press, 2016.

———. "Toward a Bestial Rhetoric." *Philosophy & Rhetoric* 44, no. 1 (2011): 81–87.

Jasinski, James. *Sourcebook on Rhetoric*. Thousand Oaks: Sage, 2001.

Kennedy, George A. "A Hoot in the Dark: The Evolution of General Rhetoric." *Philosophy & Rhetoric* 25, no. 1 (1992): 1–21

———. *Comparative Rhetoric: An Historical and Cross-Cultural Introduction*. New York: Oxford University Press, 1998.

Latour, Bruno. *Science in Action: How to Follow Scientists and Engineers through Society*. Cambridge: Harvard University Press, 1987.

Massumi, Brian. *Parables for the Virtual: Movement, Affect, Sensation*. Durham: Duke University Press, 2002.

McGee, Michael Calvin. "A Materialist's Conception of Rhetoric," in Barbara A. Biesecker and John Louis Lucaites (eds), *Rhetoric, Materiality, and Politics* (New York: Peter Lang Publishing, 2009), 17–42.

Parrish, Alex C. *Adaptive Rhetoric: Evolution, Culture, and the Art of Persuasion*. New York: Routledge, 2013.

Pflugfelder, Ehren Helmut. "Rhetoric's New Materialism: From Micro-Rhetoric to Microbrew." *Rhetoric Society Quarterly* 45, no. 5 (2015): 441–61.

Plec, Emily (eds). *Perspectives on Human–Animal Communication: Internatural Communication*. New York: Routledge, 2013.

Rickert, Thomas J. *Ambient Rhetoric: The Attunements of Rhetorical Being*. Pittsburgh: University of Pittsburgh Press, 2013.

Seegert, Natasha "Play of Sniffication: Coyotes Sing in the Margins." *Philosophy and Rhetoric* 47, no. 2 (2014): 158–78.

Sellars, Wilfrid. *Science and Metaphysics: Variations on Kantian Themes*. New York: The Humanities Press , 1968.

Shapiro, Stuart . "Caught in a Web: The Implications of Ecology for Radical Symmetry in STS," *Social Epistemology* 11, no. 1 (1997): 97–110.

Singer, Peter. *Animal Liberation*. New York: Random House, 1995.

Paul Lynch and Nathanial Rivers (eds), *Thinking with Bruno Latour in Rhetoric and Composition*. Carbondale: Southern Illinois University Press, 2015.

Waldau, Paul. *Animal Studies: An Introduction*. Oxford: Oxford University Press, 2013.

Wolfe, Tom. *The Kingdom of Speech*. New York: Little, Brown, and Company, 2016.

Part I

EXPANDING BOUNDARIES—INTERNALLY

Chapter 1

Multiple Rhetorical Animals

Motivation and Fairness in a Paradigm of Rhetoric as Emotive Consciousness

David R. Gruber

Amid the materialist turn[1] and affective turn,[2] scholars are currently reconsidering the role of the emotions in rhetoric.[3] Celeste Condit argues that pathos has been overlooked in comparison to logos and ethos, and, as a result, scholars could now "expand tools for building pathos."[4] This chapter engages new neuroscience research to attempt to do exactly that, taking inspiration from Jordynn Jack's call for a "neuroscience of rhetoric," which intends to carefully consider the potential of new brain findings to contribute to rhetorical theory.[5]

Recent work in the behavioral and cognitive neurosciences[6] suggests that brain regions related to so-called "survival circuitry" are active during certain emotional experiences involving jealousy or feelings of "unfairness," and that the circuitry involved across the brains of many animals appears structurally similar to humans.[7] The implication is not necessarily that humans and animals feel exactly the same way but, rather, that certain emotional states appear to be tied to "motivation" and perceptions of "fairness." Carefully analyzing how those key terms are employed in the neuroscience literature as well as in rhetoric's own tradition proves useful to understanding certain kinds of emotional experiences as felt bodily states generated *in the midst of* rhetorical situations involving perceptions of inequality or "unfairness"; importantly, these states are produced without necessary default to rationality and have striking neurobiological *and* behavioral commonalities across multiple animals. The neuroscientific insight offers the opportunity for ecological realignment in definitions of rhetoric, foregrounding a biological orientation that pushes beyond a human-specific and rational framework for rhetorical action; the move is similar to Thomas Rickert's realignment of "The Logos" with Chora, which attempts to push Logos away from its "overly rational character."[8] Likewise, instead of stressing a rhetoric as identification that is symbolic and conscious—although not always deliberate[9]—rhetoric can be conceptualized, instead, as an emotive form of consciousness, or a neurological orientation sprung from a brain-body engagement with the world, a trait shared among many animals.

Theorizing rhetoric in this way stands in contrast to an "always-already" interior unity or automatic "identification" with others, as Diane Davis forwards, which is a form of identification that humans consistently "fail to uphold" upon cognitively surveying the landscape and dividing themselves from Others.[10] The concept offered here, rather, begins, as Burke does, with a form of consciousness—a form generating a foundational compulsion to look out, to evaluate, to gage distance, and to feel. Although the presumption of division informing the "looking out" of living beings having such forms of consciousness may on the surface appear to rebuff a politics of ecological expansion and affective unity with all others, in contrast to Davis, this is not the case; sharing modes of conscious experience arising at least in part from parallel biological structures unites sets of living beings in a range of reactive, emotive responses within lived experience such that many common symbolic evaluations and responses are also shared. The suggestion, as a result, may be even more radical than any "pre-originary self" denied and washed away by the dominance of symbolic division.[11]

In making an expansive claim about the emergence of "rhetoric" as operative from a shared form of emotive consciousness, I recognize the conditions on which the claim is built. The situated, disciplined, constructed condition of neuroscience does not escape oversight. In particular, by addressing critical rhetorical concerns with respect to the deployment of a neuroscience literature on emotions—focusing on the key terms "motivation" and "fairness"—I seek to avoid reductive readings that separate emotions from sociocultural contours. I aim to balance a "rhetoric of neuroscience" with a "neuroscience of rhetoric"[12] as I review neuroscientific understandings of "motivation" and "fairness," situating these terms for a rhetorical audience and contrasting them with rhetoric's own elaborations of similar terms.

Yet, I also treat neuroscience research as informative and as a means toward rhetorical invention, that is, what John Muckelbauer calls "an affirmative sense of change" not mired in the "dialectical negation" of fields or ideas but focused on blending and enacting responses, seeking what lies "in-between" disciplines and perspectives.[13] Doing so, I examine the concept of "fairness" as used strategically to theorize neuroscience,[14] tracking how it operates within a broader evolutionary framework upholding a "survival" mechanism/interpretation from which these neuroscientists can then argue for the emergence of particular forms of consciousness. Likewise, I examine Aristotle, Burke, and McClelland's engagement with concepts of motive and fairness, tracking how they wield those (similar) terms to revise rhetorical theorizations. Ultimately, I stress human-animal commonalities, seeking to invent in the space between an emotive form of conscious awareness[15] and symbolic evaluations of behavior, which seem to instigate particular types of shared experiences, as, for example, in the fear displayed similarly across birds, dogs, and cats confronting aggressive movements or in the envy shared in situations of perceived inequity. In the conclusion, I extend such human-animal commonalities and shared forms of consciousness to rethinking human-centric definitions of rhetoric.

As an intervention, tying rhetoric to emotive forms of consciousness expands rhetorical thinking away from the rational and logical toward the many pathic ways of being that are not necessarily unique to a human experience of the world. If Rickert is correct in saying that "rhetoric itself is ontological, having to do with being not just knowing," then a materialist perspective recognizing how "human beings in the world *are*," as Rickert says, and seeing how they "emerge in the world" leaves us with the imperative to examine how similar we *are* to other animals.[16] Rhetoric cannot be content to revise itself as ontological formation and yet turn only to human beings. An ontological rhetoric must turn back around—utilizing the backbone shared by other vertebrates—to see the animals living right in front of us; accordingly, examination of "rhetorical attunements" must extend at least as far as those animals "emerging in the world" with biological and behavioral similitudes.[17]

Accordingly, rhetorical scholarship can and should expand its own theorizing, blending neuroscientific knowledges with the rhetorical tradition. Such a move need not give excessive or undue epistemological privilege to neuroscience but can, rather, treat neuroscience as valid on its own grounds and as useful to invention. Theoretically, then, I seek to balance what is constructed symbolically with and against an organism's biological processes. Allowing biology and symbolicity to collude, in this case, exposes many kinds of animals as always having been, at least to some extent, rhetorical beings.

THEORIZING THE EMOTIONS IN COGNITIVE NEUROSCIENCE

Three theories tend to structure cognitive investigations of the emotions. All are thoroughly detailed by applied psychologists Keith Oatley and P. N. Johnson-Laird;[18] however, the theories are briefly summarized here to bring attention to underlying commonalities relevant to interpreting "motivation" and "fairness" within the neuroscience literature.

The first theory termed the Action-Readiness Theory suggests that "emotions are built from elements that are not themselves emotions" but are "states of readiness for certain kinds of actions."[19] Nico Henri Frijda forwards the theory, suggesting that the body seeks to do something through emotions, which is not necessarily rational or intentional.[20] The body, for example, might seek to enhance engagement (joy) or impart urgency (fear) for evolutionarily beneficial reasons.[21]

The second theory termed the Core-Affect Theory imagines emotions as double-layered. Much like Brian Massumi's concept of emotions, which has circulated throughout rhetoric and critical-cultural studies likely for its resonances with Deluezian scholarship,[22] Core-Affect Theory is the generation of an "emotion" from a pre-existing "affect" state. The body, according to this theory, automatically evaluates all situations on a scale of valence (positive versus negative) weighed against cultural "emotional prototypes."[23] When the valence meets the prototype—or what Massumi might call the "semantically or semiotically ordered" affect[24]—an emotion emerges. Barrett and Bliss-Moreau forward the theory, terming affect a "psychological primitive" of an emotion.[25]

The third theory termed the Communicative Theory proposes that emotions are "evolutionary adaptations" to "signals" that "have propositional content" relating to one's own self and goals.[26] The theory sounds more cognitive and deliberate than the others but upholds that such signals can be physiologically embedded and triggered without conscious reflection. Emotions, thus, arise from a complex of "antecedents," "objects," and/or "consequences" without any one emotion necessarily having all three.[27] For example, an emotion might arise from wanting to succeed (antecedent), from considering one's partner (object), and/or witnessing two people cooperating on a task (consequence). Emotions themselves, then, are adapted bodily responses tied to the world.

Reviewing these three theories highlights striking commonalities, which prove exigent to the neuroscience conversations of motivation and fairness; the commonalities also expose how rhetoric-as-ontology—a way of *being* in the world—might be reasonably extended beyond the human form.

> First: emotions are "built." They do not exist *as such* in the brain/body.
>
> Second: emotions are felt, embodied *interactions with environmental situations*. Whether they are products of consciously, culturally ordered affect states or whether they are of general readiness states, they are intimately tied to lively environments.
>
> Third: emotions *do something to the body and others*; they change the direction or trajectory of the body and, accordingly, of social interaction.

It is from this background that current cognitive neuroscience research on the emotions engages "motivation" and "fairness." Indeed, commonalities among the aforementioned views on the emotions evidence how brain-body states and neurological structures for appraisals might make environments appear reactive and symbolic in ways that are shared across many animals. As an example, neuroscientists Elizabeth Phelps and Joseph LeDoux argue, "the amygdala has been shown to play similar roles in humans and other animals" such that humans and other animals seem to all live with a "form of emotional learning" wherein "a neutral stimulus comes to elicit defensive behavior and physiological responses after being associated with an aversive event."[28] Fear responses across humans and other animals, specifically involving repeated or environmentally conditioned fear experiences such as loud noises or electric shocks, share neural pathways and activation patterns.[29] Neurological structures, in other words, *do something* similar and something predictable to bodies having those structures and engaging similar environmental situations. Shared types of brain-bodies, perhaps not surprisingly, alter their environments or react in similar ways.

Returning briefly to definitions of rhetoric: thinking about "motivation" and "fairness" in terms of 1) being an embodied interaction with the world, 2) being built within situations, and 3) being able to do something to the body and environment allows for conceptual experimentation with the ontological dynamic informing rhetoric. Seeing emotional experience as interactional, situational, and bio-life altering provides an opportunity to reconsider rhetoric itself in terms directed now toward scientific observations about human-animal commonalities. Philosophical declarations of a rhetorical-ontology and the entwined relations that become "world-transforming" for individuals "immersed in vibrant, ecologically attuned environments"[30] can, at least to some extent, be examined biologically by those who study the body's materiality; in so doing, any vagaries swimming within nonetheless useful concepts like "attunements" or "identifications" can find further and more interdisciplinary grounding in scientific constructions of material specificity.

READING MOTIVATION AND FAIRNESS IN COGNITIVE NEUROSCIENCE

Examining how "motivation" and "fairness" are employed and entwined in much of the cognitive and affective neuroscience literature on emotions enables a comparison to rhetoric's tradition. Just as a rhetorical scholar might start with Kenneth Burke for many discussions of basic terms, in this case, starting with the well-known neuroscientist Joseph LeDoux seems apropos.

For LeDoux, "motivation" sits at the very "core of consciousness."[31] He states, "what we do in life is controlled by non-conscious processes that we only come to understand by monitoring and interpreting their expression in behavior or in other body states."[32] This "monitoring" and "interpreting" *is* consciousness, which then transforms through memory into a kind of "self-story that we use to understand the motivations and actions that arise."[33] In brief, for LeDoux, our consciousness can be described as what "arises," or perhaps, what congeals amid the "monitoring" and "interpreting" that happens in certain kinds of brains automatically.

Accordingly, LeDoux's usage of the term motivation exists on at least two levels: a specific interpretation—of one's own and others' personal motivations—and a "more basic" biological response. One is cognitive, while the other is involuntary. From LeDoux's writings, the former typically follows the latter. He states, "When a threat activates one of these hard-wired circuits ["survival circuits" that do the monitoring], the result is the establishment of a global motivational state in the organism, a condition that spreads throughout the brain and body to mobilize the organism's resources to deal with the danger."[34] Focusing here on "hard-wired circuits," or what LeDoux calls "survival circuits," situates the discourse of emotions in terms of similarities of circuitry across species whose engagement with the environment required evolutionary adaptation with commonalities.[35] This foregrounding of "circuitry" is a scientifically judicious choice wherein LeDoux is careful not to expand claims too far, that is, not to say that different animals have exactly the same emotions but only similar neurological architectures for reasons, presumably, of all growing on the planet over time. The move highlights the undeniable, observable (neuro-architectural) parallels among humans and animals, helping to explain why humans and animals might display similar behaviors when confronted, say, with a threat.

Another way to put this is to say that similar brains activate similarly in response to similar environments, and neurological activations seem often—but not always—to create some internal awareness of the changes in the brain-body that result. Such awareness, LeDoux argues, emerges from the underlying, built-in "motive state" that is a product of a biological architecture, which also contributes to the formation of consciousness. Again, here, the consciousness is an emotive, scanning consciousness that is sometimes complimented by conscious awareness and, accordingly, the production of defined emotions. As LeDoux states, "When a motive state related to danger is active, we become sensitive .... Feeling afraid only occurs in organisms that can be conscious that they are in danger."[36] LeDoux herein adopts something like the Action-Readiness Theory of emotions, which proposes that the brain-body in the production of consciousness itself does something, which is to have states of readiness for threats that can, subsequently, produce feelings of fear.[37] Having this basic readiness or emotive awareness of the environment is where the word "fairness" proves useful and prominent in a discussion of human-animal commonalities.

Fairness

Nichols defines "fairness" as detecting "norms of equal division."[38] Distinguished professor and primatologist Frans de Waal also adopts the definition, stating that his research team is interested in "why humans aren't happy with what we have, even if it's good enough, if someone else has more."[39] Building from LeDoux's "motivation"—that survival circuits necessarily evaluate threats by automatically "monitoring and "interpreting" the world in the production of consciousness—fairness becomes one specific form of interpretation; a shared form.

Indeed, like motivation, fairness retains a cognitive and involuntary dimension—a dimension of circuitry that reacts or engages with higher-level evaluative systems. In one way, then, fairness can mean responding automatically to the environment due to evolutionary development, while it might also refer to a rational situational judgment of relations. Researchers such as Sarah Brosnan and Frans de Waal offer a reason for this, arguing that the brain evolved to monitor other animals and keep them from doing better than one's self and gaining an advantage.[40] They state, "Species other than humans respond to inequity in a variety of situations. This has been particularly well documented in two species of non-human primates, capuchin monkeys and chimpanzees, as well as in domestic dogs ... these species show negative reactions to continued inequity between themselves and a social partner."[41] Another theory suggests that fairness results from an animal need to evaluate whether or not another animal wants to cooperate and/or ensures that other animals will cooperate over long spans of time.[42]

The underlying need for cooperation seems a good explanation in a particular evolutionary framework for the biological development of a "survival circuit" that causes an awareness of what we, as humans, call "fairness" (Brosnan and De Waal, 2013,1–2). In other words, "fairness" is explained, as a phenomenon, through the concept of an automatic monitoring—"motivation"—while the evolutionary development of the circuitry producing such monitoring is granted, at least in part, to the evolutionary need to evaluate cooperation, which, thus, results in levels of "fairness." The reasoning is circular. But the circular, internal bind that is evident might not be so much a problem to be solved as an indicator of the mutual codevelopmental, progressive relationship between brains and environments. This is not to shirk off circular reasoning but, rather, to recognize how complicated rhetorics arise from complex living organisms developing in a dynamic environment. Words and things, actions and narratives spin a web of loops, relapses, returns, and suggestions whose composition becomes a messily meaningful brain-body-world.

With respect to neuroscientific views of "fairness," Lind again draws a connection to social cooperation: "fairness judgments are assumed to serve as a proxy for interpersonal trust in guiding decisions about whether to behave in a cooperative manner."[43] For Lind, such evaluations are "the fundamental social dilemma" and indicate safe environments that can be explored and used.[44] Interestingly, an awareness of the brain's monitoring for cooperation (and/or non-cooperation) produces a symbolic interpretation of bodies performing in-fairness or out-of-fairness. Thus, fairness becomes both neurobiological and automatic as well as codes the symbolic and social—or, perhaps, better stated, it generates what is considered cooperative and social.

How such terms have been employed in the rhetorical tradition allows for subsequent reflection on interrelations among symbols and biology—and importantly among humans and animals. In particular, rhetorical scholars' attempts to define "motive" (as opposed to "motivation") and "fairness" expose a tendency to ignore or reject the animal and avoid any discussion of precise biological features detailed in the sciences. Digging down, so to speak, into shared mechanisms and scientific knowledges holds potential, I believe, to invigorate rhetorical understandings, which risk upholding Modern dedications to rationality and human uniqueness without adopting the animal.

READING MOTIVATION AND FAIRNESS IN THE RHETORICAL TRADITION

Motive and Motivation

Perhaps the best place to begin a discussion of motivation in the rhetorical tradition is with the root word, motive. In rhetoric, motive has meant the psychological effects of mythical images and vocabularies mediating human interactions and which form one's basic impetus for action.[45] Motive has also meant cognitive end-goal states, which are not always part of a person's conscious self-image.[46] Motive is, thus, symbolic in rhetoric while also being, to some extent, involuntary and entwined with consciousness itself. In some ways, then, motive in rhetoric has dual deployments and resembles "motivation" in the neuroscience literature. Yet, two consistent differences persist: an absence of discussion in rhetoric around biological mechanisms and little suggestion that motive and symbolic action are differently involved over time and situation.

Resolving those differences aims to craft a more material-centric rhetoric. Putting contemporary neuroscience into the picture deepens a "rhetoric of motives" by delving into the inner workings of the brain-body and avoiding vague theoretical proclamations about the body, at least as much as is possible and sensible. Biological action and conscious action can be thought together, mutually entwined in doing and meaning-making *in some specific kind of way*.

Returning to "motive" in the rhetorical tradition: the first stop, predictably, must be Kenneth Burke. His *A Rhetoric of Motives* has been called "logocentric" and obsessed with symbols[47] as well as deeply invested in materiality.[48] The consensus today seems to be that Burke did not neglect materiality; indeed, he is, for Celeste Condit, "superior to most post-structuralists" with respect to material considerations.[49] However, for Richard Engnell, rhetorical scholars are not so much engaging Burke's original meanings but are, rather, "creatively refashioning his work" to explore "what it may mean to be material creatures that also function symbolically."[50] Whatever the case, Burke grasps the inherent role of the body and the need for human cooperation within his redefinition of rhetoric as identification. In fact, Burke's starting point for the concept of "motive" resembles the neuroscientific one—a need for cooperation set against a kind of automatic, if not symbolic or symbolically informed, evaluation of others.

Burke's notion of "motive" is commonly positioned as the sum of those vocabularies situated around all of the engaged processes of identifying.[51] Any further definitional rigor regarding his idea of motive has proven "elusive."[52] Yet, in general, Burke's idea presumes, as Engnell states, a dramatistic and human-centric positioning wherein "animals move, human beings act."[53] Burke's "motive" is, thus, some human outcome, expressed symbolically, of so-called "rational" bodies and does not seem to extend specifically to human-animal commonalities of circuitry or to nonhuman animals.

Drawing from Burke's work in *Rhetoric*, Brian Crabel slightly revises Burke's position.[54] Crabel argues that rhetoric itself, for Burke, was most basically not about identification, but about sensing and bridging distance. For Crabel, the ultimate human "motive," then, would then be to deal with "distance." Humans seek to bridge distance and to find a way to "transcend the nonverbal, to create distance," a persuasion of being.[55] In Crabel's view, Burke understands rhetoric as the compulsion to see speaker and listener as separated and, thus, a need to create appeals and an ability to react to them. Inherent in the idea is a compulsion—a material, bodily, non-symbolic force driving symbolic action. However, such an idea remains in the realm of explanatory plenitudes absent specific biological mechanisms. When and how the non-symbolic and the bodily lead to "motives" or "motive-ate" symbolic action—or may do so differently across situation and/or time—is left unaddressed.

McClelland similarly conceptualizes motive in broad theoretical terms as emergent among symbols and bodies. For him, motives are similar to "concerns," but they are not necessarily conscious and may be variable in kind and strength.[56] For McClelland, a motive is a "natural incentive," but there may be different kinds of these motives for different individuals. What proves striking about McClelland's argument is its mutual dependence upon the biological "naturalness" of the body as well as the environmental situation. Similar to neuroscientific conceptions, McClelland's motive, like motivation in the neuroscience literature, arises variably from being a body in the world. McClelland's motive is both an effect of bodies (circuitry, in much neuroscience lingo) and a semiotic assessment of an environmental situation. For rhetorical scholars, though, discussion of the variability has mostly remained a task of examining vocabularies without the risk or benefit of reference to scholars studying the body scientifically.

As a result, McClelland forwards something that looks similar to the Communication Theory of emotions. He proposes motives as a kind of internal incentive or "concern," like a feeling, which arises from antecedents, consequences, or objects. Accordingly, people may hold different motives, and the involvement of such motives in human communication appears always necessary; the lingering question for the rhetorical scholar, then, given his view, is how do motives arise?

The neuroscientific understanding mapped in this chapter makes the mechanism of "motivation" inseparable from the very basic production of consciousness.[57] And here is one answer to how and why "motives" arise. Rhetorical action—the compulsion or need for bridging distance, making known, and reaching out—arises from ontology, owning or living with/in a particular form of consciousness. The point is non-trivial when animal being is considered. Put succinctly: animals can also be rhetorical beings when their forms of consciousness compel or spur them to motivations, concerns, feelings, or social incentives arising from the contextual and environmental, that is, precisely what McClelland holds as human motives in the rhetorical tradition. That some animal feelings may be quite similar to our own—anger or jealously at unequal treatment, for example—is of our emotive consciousness, and their actions, as we are moved now to understand animals better, can be viewed, at the least, as rudimentarily rhetorical.

Fairness

Fairness provides another instance in the rhetorical tradition where the symbolic and bodily interrelate and yet retain little reference to biological specificity. With that said, few detailed theorizations of "fairness" exist beyond Aristotle. Attempts at precisely interrelating symbolic evaluations of the environment and the mechanisms of the body in classical discussions of fairness remain centered on concepts of justice. In contemporary discussions, any mention of fairness usually remains focused on how a rhetor composes persuasive political rhetorics of what is "fair" or not. Briefly examining these rhetorical ruminations on fairness makes evident where and how rhetorical understandings overlap with neuroscientific ones.

Much has been made of Aristotle's turn to the concept of "fairness" as universally recognizable and useful when the laws of a State fail to address or satisfy a situation.[58] The lack of satisfaction on the part of a juror can, itself, imply a deeply embedded, perhaps enculturated, sense of "unfairness" that radiates from the chest and sweaty palms of the juror, an ideal left terribly unfulfilled. James Murphy and colleagues suggest that Aristotle understood fairness as a part of a natural order and an embodied sensibility. "The rules of equity [*epieikes*, or fairness] are embodied in one species of unwritten law"; since Aristotle "avoids tracing unwritten law to the gods," one can then conclude that Aristotle saw fairness as a condition of the psyche and something intuitive.[59] The classical concept of fairness appears even so powerful that, as Edward Harris argues, many ancient Greeks believed that an interior demand or strong feeling of *epiekikea* allowed judges to "ignore the law completely" in cases where strict applications of the law would be unfair to the case under consideration.[60] Fairness here is positioned by the Greeks as having an undeniably bodily character and is, more accurately, a compulsion or demand of human sociality.

Contemporary discussions of fairness in rhetorical scholarship touch on these same themes, that is, the unavoidability of fairness and its social importance. Yet, they display a distinctive semiotic and poststructuralist character. For example, Katherine Smith argues that fairness functions in local communities as "an idiom of exclusion as well as inclusion"[61] insofar as "the language of fairness mediates" relational strife and competing moralities.[62] In other words, "fairness" is a language strategically deployed in the British context that Smith critiques; fairness operates as an appeal left unspecified because it functions rhetorically—"Hey, consider what is fair!"—to resolve a "tension and confusion between a liberal and multi-cultural Britain."[63] As Smith notes, "'unfair judgments" about the rights of identity and representation are real possibilities, so a vague, general appeal to "fairness" tends to negotiate competing claims.[64]

Such scholarship rightfully highlights the cultural components of fairness and its flexibility. It also usefully demonstrates the constructedness of rhetorics and how local situations demand varying concepts of "fairness" such that there may be very little that remains "natural" about fairness across bodies in different times and places. In this respect, contemporary rhetorical scholarship adheres to important poststructuralist insights that must not be neglected, even as doing so might risk dismissing how the production of consciousness—or an alertness about living environments and social relations—might very well have something to say about why fairness proves salient and works rhetorically in the first place. The textual, rhetorical analysis of fairness leaves unanswered a question of why "fairness" proves a powerful solution to a difficult problem of sociality, of "unfair" situations and "unjust" judgments across time and place.

Yet, what classical notions of fairness and the contemporary rhetorical scholarship have in common is a turn back to emotive life. Emotive considerations and appeals serve to resolve difficult social circumstances and rhetorically mediate complex relationships. In this way, the focus in classical works on a "natural" sense of fairness as well as the focus on a rhetorically constructed sense of fairness call attention to a massive, muddy middle ground between what LeDoux terms a biological "motivation state" embedded in our brains and what those focused on the poststructuralist tradition might deem an enculturated, symbolic evaluation manifesting as strategic exercise.[65] Indeed, the rhetoric of fairness arises from "the fundamental social dilemma"[66]—cooperation or closing distance—which itself remains shrouded in a complex bodily dilemma: how exactly do we encounter one another in an environment and sense a distance and the need to close it? Fairness, then, expresses, as a term as much as a feeling, the enmeshment of materiality and symbolicity.

CONCLUSION: MANY RHETORICAL ANIMALS, OR NEW DIRECTIONS FOR RHETORIC AND THE EMOTIONS

Many animals are rhetorical animals, not humans alone. The conclusion may sound like a radical divergence from famous quotations about *the* symbol-using animal,[67] the political animal,[68] and the rhetorical animal; yet, the conclusion might also sound familiar due to its roots in discussions of distance, commonality, and identification composed by rhetorical scholars. More so, the rhetorical of the animal resonates as a familiarity from a repeated scientific grounding in evolutionary frameworks guiding interpretations of brain data and those related theories of human brain development. Put differently, at the heart of rhetoric resides the image of an identifying-being with a compulsion to "close distance," while at the heart of neuroscience resides the image of a being whose "monitoring" and "interpreting" *is* its production of an emotive form of consciousness driving selective behavioral responses, which are shared with many animals; amid the overlap, rhetorical animals appear in many shapes and kinds, a multiplicity across an animal range.

The dethroning of the human as rhetorical in exchange for rhetorical animals is not merely an intellectual dethroning of Modernism[69] but also a scientific dethroning of a human-centric rhetoric too narrowly focused on a human consciousness predominately defined by higher-order rational thinking. The commonalities observed across brain circuits and certain emotional states indicate just how social-symbolic situations can appear "fair" or not to many species and how those understandable feelings lead to targeted actions—to slap away, to frown, to scream, to grab, to hide—that are undeniably *rhetorical*. The act of howling and begging for food to catch the attention of humans distributing food, as chimpanzees have been observed to do, demonstrates a need for a more open definition of rhetoric.[70] The act of a dog pouting and using the same voice tones and "voice areas" of the brain as humans to express sorrow and frustration adds evidentiary intensity to the argument.[71] It is my hope that evaluations of "fairness" and its ties to a form of consciousness producing common "motivation states" expands a rhetoric of motives and pushes rhetorical thinking toward (e)motive explanations not necessarily unique to a human experience of the world.

New directions for rhetoric, as Rice and Condit have indicated, must involve further considerations of the emotions, their ranges and types, as well as how they are rhetorically constructed and deployed;[72] yet, new directions must not overlook implications for definitions of rhetoric itself. If Kenneth Burke's major contribution to the field was defining rhetoric as identification, then re-seeing rhetoric as having its neurological basis in an emotive consciousness should encourage future definitional commentary to address the particularities of the bodily, to closely consider nonrational environmental monitoring and interpreting, and to rethink many animals as often acting rhetorically as well.

The discussion in this chapter also implies that neuroscience can be more than a strange outsider to rhetoric or positioned simply as another interesting set of texts to critique. A turn toward materiality and affect as well as toward the animal requires that the study of the brain and biological development be respected and engaged. Exercising one's own rhetorical lens is, of course, a necessity when dealing with any disciplined text; Jordynn Jack makes this point clearly when asserting that scholars must toggle back and forth between looking with a neuroscientific and rhetorical lens.[73] However, that said, when definitions of rhetoric link to scientific observations and ideas of consciousness, then rhetorical theorists have an opportunity to re-evaluate their theories and imagine new stances. Affirming many animals as rhetorical animals becomes possible and relevant exactly because of the linkages between the rhetorical tradition and neuroscience. New ideas emerge from sharing/shared examinations of distance, identification, situation, and persuasion. At those intersections, we are forced to again confront what the animal means to being human because we see that we are, as Jacques Derrida once said, already inextricably caught "in an 'animal' relation."[74]

NOTES

1. See: Diana Coole and Samantha Frost, *New materialisms: Ontology, agency, and politics* (Durham: Duke University Press, 2010); Alex Parrish, C., *Adaptive*

rhetoric: Evolution, culture, and the art of persuasion (New York: Routledge, 2013); Thomas Rickert, *Ambient rhetoric: The attunements of rhetorical being* (Pittsburgh: University of Pittsburgh Press, 2013).

2. See: Jenny Edbauer Rice, "The new 'new': Making a case for critical affect studies," *Quarterly Journal of Speech* 94, no. 2 (2009): 200–212; Joddy Murray, *Non-discursive rhetoric: Image and affect in multimodal composition* (Albany: SUNY Press, 2009); Melissa Gregg and Gregory Seigworth, *The affect theory reader* (Durham: Duke University Press, 2010); Jeff Pruchnic and Kim Lacey, "The future of forgetting: Rhetoric, memory, affect," *Rhetoric Society Quarterly* 41, no. 5 (2011): 472–494; David Gruber, "The (Digital) Majesty of all under heaven: Affective constitutive rhetoric at the Hong Kong History Museum of History's terracotta warrior exhibit," *Rhetoric Society Quarterly* 44, no. 2 (2014): 148–167.

3. See: Brian Ogren, "Aristotle's rhetoric and the cognition of being: Human emotions and the rational–irrational dialectic," *Minerva—An Internet Journal of Philosophy* 8 (2004): 1–19; Daniel Gross, *Secret history of emotions: From Aristotle's rhetoric to modern brain science* (Chicago: University of Chicago Press, 2006); Lea Povozhaev, "Metaphor as emotive change: A triangulated approach to thought, language, and emotion relatable to Aristotelian sensate perception," *Rhetoric Review* 32, no. 1 (2013): 44–63; Erin Rand, J., "Bad feelings in public: Rhetoric, affect, and emotion," *Rhetoric & Public Affairs* 18, no. 1 (2015): 161–175.

4. Celeste Condit, "Pathos in criticism: Edwin Black's communism as cancer metaphor," *Quarterly Journal of Speech* 99, no. 1 (2013): 1.

5. Jordynn Jack, "What are neurorhetorics?" *Rhetoric Society Quarterly* 40, no. 5 (2010): 405–410.

6. See: David G. Amaral, "The amygdala, social behavior, and danger detection," *Annals of the New York Academy of Sciences* 1000 (2003): 337–347; Elizabeth A. Phelps, "Emotion and cognition: Insights from studies of the human amygdala," *Annual Review of Psychology* 57 (2006): 27–53; Shaun Nichols, "Emotions, norms, and the genealogy of fairness," *Politics, Philosophy & Economics* 9, no. 3 (2010): 275–296.

7. Joseph E. LeDoux, "Feelings: What are they & how does the brain make them?" *Dædalus, Journal of the American Academy of Arts and Sciences* 144, no. 1 (2015c): 96–114.

8. Thomas Rickert, "Toward the Chora: Kristeva, Derrida, and Ulmer on emplaced invention," *Philosophy and Rhetoric* 40, no. 3 (2007): 251–273.

9. The exception here is conversations of Burkean identification as always-already affective. See: Diane Davis, "Identification: Burke and Freud on who you are," *Rhetoric Society Quarterly* 32, no. 2 (2008): 123–147; however, Burke himself, I believe, did not conceptualize identification in precisely this way. For discussion, see: David Gruber, "The neuroscience of rhetoric and the meaning of identification," in *Neurorhetorics*, ed. Jordynn Jack (New York: Routledge, 2013), 37–52.

10. Davis, "Identification," 144.

11. Diane Davis, *Inessential solidarity: Rhetoric and foreigner relations* (Pittsburgh: University of Pittsburgh Press, 2010), 6–7.

12. Jack, "Neurorhetorics," 406.

13. John Muckelbauer, *The future of invention: Rhetoric, postmodernism and the problem of change* (Albany: SUNY Press, 2008), 5–13.
14. See: Sarah F. Brosnan and Frans B. M. De Waal, "Fairness in animals: Where to from here?" *Social Justice Research* 25, no. 3 (2012): 336–351; Sarah F. Brosnan and Frans B. M. De Waal, "Evolution of response to (un)fairness," *Science* 346, no. 6207 (2013): 1–6; Joseph E. LeDoux, "Rethinking the emotional brain," *Neuron* 73.4 (2012): 653–676; LeDoux, "Feelings," 96–114; Nichols, "Emotions," 275–296.
15. Since consciousness is commonly defined as being conscious of some X object, I use the terms *conscious awareness* and *consciousness* interchangeably here.
16. Thomas Rickert, *Ambient rhetoric: The attunements of rhetorical being* (Pittsburgh: University of Pittsburgh Press, 2013), xv.
17. Ibid.
18. Keith Oatley and Philip N. Johnson-Laird, "Cognitive approaches to emotions," *Trends in Cognitive Neuroscience* 18, no. 3 (2014): 134–140.
19. Oatley and Johnson-Laird, "Cognitive," 134.
20. Nico H. Frijda, *The emotions* (Cambridge: Cambridge University Press, 1986); Nico H. Frijda, *The laws of emotion* (New York: Erlbaum Press, 2007).
21. Oatley and Johnson-Laird, "Cognitive," 134.
22. Brian Massumi, *Parables of the virtual: Movement, affect, sensation* (Durham: Duke University Press, 1995).
23. Oatley and Johnson-Laird, "Cognitive," 135.
24. Massumi, *Parables,* 85.
25. Lisa F. Barrett and Eliza Bliss-Moreau, "Affect as psychological primitive," *Advanced Experimental Social Psychology* 41 (2009): 167–212.
26. Oatley and Johnson-Laird, "Cognitive," 135.
27. Keith Oatley and Philip N. Johnson-Laird, "Basic emotions in social relationships, reasoning, and psychological illnesses," *Emotion Review* 3 (2011): 424–433; See also: Oatley and Johnson-Laird, "Cognitive," 135.
28. Elizabeth A. Phelps and Joseph E. LeDoux, "Contributions of the amygdala to emotion processing: From animal models to human behavior," *Neuron* 48, no. 2 (2005): 175.
29. Phelps and LeDoux, "Contributions," 176–178.
30. Rickert, *Ambient*, xvi.
31. LeDoux, "Feelings," 101.
32. Ibid.
33. Ibid.
34. LeDoux, "Feelings," 102.
35. LeDoux, "Rethinking," 653.
36. LeDoux, "Feelings," 102.
37. Frijda, *The laws of emotion,* 4.
38. Nichols, "Emotions, norms," 275.
39. "Human sense of fairness evolved to favor long-term competition," *Psy.org* (September 18, 2014), para 4, a news article drawing from Sarah F. Brosnan and Frans B. M. de Waal, "Evolution of responses to (un)fairness," *Science*, October 17, 2014.
40. Brosnan and De Waal, "Evolution," 337; "Fairness," 1.

41. Brosnan and De Waal, "Evolution," 337.

42. Brosnan and De Waal, "Fairness," 1–2.

43. Allen E. Lind, "Fairness heuristic theory: Justice judgments as pivotal cognitions in organizational relations," in *Advances in Organizational Justice*, ed. J. Greenberg and R. Cropanzano (Palo Alto, Stanford University Press, 2001), 56.

44. Allen E. Lind, "Social conflict and social justice," *Inaugural oration for the Leiden University Fund Chair in Social Conflict* (Presentation to the Faculty of the University, Leiden, the Netherlands, June, 1995), para 17.

45. Kenneth Burke, *A rhetoric of motives* (Berkeley: University of California Press, 1969), 202–224.

46. David C. McClelland, *Human motivation* (Cambridge: Cambridge University Press, 1987), 543.

47. James W. Chesebro, "Extensions of the Burkeian System," *Quarterly Journal of Speech* 78, no. 3 (1992): 358.

48. Debra Hawhee, *Moving Bodies: Kenneth Burke at the edges of language* (Columbia: University of South Carolina Press, 2009).

49. Celeste Condit, "Framing Kenneth Burke: Sad tragedy or comic dance?" *The Quarterly Journal of Speech* 80, no. 1 (1994): 80.

50. Richard A. Engnell, "Materiality, symbolicity, and the rhetoric of order: Dialectical biologism as motive in Burke," *Western Journal of Communication* 62, no. 1 (1998): 2.

51. James Jasinski, *Sourcebook on rhetoric* (Thousand Oaks: Sage, 2001), 369.

52. Ibid.

53. Engnell "Materiality," 5.

54. Bryan Crabel, "Distance as ultimate motive: A dialectical interpretation of a 'rhetoric of motives,'" *Rhetoric Society Quarterly* 39, no. 3 (2009): 213–239.

55. Crabel "Distance," 236.

56. McClelland, *Human motivation,* 591–593.

57. LeDoux, "Feelings," 96–100.

58. See: Edward Harris, "How strictly did Athenian courts apply the law? The role of Epieikeia," *Bulletin of the Institute of Classical Studies* (June 2013): 27–48: James J. Murphy, Richard J. Katula, and Michael Hoppmann, *A synoptic history of classical rhetoric* (New York: Routledge, 2014); John D. Staines, *Mary's tragedy and public rhetoric in an age of reform and revolution* (Burlington: Ashgate Publishing, 2009), 15; Ann Ward, "Justice as Economics in Aristotle's Nicomachean Ethics," *Canadian Political Science Review* 4, no. 1 (2010): 1–11.

59. Murphy et al., *A synoptic history,* 82.

60. Harris, "How strictly," 27–28.

61. Katherine Smith, "Rhetoric and the everyday—Fairness as rhetorical force and the micro-politics of intentionality in a North Manchester town," in *Rhetoric, Politics and Society*, ed. Alan Finlayson, James Martin, and Kendall R. Phillips (London: Palgrave McMillan, 2014), 171.

62. Smith, "Rhetoric," 160.

63. Ibid, 168.

64. Ibid.

65. LeDoux, "Rethinking," 653; Also see: Joseph E. LeDoux, "'Inside out' draws from Darwin's theory of the nature of emotions: I think he got it wrong," *Salon*, July 20, 2015.

66. Lind, "Social conflict," para 17.

67. Burke, *Rhetoric*, 16.

68. Aristotle, *Politics, Book 1* (Cambridge: Harvard University Press), b1, sec. 1253.

69. Katherine Hayles suggests a move to undo the human and/or resituate what it means to be human is a reaction to Modern narratives. See: Katherine N. Hayles, *How we became posthuman: Virtual bodies in cybernetics, literature, and informatics* (Chicago: University of Chicago Press, 1999).

70. William D. Hopkins, Jared P. Taglialatela, and David A. Leavens, "Chimpanzees differentially produce novel vocalizations to capture the attention of a human," *Animal Behavior* 73, no. 2 (2007): 281–286.

71. Attila Andics, Marta Gácsi, Tamas Faragó, Anna Kis, and Adam Miklósi, "Voice-sensitive regions in the dog and human brain are revealed by comparative fMRI," *Current Biology* 24, no. 2 (2014): 574–578.

72. Rice, "The new," 200–201; Condit, "Pathos," 1–2.

73. Jack, "Neurorhetorics," 406–407.

74. Jacques Derrida, *The animal that therefore I am* (New York: Fordham University Press, 2008), 160.

BIBLIOGRAPHY

Amaral, David G. "The amygdala, social behavior, and danger detection," *Annals of the New York Academy of Sciences* 1000 (2003): 337–347.

Andics, Attila, Gácsi, Marta, Faragó, Tamas, Kis, Anna, and Miklósi, Adam. "Voice-sensitive regions in the dog and human brain are revealed by comparative fMRI," *Current Biology* 24, no. 2 (2014): 574–578.

Aristotle. *Politics, Book 1*. Cambridge: Harvard University Press: b1, sec. 1253. http://www.perseus.tufts.edu/hopper/text?doc=Perseus%3Atext%3A1999.01.0058%3Abook%3D1%3Asection%3D1253a.

Barrett, L. F. and Bliss-Moreau, E. "Affect as psychological primitive," *Advanced Experimental Social Psychology* 41 (2009): 167–212.

Brosnan, Sarah F. and DeWaal, Frans B. M. "Fairness in animals: Where to from here?" *Social Justice Research* 25, no. 3 (2012): 336–351.

Brosnan, Sarah F. and DeWaal, Frans B. M. "Evolution of response to (un)fairness," *Science* 346, no. 6207 (2013): 1–6.

Burke, Kenneth. *A Rhetoric of Motives*. Berkeley: University of California Press, 1969.

Chesebro, James W. "Extensions of the Burkeian System," *Quarterly Journal of Speech* 78, no. 3 (1992): 356–368.

Condit, Celeste. "Framing Kenneth Burke: Sad tragedy or comic dance?" *The Quarterly Journal of Speech* 80, no. 1 (1994): 77–82.

Condit, Celeste. "Pathos in criticism: Edwin Black's communism as cancer metaphor," *Quarterly Journal of Speech* 99, no. 1 (2013): 1–26.

Coole, Diana and Frost, Samantha. *New Materialisms: Ontology, Agency, and Politics*. Durham: Duke University Press, 2010.

Crabel, Bryan. "Distance as ultimate motive: A dialectical interpretation of a 'Rhetoric of Motives,'" *Rhetoric Society Quarterly* 39, no. 3 (2009): 213–239.

Derrida, Jacques. *The Animal that Therefore I Am*. New York: Fordham University Press, 2008.

Engnell, Richard A. "Materiality, symbolicity, and the rhetoric of order: Dialectical biologism as motive in Burke," *Western Journal of Communication* 62, no. 1 (1998): 1–25.

"Evolution of responses to unfairness." *Woodruff Health Sciences Center, Emory University*. September 14, 2014, http://news.emory.edu/stories/2014/09/de_waal_science_fairness/campus.html.

Frijda, Nico H. *The Emotions*. Cambridge: Cambridge University Press, 1986.

Frijda, Nico H. *The Laws of Emotion*. New York: Erlbaum Press, 2007.

Gregg, Melissa and Seigworth, Gregory. *The Affect Theory Reader*. Durham: Duke UP, 2010.

Gross, Daniel. *Secret History of Emotions: From Aristotle's Rhetoric to Modern Brain Science*. Chicago: University of Chicago Press, 2006.

Gruber, David. "The (Digital) majesty of all under heaven: Affective constitutive rhetoric at the Hong Kong History Museum of History's terracotta warrior exhibit," *Rhetoric Society Quarterly* 44, no. 2 (2014): 148–167.

Harris, Edward. "How strictly did Athenian courts apply the law? The role of Epieikeia," *Bulletin of the Institute of Classical Studies* (2013): 27–48.

Hawhee, Debra. *Moving Bodies: Kenneth Burke at the Edges of Language*. Columbia: University of South Carolina Press, 2009.

Hopkins, William D., Taglialatela, Jared P., and Leavens, David A. "Chimpanzees differentially produce novel vocalizations to capture the attention of a human," *Animal Behavior* 73, no. 2 (2007): 281–286.

"Human sense of fairness evolved to favor long-term cooperation," *Pys.org*, September 18, 2014. Accessed February 20, 2016, http://phys.org/news/2014-09-human-fairness-evolved-favor-long-term.html.

Jack, Jordynn. "What are Neurorhetorics?" *Rhetoric Society Quarterly* 40, no. 5 (2010): 405–410.

Jasinski, James. *Sourcebook on Rhetoric*. Thousand Oaks: Sage, 2001.

Konnikova, Maria. "You have no idea what happened." *The New Yorker*, February 4, 2015, http://www.newyorker.com/science/maria-konnikova/idea-happened-memory-recollection.

LeDoux, Joseph E. "Rethinking the emotional brain," *Neuron* 73, no. 4 (2012): 653–676.

LeDoux, Joseph E. "'Inside out' draws from Darwin's theory of the nature of emotions: I think he got it wrong," *Salon*, July 20, 2015, http://www.salon.com/2015/07/19/inside_outs_psychology_of_emotion_is_based_on_the_darwinian_tradition_but_thats_just_one_view/.

LeDoux, Joseph E. "'Run, hide, fight' is not how our brains work," *The New York Times*, December 18, 2015.

LeDoux, Joseph E. "Feelings: What are they & how does the brain make them?" *Dædalus, Journal of the American Academy of Arts and Sciences* 144, no. 1 (2015): 96–114.

Lind, Allen E. "Social conflict and social justice." *Inaugural oration for the Leiden University Fund Chair in Social Conflict. Presentation to the Faculty of the University, Leiden, the Netherlands, June, 1995,* http://people.duke.edu/~alind/INAUG.html.

Lind, Allen E. "Fairness heuristic theory: Justice judgments as pivotal cognitions in organizational relations." In *Advances in Organizational Justice*, edited by Jerald Greenberg and Russell Cropanzano, 56–88. Palo Alto: Stanford University Press, 2001.

Lind, Allen E. and van de Bos, Kees. "When fairness works: Toward a general theory of uncertainty management," *Research in Organizational Behavior* 24, (2002): 181–223.

Massumi, Brian. *Parables for the Virtual: Movement, Affect, Sensation.* Durham: Duke University Press, 1995.

McClelland, David C. *Human Motivation.* Cambridge: Cambridge University Press, 1987.

Muckelbauer, John. *The Future of Invention: Rhetoric, Postmodernism, and the Problem of Change.* Albany: SUNY Press, 2008.

Murray, Joddy. *Non-discursive Rhetoric: Image and Affect in Multimodal Composition.* Albany: SUNY Press, 2009.

Murphy, James J., Katula, Richard J., and Hoppmann, Michael. *A Synoptic History of Classical Rhetoric.* 4th ed. New York: Routledge, 2014.

Nichols, Shaun. "Emotions, norms, and the genealogy of fairness," *Politics, Philosophy & Economics* 9, no. 3 (2010): 275–296.

Oatley, Keith and Johnson-Laird, P. N. "Basic emotions in social relationships, reasoning, and psychological illnesses," *Emotion Review* 3 (2011): 424–433.

Oatley, Keith and Johnson-Laird, P. N. "Cognitive approaches to emotions," *Trends in Cognitive Neuroscience* 18, no. 3 (2014): 134–140.

Ogren, Brian. "Aristotle's rhetoric and the cognition of being: Human emotions and the rational–irrational dialectic," *Minerva—An Internet Journal of Philosophy* 8 (2004): 1–19.

Parrish, Alex C. *Adaptive Rhetoric: Evolution, Culture, and the Art of Persuasion.* New York: Routledge, 2013.

Phelps, Elizabeth A. and LeDoux, Joseph E. "Contributions of the amygdala to emotion processing: From animal models to human behavior," *Neuron* 48, no. 2 (2005): 175–187.

Phelps, Elizabeth A. "Emotion and cognition: Insights from studies of the human amygdala," *Annual Review of Psychology* 57 (2006): 27–53.

Povozhaev, Lea. "Metaphor as emotive change: A triangulated approach to thought, language, and emotion relatable to Aristotelian sensate perception," *Rhetoric Review* 32, no. 1 (2013): 44–63.

Pruchnic, Jeff and Lacey, Kim. "The future of forgetting: Rhetoric, memory, affect," *Rhetoric Society Quarterly* 41, no. 5 (2011): 472–494.

Rand, Erin J. "Bad feelings in public: Rhetoric, affect, and emotion," *Rhetoric & Public Affairs* 18, no. 1 (2015): 161–175.

Rice, Jenny Edbauer. "The new 'new': Making a case for critical affect studies," *Quarterly Journal of Speech* 94, no. 2 (2009): 200–212.

Rickert, Thomas. *Ambient Rhetoric: The Attunements of Rhetorical Being*. Pittsburgh: University of Pittsburgh Press, 2013.

Smith, Katherine. "Rhetoric and the everyday—Fairness as rhetorical force and the micro-politics of intentionality in a North Manchester town." In *Rhetoric, Politics and Society*, edited by Alan Finlayson, James Martin, and Kendall R. Phillips, 160–172. London: Palgrave McMillan, 2014.

Staines, John D. *Mary's Tragedy and Public Rhetoric in an Age of Reform and Revolution.* Burlington: Ashgate Publishing, 2009, 15.

Walsh, Jennifer. "How animal and human emotions are different," *Livescience*, February 29, 2012, http://www.livescience.com/18750-animal-human-emotions-fears.html.

Ward, Ann. "Justice as economics in Aristotle's Nicomachean ethics," *Canadian Political Science Review* 4, no. 1 (2010): 1–11.

Chapter 2

A Humanimal Rhetorics of Biological Materiality

Hayley Zertuche

This chapter explores the undecidable, abyssal depths between binary terms like "Life" and "Death" and "Human" and "Animal" that have been taken as self-evident in Western philosophy and rhetoric. What follows will be an attempt to entangle rhetoric and rhetorical studies in the kind of "wet pivot" Bhanu Kapil poetically describes. My argument for rhetorical studies takes issue with the clean definitions we refer to when we say "Human" and when we say "Animal." Despite complexities continually blooming within behavioral ecology and evolution, we cannot shake the singular "Animal"—just as we cling to the comfortable and clean "Human." These categories, which we affirm every time we use them, presume a distinction that simultaneously divides life into that which matters to rhetorical studies and that which does not. Therefore, I seek to introduce rhetorical studies to a transcorporeal[1] dis*ease* of the human as *humanimal*: of the human as always already animal, as matter, and as *rhetorical life*. In contrast to the transcendent, disembodied thinking-therefore-being subject, humanimal indicates an enfolding material biological reality—an agential reality[2]—in which we are always creatively becoming.

In "Domesticating Animal Theory," John Muckelbauer jokingly refers to "animality" as "the new fall line of theory ware" emerging within a larger ecological movement in the humanities.[3] It is indeed difficult to distill the practical exigencies for considering animals and the larger nonhuman world anew within all fields. After all, we are living in an era now ominously referred to as the "Anthropocene," in which practices of human exceptionalism are realizing devastation on a massive scale, perhaps even propelling us toward a sixth massive global extinction.[4] In *History and Its Limits: Human, Animal Violence*, Dominick LaCapra examines the persistence in Western culture for seeking a decisive criterion that separates human and animal and justifies such human exceptionalism. These diverse criteria have ranged from soul, spirit, and nobility to language and even laughter.[5] Although many of the behavioral and biological "golden" criteria have been largely discredited over time, LaCapra remarks that the "bewildering heterogeneity of seemingly endless criteria used to arrive at the same conclusive divide attests to the force of the desire for that divide or radical separation."[6] Precisely what is at stake in seeking the "conceptual Grand Canyon that divides into two the deceptively massive categories of human and animal?"[7]

In *The Animal That Therefore I Am*, Jacques Derrida pursues a similar examination and stance that seeks to complicate a clean human/animal distinction. In Derrida's view, while the question of whether a discontinuity exists between human and animal is not debatable,

> The discussion becomes interesting once, instead of asking whether or not there is a limit that produces discontinuity, one attempts to think what a limit becomes once it is abyssal, once the frontier no longer forms a single indivisible line but more than one internally divided line; once, as a result, it can no longer be traced, objectified, or counted as single and indivisible. What are the edges of a limit that grows and multiplies by feeding on an abyss?[8]

For Derrida, human and animal exist rather as "relations … at once intertwined and abyssal, and they can never be totally objectified."[9] And yet the clean and easy human/animal distinction persists, allowing justification for "all manner of atrocities,"[10] including the staggering rate at which (nonhuman) animals are killed each year for consumption (a number in the billions, surpassing even the global human population).[11]

The abyssal rupture between "Human" and "Animal" haunts rhetoric and rhetorical studies today. As Diane Davis points out in "Autozoography: Notes Toward a Rhetoricity of the Living," rhetorical studies has been a discipline that, "despite everything," traditionally presumes the human with little-to-no question.[12] George Kennedy's seminal essay "A Hoot in the Dark: The Evolution of General Rhetoric," which provides evolution as a gradient against which to read rhetoric, plunged rhetoric into the dark, abyssal depths of these specters for the first time in 1991 when he suggested, through a number of theses, that rhetoric is, in fact, prior to speech and even intent. Kennedy even goes as far as to suggest that, as "an energy existing in life" and therefore necessarily prior to life itself, rhetoric may be a "special case of the energy of all physics as known from subatomic particles."[13]

Evolution should not be a surprising choice with which to engage the problematic presumptions of rhetorical ability in humans and animals. In *Animal Wise: The Thoughts and Emotions of Our Fellow Creatures*, Virginia Morell explains how Darwin's theory of evolution "presented a solution to

[the] impasse" between "philosophers and theologians who saw humans as the product of a special creation and therefore completely separate from other animals" on one side and "thinkers who argued that there were sufficient similarities between people and animals" on the other.[14] Darwin's evolutionary biology submits compelling evidence that humans *are also* animals, thus collapsing, in quite a revolutionary fashion, the massive gulf between what is (often still) passionately viewed as two radically opposed ontologies. In other words, evolution traces the (divergent and convergent) entanglement of biological life. In "Darwin and Feminism," Elizabeth Grosz echoes this conclusion when she suggests evolution allows "a way of reconceptualizing the relations between the natural and the social, between the biological and the cultural, outside the dichotomous structure in which these terms are currently enmeshed."[15]

Evolution and its various mechanisms are indeed unique, because they allow us to see the histories and the complexities of how life is interwoven. Individuals and populations are condensations of this process, forming entangled *umwelten*.[16] Evolution is an enactment between and within species and their environments. Characteristics we are able to witness around us have emerged out of this very real process of matter-ing, and the examples of our material entanglements with all of life are as remarkable as they are vast and numerous. We are symbiotically riddled with microorganisms throughout our bodies such that we would not survive without them. The toxins in our bodies tell stories about where we were born and where we have been. Mothers have been found to retain DNA in their bodies from their children long after giving birth. We are composed of multitudes. These material entanglements, in which life is born in foliated patterns of difference, blooming out and folding back upon itself, are traced with evolution. And while selection processes are essentially "mindless," I intend to show how rhetoric manifests itself in these movements of life.

MINDS AND BEHAVIOR

Alex Parrish's 2014 *Adaptive Rhetoric: Evolution, Culture, and the Art of Persuasion* revisits Kennedy's theory of a general rhetoric with additional insights from evolutionary biology. Relying upon cognitive science and behavioral ecology, Parrish establishes a more thorough "biocultural paradigm" for investigating the origins and forces of rhetoric. He argues: "To better understand the art of persuasion, rhetoricians need to closely examine both cultural and biological influences on human behavior, which requires us to locate the evolved mind as the production point of behavior and the nexus where culture and evolution meet."[17] Mind and behavior are critical points of qualification when arguing for this adaptive rhetoric, for Parrish seeks to extend the classical understanding of rhetoric as a discursive, situated, and intentional art to animals with scientifically supported abilities to engage in just that. Therefore, such qualifying animals must demonstrate, among other things, 1) theory of mind as well as 2) rhetorical intention. That is, can a particular animal judge that a fellow animal has a mind that can *be persuaded*? And does this animal engage in an intentional behavior in order to persuade? These factors are, indeed, the most compelling in an argument for extending rhetoric to certain animals.

Theory of mind is an interesting concept within evolutionary psychology, and there are many animals that arguably demonstrate this ability, not least of which are dolphins and elephants—two species among others who can recognize themselves as individuals—as *self*—in mirror tests[18]. The scrub jay's food caching and recaching behavior is a striking (possible) example of theory of mind in corvids. Scrub jays seem to anticipate the likely thieving behavior of other jays and will return to and rebury food stashes they previously buried when in the presence of other jays.[19] Behavior like this, which seems to demonstrate knowledge of an other's thoughts, is compelling; but the problem is what gets lost when something like theory of mind, an admittedly significant concept, is selected as a threshold for definitions. It does not separate humans and animals, but it separates the "haves" of rhetorical ability from the "have-nots." For example, a study published about scrub jays' caching behavior in 2012 persuasively argues that such "sophisticated social cognition" as theory of mind is not a necessary cause of food recaching, citing mere stress as another possibility.[20] What, then, are we to make of the scrub jays' rhetorical abilities, given Parrish's definition? This study, which argues against theory of mind in scrub jays based on a witnessed behavior and its link to stress, can also be imagined to severely limit the rhetorical behaviors within the human species.[21] Perhaps a true rhetorical behavior (one limited to intention and theory of mind) would be limited based on the context in which it occurs (those initiated by stress being excluded), but then who decides the status of a behavior's intentionality and theory-of-mind enactment? Often we are unaware of even our own intentions.

Cuts and definitions are a necessary part of making sense of the world. Indeed, according to quantum physicist Karen Barad, such cuts constitute the fundamental units of reality from which we cannot be disentangled. And these decisions have important ethical implications; Barad describes an agential realism in which "ethicality is part of the fabric of the world."[22] It is essential, then, that we follow this reliance upon mind and intention to its inevitable ethical conclusions. Daniel Dennett is well known for his functionalist approach to theories of mind with insights from cognitive science. He is also cited by Parrish. As Cary Wolfe calls to our attention in *What is*

Posthumanism? Dennett's approach to animal minds is problematic when he becomes trapped in the same Cartesianism he seeks to reject. The ability to host representations of thoughts within a "Cartesian theater" of the mind from which intention and consciousness can arise results in a hierarchy of ability and capacity for suffering in which "different forms of being human in the world are rewritten … in terms of a homogenous Cartesian ideal," which is dire for humans who are not of the minded ideal, and in which "nonhuman beings, in all their diversity, are now rendered not as fully complete forms of life … but rather as diminished or crippled versions of the fantasy figure called the human."[23] Such a conclusion about *who can suffer* obviously has profound implications and illustrates how careful we need to be when making these cuts and exclusions. It is essential to investigate the assumptions responsible for these exclusions, since they have such a grounding effect on how we interpret reality and make meaning. As Cary Wolfe urges, "even though thinking about the consciousness, intelligence, and emotional and mental lives of nonhuman animals in terms of their linguistic abilities has historically been a crucial means for getting such questions on the table *at all*, it may not be the best way, and it certainly is not the only way, of approaching these questions."[24] Rather, perhaps we should be interested in animals for their "very different way[s] of being in the world that calls on us to rethink … what we mean by 'person,' 'mind,' 'consciousness'—that entire cluster of terms and the ethical implications that flow from them."[25]

What other way might we approach animals in rhetoric? In *Ambient Rhetoric: The Attunements of Rhetorical Being*, Thomas Rickert describes a rhetoric without prescribed borders that can encompass narrower versions "without in any way being limited to [them]."[26] Rickert's ambient rhetoric that is "dispersed, embodied, and embedded" and is "itself ontological" informs a humanimal rhetorics as well.[27] More restrictive definitions, that presume intent as a (human or animal) requirement, "no longer suffice to determine what is rhetoric and what is not."[28] The "nebulous concept" of mind as the origin of rhetoric similarly betrays the extent to which minds are embodied and dispersed (and entangled) in the world.[29] These entanglements extend not only to our ambient surroundings and other animals; Michelle Ballif suggests they extend to the dead and dying as well. Ballif explains how rhetoric, being an address between self and other—an address that is always already in mourning for the death of the other—constitutes rhetoric as "an ethical rather than epistemological enterprise."[30] Theory of Mind, intention, and—slipping further down this problematic slope, as Dennett's discriminations further illustrate—abilities of suffering limits the scope of what kinds of life matter to rhetorical studies.

Rhetorical theorists like Davis, Rickert, and Ballif become even more exciting for a humanimal approach to rhetoric when they are folded into fleshy contours of new material feminists like Karen Barad, Donna Haraway, Elizabeth Grosz, and Stacy Alaimo. Such theorists have returned to bodies (and their inherent biology or quantum properties) as dynamic and changing matter rather than determinant, static, and immutable, and they do so with the help of the biological and quantum sciences. These theories emphasize relationality, engagements, and entangled performances of materiality rather than individuals and their "separately determined properties."[31] Indeed, Barad argues that static qualities of an individual do not exist even at the most fundamental level of reality, and it is Barad's unique and radical insights from quantum physics that provide perhaps the most helpful departure from an approach like Dennett's, which relies upon reflection and representation in the mind for making meaning that matters in any ethical capacity. In its place, Barad puts forth the quantum phenomenon of diffraction as a methodology for engaging with the world. Rather than presuming a decisive cut—an abyssal space, if you will—between image and reality like reflection does (which necessitates representation of that reality), Barad explains how the wave patterns of diffraction create overlaps and superpositions—essentially "ontologically indeterminate states"—inherent to the wave-particle paradox that has made "the downfall of classical metaphysics explicit."[32] The special "weird" quality of quantum phenomena that throws classic Newtonian physics into an uproar *is* entanglement—and it is not something anyone can witness by standing outside, at a distance. We are each enfolded within these ontological states, cocreating reality as we participate in it. Entanglement denies the idea of static, unchanging individuals or states: "What often appears as separate entities (and separate sets of concerns) with sharp edges does not actually entail a relation of absolute exteriority at all."[33] Rather, the most primary ontological unit on a quantum level is this entangled phenomenon and its "*dynamic* relationality to the other."[34] Furthermore, while diffraction is useful in its attunement to difference, these diffractive apparatuses "highlight, exhibit, and make evident the entangled structure of the changing and contingent ontology of the world, including the ontology of knowing," for, as Barad explains, diffraction "is itself an entangled phenomenon."[35] In other words, quantum entanglement and the physical phenomenon of diffraction make manifest a reality in which all of matter engages in an ongoing, dynamic ontological relationality.

What does a methodology of diffraction look like when applied to rhetoric? What once appeared as clear and definitive boundaries suddenly become uncertain *by nature*. For example, Parrish offers a seemingly simple example of intention for rhetorical ability with the peacock. According to Parrish, peacocks may have large, brightly colored tails that participate in the persuasive courtship, but these feathers are not rhetorical until they are intentionally put into action with a characteristic mating behavior. In other words, the tail itself is not rhetorical until it is intentionally displayed for that purpose. But the methodology of diffraction leads me to propose that it is precisely here—in this complex evolutionary movement that shapes phenotypes with the dynamic materiality of genes, of ecologies, of peahen bodies and choices, of a temporal emergence in which these varied and diverse factors become entangled and changed—that we can witness a creative *becoming* of rhetoric. It is in the formation, the continuous becomings of bodies as they inter and intra-act, that we see a dynamic, adaptive, but also performative rhetoric of *agential* matter. To view these physical traits of biological bodies as static and immutable as they appear to our eyes, then, is deceiving. These morphologies and physiologies bear visible traces of the movement of rhetoric "in" evolution, as Kennedy suggests, and therefore within the fleshy contours and material folds of biological—rhetorical—life. Parrish does a great job of demonstrating a careful and thorough reading of rhetoric as adaptive. Intention and theory of mind are well supported with scientific study and well theorized as applicable to rhetoric in the capacity that Parrish frames it. However, I suggest that these convenient and clean distinctions are actually uneasy lines within which much more entangled rhetorical relationalities are at work. Parrish's adaptive rhetoric describes a behavioral ability arising from special kinds of mind. A humanimal rhetorics, with its diffractive methodology, reaches for more.

BODIES AND BODYMINDS

Referring to our human animal beings as *humanimal* is appropriately awkward, as it recalls our biological materiality, indicating the inherent difficulty in distinguishing between subject-object positions of human/animal. Humanimal also inherits its hybridity from a number of similar ambitious terms, such as Donna Haraway's well-known "natureculture" and Wendy Wheeler's "bodymind."[36] As Jane Bennett describes in *Vibrant Matter*, that which is frequently referred to as the object of exteriority, the environment, "is actually inside human minds and bodies" in such a way that we are "inextricably bound" ontologically, epistemologically, and ethically with all of matter.[37] Suddenly, our presumed human form, delineated by a smooth and constant exteriority of skin and flesh that holds the world—Nature or environment—at bay, is plunged back into the dark and nascent unknowability in which it first forms. Limbs become almost indiscernible as we strain to make out budding contours. Life divides and folds upon itself while being further enfolded within life. The humanimal exists within the mysterious darkness of these originary depths.

In "Natural Convers(at)ions: Or, What if Culture Was Really Nature All Along?" Vicki Kirby succinctly captures the issue of the human/animal, nature/culture, mind/body constellation of oppositions that constitutes the general concern of much of new material feminist theorists:

> The most important [assumption] is the assertion that humanness is profoundly unnatural. The abstracting technology of language, intelligence, and creative invention is separated from the body of the material world, indeed, from the material body of human animality. [...] If we translate the separation of culture from nature into the mind/body split, it seems that Cartesian subject can admit that s/he has a body (that *attaches* to the self), and yet s/he is somehow able to sustain the belief that *s/he is not this body*. This denial is necessary because to contest the latter and all its possible consequences would at least suggest that it might be in the nature of the biological body to argue, to reinvent, and rewrite itself—to cogitate.[38]

An emphasis on the body as active and inseparable from mind also finds overwhelming support from the biological sciences. Recent scholarship within the burgeoning subfield of biosemiotics similarly complicates any distinction between mind/body or mind/matter with some scholars even claiming body is itself synonymous with mind[39] to such an extent that Wendy Wheeler fuses the two into bodymind.[40]

The appropriateness of a hybrid term such as Wheeler's bodymind is illustrated by a species of brittlestar. *Ophiocoma wendti* is an invertebrate related to starfish. This remarkable organism has no brain and, therefore, no chance of receiving consideration for the coveted theory of mind. Instead, its body is covered in thousands of dome-shaped crystals that optimize and focus light onto its diffuse nervous system, ultimately creating an embodied compound eye with a 360-degree view of its surroundings.[41] According to Barad, "Brittlestars challenge not only disembodied epistemologies but also traditional, and indeed many nontraditional, notions of embodiment. Bodies are not situated *in* the world; they are part *of* the world."[42] For Barad, the brittlestar is an apt example of her agential realist philosophy and the onto-epistemo-ethical entanglements implied: "Parts of the world are always intra-acting with other parts of the world, and it is through specific intra-actions that a differential sense of being—with boundaries, properties, cause and effect—is enacted in the ongoing ebb and flow of agency."[43] The brittlestar's "ongoing reconfigurings of its bodily boundaries … through which the agential cut between 'self' and 'other' (e.g., 'surrounding environment') is differentially enacted"[44] suggests an ability Diane Davis identifies as the primary criterion with which rhetorical studies bestows upon the human the exclusive right of the rhetorically able: the ability of marking "I." Davis explains: "Rhetorical acts are held to be distinctive human acts inasmuch as they're grounded not so much in the power to speak (certain birds can do that) but in a more originary capability of self-reflexivity."[45] Davis goes on to argue that this ability is not simply situated with humans but is a "preoriginary rhetoricity" that exists as "the very condition for the identity and functioning of *any* living being."[46]

As Davis further states, "This rhetoricity cannot be innate because it cannot *not* be relational."[47] Symbiotic relationalities are continually occurring within any one "individual." These include everything from relatively benign whale barnacles to beneficial humanimal intestinal bacteria to the often horrific examples of parasitism, such as the crab parasite *Sacculina*, which turns its host into what Carl Zimmer calls a "walking corpse"—truly a foil of any easy distinction between life and death for the unfortunate crab.[48] The list of our material mutliplicities is endless.

As examples of a humanimal rhetorics that is "in" biological evolution—that is entangled, enfolded, invaginated in the becoming and enactment of life itself—I will next trace its movements in select entanglements of sex and symbiosis—relationalities that do much to confuse any clear distinctions between behaviors and bodies.

SEX AND SYMBIOSIS

Perhaps the most appropriate area for seeking to trace these dynamic rhetorical becomings in our entangled materiality is in the movement of DNA. This is where the magic happens, so to speak, as the very earliest form of this process is responsible for life's origins.[49] In *Nature's Nether Regions*, Menno Schilthuizen offers examples of the stunning diversity that exists to exchange genetic information in every biological organism, from bacteria, which swap DNA among themselves or get stray bits from their environment and then reproduce asexually by budding off little clones of themselves, to corals that "release their eggs and sperm into the waters and hope for the best."[50] But Schilthuizen explains how the "evolutionary play" of sexual reproduction makes genitalia in particular "the best body parts to illustrate the power of evolution."[51]

One striking example for witnessing this fascinating material relationality of sexual reproduction is in the genital evolution of ducks. In 2009, the rather strange and unexpected morphologies of duck genitalia made a big splash in popular culture. Ed Young explains Yale researcher Patricia Brennan's study of this extraordinary duck sexual reproduction in *National Geographic*:

> Many ducks form bonds between males and females that last for a whole mating season. But rival males often violently force themselves onto females. To gain the edge in these conflicts, drakes have evolved large corkscrew phalluses, lined with ridges and backward-pointing spines, which allow them to deposit their sperm further into a female than their rivals.[52]

While these "corkscrew" duck penises are indeed strange, the duck vaginas that have co-evolved with these penises are even more interesting, and this is the focus of Brennan's research.[53] As "organic chastity belts," duck vaginas are described as coiling clockwise (in the opposite direction of the penis) and being something like a maze in order to make successful fertilization more difficult for the aggressive rival ducks.[54] Indeed, Schilthuizen explains the results of Brennan et al.'s research on sixteen duck species, "which showed that the complexity of penis and vagina went hand in hand: whenever the penis was large and curly, so was the vagina," and "the species in which rape was rife were also the ones with the most complex genitalia."[55] The evolution of duck genitalia is truly a story of at least "competing and cooperating interests" such as those Parrish describes as central to his requirement for intention, though the ducks' interests are also manifesting morphologically rather than purely behaviorally.

Female duck behavior also works in concert with the convoluted morphology of her genitalia: if sex with a particular drake is unwanted, she does not do him any favors in navigating her labyrinthine vagina like she does in consensual encounters. If a female wishes to engage in a sexual encounter, she will "strike a pose that signals her receptiveness, keeping her body level and lifting her tail feathers high" while further "repeatedly contract[ing] the walls of her genital tract, relaxing them for long enough for favoured suitors to achieve full penetration."[56] In contrast, and perhaps more pointedly, Schilthuizen further explains: "by flexing the muscles in her vagina wall, the female would be able to block a rapist male from properly inflating his countercoiled penis into her vagina."[57] The co-evolution in this, what Carl Zimmer calls, "war of the sexes" in ducks is an extraordinary example of the difficulty in drawing a line between persuasive, intentional behavior and persuasive physiology.[58] The female duck's behavior in thwarting or helping her suitors cannot be severed from the contours and contractions of her vagina. The gradual co-evolution of both male and female ducks' bodyminds is a delicate and complex movement of entangled persuasion.

If we continue to look into the bizarre sexual entanglements manifested in animal bodyminds as condensations of evolutionary enactments, there is much to see. Sperm is particularly fascinating, as it has been found to contain a surprising number of proteins in large quantities. In a creature as small as the banana fly, an amazing 133 seminal proteins have been found that apparently have a dramatic effect on females in "a kind of neuropsychological manipulation" by "hijack[ing] a female's hormonal system by shutting down her sex drive."[59] Humanimals are not exempt from this startling behavior, as sperm cells have been found floating around outside of expected areas as well.[60] How these wandering sperm cells may affect women in ways other than fertilization is not fully understood, though past studies suggest that women who have unprotected sex have fewer symptoms of depression and, while pregnant, are also less likely to develop preeclampsia.[61] Humanimal male sperm cells have also been found to contain surprising hormone compounds that both stimulate ovulation processes and also maintain pregnancy. Scientists theorize these seminal compounds may be a "counter-strategy to concealed ovulation in women," to which I will return shortly.[62]

Perhaps even more fascinating than sperm and its mysterious components is how female bodyminds *react to* the sperm or to sex itself, from full out rejection of sperm to sperm storage to the developmental process of young and all the strange divergences in between. In mammals, including humanimals, females have been found to retain sperm much more successfully following an orgasm, which seems to create a low pressure vacuum in the uterus.[63] Some female species have also been found to expel some or all of a sperm "deposit" while others may hang onto sperm for several months or even *years*, which occurs in species as different as snakes and ants.[64] Females like various insects with "sperm pouches" and turtles with "countless miniscule sperm-harboring tubes all along their oviduct" can even store sperm from *multiple* males for future use.[65] These are examples of what senior staff scientist at the Smithsonian and behavioral ecologist William G. Eberhard has coined as "cryptic female choice." The forms cryptic female choice can take—like the "Bruce Effect" of induced abortion in pregnant mice when a new male is smelled or touched, or in pregnant geladas when a new male becomes alpha—are staggering in diversity and can be enacted at any stage of mating.[66]

The chemical communication that enfolds bodies and behaviors in this process of evolutionary becoming is complex and mysterious. Even in humanimals, the processes are not completely understood. An intriguing example of the degree to which chemical communication can influence something like attraction is demonstrated in a scientific article in the journal *Evolution and Human Behavior* titled "Ovulatory Cycle Effects on Tip Earnings By Lap Dancers: Economic Evidence of Estrus?" The title pretty much says it all. As the authors explain, it has been theorized that estrus was lost at some point in humanimal evolution. Estrus, or "increased female sexual receptivity, proceptivity, selectivity, and attractiveness" immediately preceding ovulation during which time a female is most likely to conceive, has been found in most mammals, but little evidence indicates its presence in female humanimals, perhaps, as scientists have speculated, to "hide" ovulation.[67] Hidden ovulation may be evolutionarily advantageous if it encourages mates to stick around longer, thus providing "male provisioning and paternal care" for longer periods of time.[68] However, Geoffrey Miller et al. have uncovered support for subtle indications of estrus in normally ovulating women (well, at least subtler than "bright red swollen behinds" in female chimps).[69] In this provocative study, researchers measured the direct effect of perceived female attractiveness on their nightly earnings and plotted this across menstrual cycle schedules. Lap dancers are particularly appropriate for a study of this kind, as the researchers explain, because of the intimate multisensory contact between the women and men:

> In each lap dance … the topless female dancer sits on the man's lap, either facing away from him (to display her buttocks, back, and hair) or facing him (either leaning back to display her breasts, and to make conversation and eye contact, or leaning forward to whisper in his ear). Lap dances typically entail intense rhythmic contact between the female pelvis and the clothed male penis.[70]

For those otherwise unfamiliar, the description of this sexually charged phenomenon sounds much like a mating ritual one would expect to find in other animals. Clearly any kind of unregistered chemical persuasion, if it exists, could be picked up by males' bodyminds in such an intensely intimate scenario, which lasts anywhere from three minutes to over an hour. Results of the study indicate that some type of estrus-induced persuasion probably exists. Not only did women in the estrus phase of their cycle earn significantly more tips per shift, normally cycling women made more overall than noncycling women who were taking some form of chemical contraception.[71] Therefore, estrus in humanimals may exist with subtle persuasive cues about when fertilization is most likely to occur.

In the evolutionary becomings of sexual reproduction, bodies and behavior are mutually enfolded in a complex rhetorical dance, just as they are in other movements of biological evolution. Rather than simply relying on observable behaviors of creatures that typically come to mind(/body) with the all-encompassing label "Animal" in order to make a decisive cut between beings and abilities, it is important to examine the strange, cryptic corridors of embodied becomings in places humanimals may not typically look.[72] But even the observable displays of sexual performativity illustrate the complexities that make mind and body, human and animal, intentional rhetoric and unintentional behavior endlessly problematic for the purposes of definition.

CONCLUSION: A RETURN TO ORIGINS

As an example of symbiosis as well as material origins, the ontological status of a humanimal mother and fetus[73] is one of intimate entanglement and mutual chemical persuasion. A developing fetus not only consists of the biological material of its parents, it is suffused with maternal chemical communication, nourishment, and protection. The contours of a fetus unfold and invaginate as nascent life emerges: budding limbs and tail, dark eye spots, folds of skin, the circulation of blood separate from maternal blood. In humanimal pregnancy and birth, a woman's placenta is the material threshold of becoming that separates as well as joins. The placenta transfers critical compounds from the mother's blood that feed and promote this flourishing of life within. The placenta is an ontological embrace.

A developing fetus is in dynamic relation with the mother as well, rather than simply passive cargo. An embryo helps initiate its own implantation in the mother's uterine wall, blocks chemicals that signal the shedding of the nutrient-rich uterine lining that is prepared throughout each menstrual cycle, creates chemical compounds that signal for continued production of the uterine lining, suppresses maternal immune reaction, and even "contributes signals that promote the onset of labor."[74] Following birth, this newly emerged life continues biochemical dances within its tissues, as in the downregulation of his or her own immune system to welcome in a vast array of the world's microbiota, beginning a lifelong symbiotic relationality in what Kramer and Bressan call "an ancient, stable, fundamental component of the human superorganism."[75] The elegant maneuvering of these chemical processes have emerged prior to "human"—just as with sex and symbiosis, the ability to create another life from one's own materiality in a dynamic biological relation connects us to, entangles us with, *makes us* animal: humanimal. I argue it is in this ongoing flux of movements, of the becoming-entanglement in which matter and energy are expressed—this tunneling of life into originary and symbiotic alignments—that rhetoric suffuses these negotiations of bodies with dynamic relationality.

NOTES

1. In *Bodily Natures: Science, Environment, and the Material Self* (Bloomington: Indiana University Press, 2010), Stacy Alaimo creates the term "transcorporeal" to describe her feminist new material feminism by indicating a thorough permeability and shared materiality across otherwise presumed bodily frontiers.

2. "Agential reality" arises from Karen Barad, most thoroughly explained in *Meeting the Universe Halfway: Quantum Physics and the Entanglement of Matter and Meaning* (Durham: Duke University Press, 2007). Agential reality suggests an ongoing, dynamic relationality of matter that is more thoroughly explained later in the present chapter.

3. John Muckelbauer, "Domesticating Animal Theory," *Philosophy and Rhetoric* 44.1 (2011): 97.

4. See Elizabeth Kolbert's *The Sixth Extinction: An Unnatural History* (New York: Henry Holt & Company, 2014).

5. Dominick LaCapra, *History and Its Limits: Human, Animal Violence* (Ithaca: Cornell University Press, 2009): 157.

6. Ibid.

7. Ibid., 150.

8. Jacques Derrida, *The Animal That Therefore I Am*. Ed. Marie-Louise Mallet. Trans. David Willis (New York: Fordham University Press, 2008): 30–31.

9. Ibid., 3.

10. Ibid., 26.

11. A more precise statistic for animals killed annually is difficult to determine, with numbers ranging from 9 billion "land" animals to 56 billion "farm" animals to just simply 150 billion animals slaughtered. According to the US Department of Agriculture (USDA), nearly 10.2 billion land animals were killed in the United States in 2010. USDA National Agricultural Statistics Service, "Livestock Slaughter 2010 Summary."

12. Diane Davis, "Autozoography: Notes Toward a Rhetoricity of the Living," *Philosophy and Rhetoric* 44.1 (2011): 535.

13. George Kennedy, "A Hoot in the Dark: The Evolution of General Rhetoric," *Philosophy and Rhetoric* 25.1 (1992): 13.

14. Virginia Morrell, *Animal Wise: How We Know Animals Think and Feel* (New York: Crown Publishers, 2013): 10.

15. Elizabeth Grosz, "Darwin and Feminism." In *Material Feminisms*. Eds. Stacy Alaimo and Susan Hekman (Bloomington: Indiana University Press, 2008): 43.

16. Von Uexküll's *umwelt* characterizes the entangled ontologies of organisms and how each *umwelt* cuts differently. This is similar to Barad's entangled phenomena. Grosz writes of Uexküll: "he advocates an extreme perspectivism in which objects are not autonomous or independent sets of qualities and quantities, but opportunities for engagement that offer themselves in particular ways to particular organs and remain otherwise indiscernible." See Elizabeth Grosz, *Chaos, Territory, Art: Deleuze and the Framing of the Earth* (New York: Columbia University Press, 2008): 41. In Uexküll's words: "Every object becomes something completely different on entering a different Umwelt." See Jacob von Uexküll, *A Foray into the Worlds of Animals and Humans* (Minneapolis: University of Minnesota Press, 2012): 108.

17. Alex Parrish, *Adaptive Rhetoric: Evolution, Culture, and the Art of Persuasion* (New York: Routledge, 2014): 2.

18. Morrell, *Animal Wise*.

19. Elske van der Vaart, Rineke Verbrugge, and Charlotte K. Hemelrijk, "Corvid Re-Caching Without 'Theory of Mind': A Model," *PLOS ONE* 7.3 (2012), journals.plos.org/plosone/article?id=10.1371/journal.pone.0032904.

20. Ibid.

21. Following this trajectory, it becomes increasingly clear how such psychological criteria for rhetoric are themselves problematic. Even in humans, it may not be clear when "theory of mind" is being enacted. Such criteria would most certainly foreclose the possibility of ambient rhetoric, more generally, as well as much visual rhetoric. This further illuminates the differences between an adaptive rhetoric and a humanimal rhetorics that affirms persuasion created and received by all the senses.

22. Barad, *Meeting the Universe Halfway*, 182.
23. Cary Wolfe, *What is Posthumanism?* (Minneapolis: University of Minnesota Press, 2009): 45.
24. Ibid., 46–47.
25. Ibid., 47.
26. Thomas Rickert, *Ambient Rhetoric* (Pittsburgh: University of Pittsburgh Press, 2013): 35.
27. Ibid., xv.
28. Ibid., 36.
29. Ibid., 40–43.
30. Michelle Ballif, "Regarding the Dead," *Philosophy and Rhetoric* 47.4 (2014): 466.
31. Barad, *Meeting the Universe Halfway*, 55.
32. Ibid., 265, 72.
33. Ibid., 93.
34. Ibid.
35. Ibid., 73.
36. Wendy Wheeler, *The Whole Creature: Complexity, Biosemiotics and the Evolution of Culture* (London: Lawrence and Wishart, 2006): 4.
37. Jane Bennett, *Vibrant Matter: A Political Ecology of Things* (Durham: Duke University Press, 2010): 116.
38. Vicki Kirby, "Natural Convers(at)ions: Or, What if Culture Was Really Nature All Along?" In *Material Feminisms*. Eds. Stacy Alaimo and Susan Hekman (Bloomington: Indiana University Press, 2008): 220–221, her emphasis.
39. The academic journal *Biosemiotics* 6.2 is a special issue on origins of mind. See especially Yoshimi Kawade, "The Origin of Mind: The Mind–Matter Continuity Thesis," 367–378; and Mario Villalobos, "Autopoiesis, Life, Mind, and Cognition: Bases for a Proper Naturalistic Continuity," 379–391.
40. Wheeler, *The Whole Creature,* 18.
41. Ibid., 369–371.
42. Ibid., 376.
43. Ibid., 338.
44. Ibid., 376.
45. Davis, "Autozoography," 535.
46. Ibid., 546.
47. Ibid., 548, her emphasis.
48. Carl Zimmer, *Parasite Rex* (New York: The Free Press, 2000): 159.
49. The evolution of life's origins is still a mystery. Many interesting developments in this theoretical research are currently underway, including geologist Mike Russell's idea that a natural proton gradient created near hydrothermal vents in the ocean created the originary scene of cellular chemiosmosis. See Tim Requarth's "Our Chemical Eden," *Aeon* (January 11, 2016).
50. Menno Schilthuizen, *Nature's Nether Regions: What the Sex Lives of Bugs, Birds, and Beasts Tell Us About Evolution, Biodiversity, and Ourselves* (New York: Penguin, 2014): 10.

51. Ibid., 5. Schilthuizen explains that there is a vast array of diversity in how, precisely, DNA gets shuffled around. Genitalia are organs used for the very specific process of "internal fertilization," 18.

52. Ed Young, "Ballistic Penises and Corkscrew Vaginas—The Sexual Battles of Ducks," *National Geographic* (December 22, 2009): Accessed June 2014.

53. Prior to Brennan's research, the majority of biologists and researchers of duck reproduction had only focused on the more observable and extraordinary male genitalia. This brings up an important bias in scientific research (and all the assumptions that stem from it) for focusing almost exclusively on males. For further discussion and research of this phenomenon, see Erika L. Milam's *Looking for a Few Good Males: Female Choice in Evolutionary Biology* (John Hopkins University Press, 2011).

54. Young, "Ballistic Penises and Corkscrew Vaginas."

55. Schilthuizen, *Nature's Nether Regions*, 128.

56. Young, "Ballistic Penises and Corkscrew Vaginas."

57. Schilthuizen, *Nature's Nether Regions*, 128.

58. Carl Zimmer, "In Ducks, War of the Sexes Plays Out in the Evolution of Genitalia," *New York Times: Science* (May 1, 2007): Accessed June 2014.

59. Schilthuizen, *Nature's Nether Regions*, 146.

60. Ibid., 134.

61. Ibid., 147–148.

62. Alison Motluk, "The Secret Life of Semen," *The New Scientist*, Press Release (August 2, 2006).

63. Schilthuizen, *Nature's Nether Regions*, 79.

64. Ibid., 60, 82–83.

65. Ibid., 83.

66. Ibid., 87.

67. Geoffrey Miller, Joshua M. Tybur, and Brent D. Jordan, "Ovulatory Cycle Effects on Tip Earnings by Lap Dancers: Economic Evidence for Human Estrus?" *Evolution and Human Behavior* 28 (2007): 375.

68. Ibid., 375.

69. Motluk, "The Secret Life of Semen."

70. Miller, Tybur, Jordan, "Ovulatory Cycle Effects," 376–377.

71. Ibid., 379.

72. Derrida comments at length on the absurd singular of "Animal" in *The Animal That Therefore I Am*: "Confined within this catch-all concept, within this vast encampment of the animal, in this general singular, within the strict enclosure of this definite article … are all the living things that man does not recognize as his fellows, his neighbors or his brothers. And that is so despite the infinite space that separates the lizard from the dog, the protozoon from the dolphin, the shark from the lamb, the parrot from the chimpanzee, the camel from the eagle, the squirrel from the tiger, the elephant from the cat, the ant from the silkworm, or the hedgehog from the echidna," 34.

73. In embryology, the developing humanimal is initially referred to as an "embryo" until the eleventh week, after which it finally gains the designation of fetus.

74. Gerard M. Dileo, "Anatomy of a Fetus: The Placenta," *Babble* (2011).

75. Kramer and Bressan, "Humans as Superorganisms," 468.

BIBLIOGRAPHY

Alaimo, Stacy. *Bodily Natures: Science, Environment, and the Material Self.* Bloomington: Indiana University Press, 2010.

Alaimo, Stacy and Susan Hekman. *Material Feminisms.* Bloomington: Indiana University Press, 2008.

Ballif, Michelle. "Regarding the Dead." *Philosophy and Rhetoric* 47.4 (2014): 455–471.

Barad, Karen. *Meeting the Universe Halfway: Quantum Physics and the Entanglement of Matter and Meaning.* Durham: Duke University Press, 2007.

Bennett, Jane. *Vibrant Matter: A Political Ecology of Things.* Durham: Duke University Press, 2010.

Davis, Diane. "Autozoography: Notes Toward a Rhetoricity of the Living." *Philosophy and Rhetoric* 47.4 (2014): 533–553.

Derrida, Jacques. *The Animal That Therefore I Am*, edited by Marie-Louise Mallet. Translated by David Willis. New York: Fordham University Press, 2008.

Dileo, Gerard M. "Anatomy of a Fetus: The Placenta." *Babble* (2011).

Grosz, Elizabeth. *Chaos, Territory, Art: Deleuze and the Framing of the Earth.* New York: Columbia University Press, 2008.

———. "Darwin and Feminism." In *Material Feminisms*, edited by Stacy Alaimo and Susan Hekman, 23–51. Bloomington: Indiana University Press, 2008.

———. *Volatile Bodies: Toward a Corporeal Feminism.* Bloomington: Indiana University Press, 1994.

Haraway, Donna. *When Species Meet.* Minneapolis: University of Minnesota Press, 2007.

Kapil, Bhanu. *Humanimal: A Project for Future Children.* Berkeley: Kelsey Street Press, 2009.

Kawade, Yoshimi. "The Origin of Mind: The Mind–Matter Continuity Thesis." *Biosemiotics* 6.3 (2013): 367–378.

Kirby, Vicki. "Natural Convers(at)ions: Or, What if Culture Was Really Nature All Along?" In *Material Feminisms*, edited by Stacy Alaimo and Susan Hekman, 214–236. Bloomington: Indiana University Press, 2008.

Kennedy, George. "A Hoot in the Dark: The Evolution of General Rhetoric." *Philosophy and Rhetoric* 25.1 (1992): 1–21.

Kolbert, Elizabeth. *The Sixth Extinction: An Unnatural History.* New York: Henry Holt & Company, 2014.

Kramer, Peter and Paola Bressan. "Humans as Superorganisms: How Microbes, Viruses, Imprinted Genes, and Other Selfish Entities Shape Our Behavior." *Perspectives on Psychological Science* 10.4 (2015): 464–481.

LaCapra, Dominick. *History and Its Limits: Human, Animal Violence.* Ithaca: Cornell University Press, 2009.

Milam, Erika L. *Looking for a Few Good Males: Female Choice in Evolutionary Biology.* John Hopkins University Press, 2011.

Miller, Geoffrey, Joshua M. Tybur, and Brent D. Jordan, "Ovulatory Cycle Effects on Tip Earnings by Lap Dancers: Economic Evidence for Human Estrus?" *Evolution and Human Behavior* 28 (2007): 375–381.

Morrell, Virginia. *Animal Wise: How We Know Animals Think and Feel*. New York: Crown Publishers, 2013.
Motluk, Alison. "The Secret Life of Semen," *The New Scientist*, Press Release (August 2, 2006).
Muckelbauer, John. "Domesticating Animal Theory." *Philosophy and Rhetoric* 44.1 (2011): 95–100.
Parrish, Alex. *Adaptive Rhetoric: Evolution, Culture, and the Art of Persuasion*. New York: Routledge, 2014.
Requarth, Tim. "Our Chemical Eden." *Aeon* (January 11, 2016).
Rickert, Thomas. *Ambient Rhetoric*. Pittsburgh: University of Pittsburgh Press, 2013.
Schilthuizen, Menno. *Nature's Nether Regions: What the Sex Lives of Bugs, Birds, and Beasts Tell Us About Evolution, Biodiversity, and Ourselves*. New York: Penguin, 2014.
USDA National Agricultural Statistics Service. "Livestock Slaughter 2010 Summary."
Van der Vaart, Elske, Rineke Verbrugge, and Charlotte K. Hemelrijk, "Corvid Re-Caching Without 'Theory of Mind': A Model." *PLoS One* 7.3 (2012): e32904.
Villalobos, Mario. "Autopoiesis, Life, Mind, and Cognition: Bases for a Proper Naturalistic Continuity." *Biosemiotics* 6.3 (2013): 379–391.
Von Uexküll, Jacob. *A Foray into the Worlds of Animals and Humans*. Minneapolis: University of Minnesota Press, 2012.
Wheeler, Wendy. *The Whole Creature: Complexity, Biosemiotics and the Evolution of Culture*. London: Lawrence and Wishart, 2006.
Wolfe, Cary. *What is Posthumanism?* Minneapolis: University of Minnesota Press, 2009.
Young, Ed. "Ballistic Penises and Corkscrew Vaginas—The Sexual Battles of Ducks." *National Geographic* (December 22, 2009).
Zimmer, Carl. "In Ducks, War of the Sexes Plays Out in the Evolution of Genitalia." *New York Times: Science* (May 1, 2007).
———. *Parasite Rex*. New York: The Free Press, 2000.

Chapter 3

Let's Listen with Our Feet

Animals, Neurodivergence, Vulnerability, and Haptic Rhetoricity

Kelin Loe

The "anthro" of anthropocentrism isn't open to all humanity equally.[1] When a human animal is being anthropocentric, we can't assume that the line he draws between human and nonhuman animals includes those human animals other to him. Consider the complexly woven traces the label "dumb" historically leaves behind. Speechless animals. Disabled humans. A history of dehumanizing racial groups through claims of a lack of intelligence. The combination of these forces acting against the life of human and nonhuman animals every time "dumb" is reiterated.

In *The Animal That Therefore I Am*, Jacques Derrida shows the history of Aristotle and continental philosophers (Descartes, Kant, Levinas, Lacan, and Heidegger) defining "human" against "animal." To be human is to think, reason, respond, use language, to recognize a self. To be animal is to operate on instinct, react, be a dumb automaton with no sense of self. And, of course, we know that history to involve the image of the human as an individual, able-bodied, white-European, monied, middle-aged man who owns property. As Agamben and others have shown, accompanying this history is a pattern of animalizing[2] human animals, degrading the life—from breath to life potential—of another human animal for the sake of those human animals in power.[3] Animalization is a justification of oppression. I am working toward habilitating the animalized being by changing the status quo to bring this being the conditions to live with health and full potential—and the societal recognition that this being has the right to do so. I use the word "habilitate" because we can't *re*habilitate, *re*store, *re*constitute, *re*vitalize those that haven't yet been consistently habilitated, stored, constituted, vitalized in their own right in Western imaginations. In my work, *to habilitate the animal is to habilitate the animalized in tandem.* To fight against this delineating degradation is to radically expand white-Western, normative, hegemonic notions of ontology—to pull them open wide enough to involve the experiences of all living beings, whether we can understand these experiences consciously, physically, or not at all. As Derrida concludes, this means pluralizing ontologies,[4] taking up what is outside of official consciousness, what is not only difficult to articulate but inarticulable. This means theorizing extralinguistic rhetorics, rhetorics of relation and sensation.[5]

What I propose here is a beginning to resisting anthropocentrism in order to habilitate both nonhuman and animalized human animals through rhetorical theories. It is a presentation of what I call haptic rhetoricity. I take "haptics" to mean sensations involving touch and feeling originating both inside and outside the skin. I'm following Diane Davis's notion of rhetoricity, the "affect*ability* or persuad*ability*" that comes with "the *exposedness* of corporeal existence."[6] Davis's rhetoricity takes root in Derrida's rejection of the foundational anthropocentrism of continental philosophy. I find that haptics can be a vocabulary for vulnerability, the experience of the exposedness that constitutes rhetoricity. If haptics are extralinguistic, what grounds can rhetorical theorists stand on to understand them? I will explore the communicative potentials of haptics in the bodies of animals (nonhuman and neurodivergent human animals). Through reading the work of Helen Keller and the research of both neurotypical and neurodivergent animal scientists, rhetorical theorists, and cultural historians, I've sifted haptics into four modes: thermodynamics, texture-pressure, vibration, and pain. These modes each provide haptic vocabularies for the vulnerable experience of rhetoricity. Before exploring the haptic modes, I will show the anti-anthropocentric, anti-ableist, and sensory roots of Davis's rhetoricity.

ANIMAL RHETORICITY

With so much work in the echo of George Kennedy's rhetoric-as-energy, rhetorical theories have clearly rounded the animal turn. And so far, recent work has focused on what rhetoric makes nonhuman animals *able* to do, such as the ability of a nonhuman animal to intentionally communicate[7] and the ability to participate in the rhetorics of human-designed spaces.[8] These approaches place an onus on *ability* to habilitate a being in rhetoric, which I consider an ablest and anthropocentric project. As Cynthia Lewiecki-Wilson works to incorporate disabled rhetors who cannot engage in verbal discourse, watch her draw this line: "we often demand some verbal response from an Other as proof of their humanness."[9] Humanity lines up behind the verbal line. If you are not *able* to engage, your humanness is at stake, and to be nonhuman, of course, is to lose everything. Jenell Johnson points out that "personhood has been defined by ability," causing nonhuman animals, disabled people, children, and seniors to lose their human status.[10] When Cary Wolfe calls Derrida's *The Animal That Therefore I Am* the most important event in the history of animal studies, he points out that Derrida finds this pattern of defining animal through ability anthropocentric.[11]

If we consider the divide between human and nonhuman animals through ability, we only, as Derrida says, breed questions "concerning *power* ... a power that consists in having such and such a faculty, thus such and such a capability, as an essential attribute."[12] This type of questioning continues to foist upon a being the labor of proving its abilities, particularly language. It constructs the possibility of deficit, allowing a being to be animalizable. In the nineteenth century, Jeremy Bentham asks *can animals suffer?* and Derrida flips the question.[13] No longer are we hunting for abilities to determine who counts. Derrida rewords it: "'Can they *not be able?*' And what of this inability [impouvoir]? What of the vulnerability felt on the basis of this inability?"[14] The ability to be unable, the power to be powerless—vulnerability—is what human and nonhuman animals share.

Diane Davis shows the roots of rhetoric in vulnerability. Davis theorizes that rhetoricity is "the already relational condition for *any* living being who presumes to recognize and refer to him- or herself," and she includes nonhuman animals in those presuming.[15] While this presumption to recognize oneself seems like an essential ability, Davis complicates it. Rhetoricity is not "innate because it cannot *not* be relational."[16] You, a being, cannot recognize yourself without first understanding that there is something *in relation to you*. This relationality is "underivable" and "irreducible,"[17] and it is rooted in mortal, vulnerable corporeality. It isn't a "specific genetic make up," but the general "exposedness of corporeal existence"—vulnerability—which allows for rhetoricity.[18] This is where Davis pulls open rhetoricity to allow nonhuman animals to participate: "To be affectable, persuadable," to be implicated in rhetoricity, "is to be always already affected, persuaded, which means: always already *responsive*."[19] To her, in pointed opposition to Descartes, because nonhuman animals are vulnerable, affectable, persuadable, they are always already *responding*, not simply reacting. Though she uses the words affect*able* and persuad*able*, her emphasis isn't on ability, but on the "always already" nature of being affected or persuaded. It isn't your ability to be responding, it's that you are already.

Rhetoricity works at the level of the trace, a relational experience anterior to language. To recognize myself is to feel something other in relation to me—this feeling is Derrida's notion of the trace.[20] Without encountering that "trace of differentiation," there would be "no need or possibility for self-reference."[21] Davis explains, the trace is the "entwinement of the other in the same," the sense of otherness that builds and marks the self. Traces are stimuli that indicate the borders of what I recognize as myself. To "mark" myself, my difference, "off from the other ..., [I] must both leave and trace traces, address and respond."[22] To recognize and show your difference from an other outside yourself, you feel, follow, and leave traces. To feel a trace is to be vulnerable to danger: "the trace is…what delivers me over 'with no possible defense' to the other: to whomever or whatever might 'do something with me and my remains. ...'"[23] By leaving a trace, you develop pathways to yourself that others can follow. The vulnerability here is multivalent: you feel the trace, you trace traces, another being feels your trace, another being traces you, you are at the mercy of the tracer, another tracer is at the mercy of you. To feel a trace must be simultaneously thrilling, camaraderous, fulfilling, dangerous, isolating, and terrifying.

If you are imagining nonhuman hunting in Davis and Derrida's depiction of the trace, know Derrida does too, and he paints it richly with sensation. He even interrupts himself to ask "why this zone of sensibility," smell specifically, "is so neglected ... in philosophy and in the arts."[24] I extend his question to that of the "lower senses" (olfactory, gustatory, haptic) in general below, and in my daily life. I manage a set of neurodivergences, one of which is sensory acuity. It is the experience of broader and more specific sensory awareness than the neurotypical human animal. The "lower senses" are louder for me. So, I wonder: can I physically sense rhetoricity? While Davis theorizes the rhetoricity of nonhuman animals through the vulnerability-based, non-ableist notion of the trace, the larger conversation hasn't yet caught up with her. I'd like to join her in theorizing rhetoricity, and to bring rhetoricity into the realm of sensation.

I ask, particularly with the trace, what does it feel like? Derrida asks, "the animal in whose tracks therefore I am (following), and who picks up traces, who is it?"[25] Derrida says to follow, to be after, to be alongside the animal "expresses a certain order of *being-huddled-together*," "the being-pressed, ...being-under-pressure, compressed, impressed, repressed, pressed-against according to" the present structures of power. He points out that the words *près* (near, close) and *après* (after) derive from *pressu*, etymologically joining the notion of pressing against with the notion of chasing after, following after.[26] For Derrida, pressure, a haptic mode, circulates traces.

Despite this image of pressure-organized huddle, I'm reminded of a torturous lack of a huddle, something Sgt. Bowe Bergdahl says in an interview with documentary filmmaker Mark Boal.[27] The Taliban held Bergdahl for nearly five years, largely in isolation (both physically forced and due to the Pashto-English language barrier). In their conversation, Bergdahl describes the specific pain of isolation, in a lightless, basement room. He didn't speak the language of his captors, and making noise wasn't worth the risk. Bergdahl says: "I would wake up not even remembering, like, what I was."

He continues: "I couldn't see my hands, I couldn't do anything. The only thing I could do is, like, touch my face." Being *held* in isolated captivity (an oxymoron humming with the same intense pain, proximity, and distance in the discussions of haptics below), he has only one self-referential affordance, and it is haptic. When he mentions it, he switches from past to present tense. This change emphasizes the intensity, the corporeality, the already, and the *haptic* nature of the trace. The vulnerability of rhetoricity and the pressures of the traces that constitute it—these all point to haptics being a primary mode of animal rhetoricity.

The physicality of haptics resonates with the trace. We shut eyes and mouths. We plug ears and noses. We hold our breath. But we cannot occlude the skin and sensations of the internal organs—we are *always* sensing through them. And the highly complex capacity of touch is nearly impossible to lose entirely, making it the sense most commonly shared in human bodies.[28] Haptic beings are vulnerable to thermodynamics, texture-pressure, vibration, and pain *always*. As Helen Keller reminds us, people who can see and hear forget the rest: "They forget that my whole body is alive to the conditions about me. The rumble and roar of the city smite the nerves of my face, and I feel the ceaseless tramp of an unseen multitude."[29] As with feeling the trace, to feel a haptic stimulant is multivalent: when I touch my hands palm-to-palm, I feel pressure, apply pressure, receive pressure.

Studying extralinguistic sensation, particularly haptic sensation, is an act of resistance, and studying it is my conscious choice to resist current power structures. Sensation lingers in Descartes' problematic line between human and nonhuman animal, and touch plays a unique role in oppression-by-animalization. One way of denigrating both human and nonhuman bodies is to associate them with what was (and is) thought of as the so-called "lower senses": smell, taste, and touch. Human societies have organized around the so-called "higher senses": sight and sound. (Perhaps this answers Derrida's question about the lack of sensory consideration in the humanities). Nineteenth-century Europe saw the crystallization of this hierarchy. It is perhaps best put in the (horrible) terms of a historian (of the time) who created a "sensory scale of 'races.'" The European man at the top: the "'eye-man'"; the African man at the bottom: the "'skin-man.'" Societies that operated through touch were thought to lack prowess in thought.[30] Due to their relationship with haptics, disabled individuals, women, and peasants were thought of as "base" or inferior.[31] Touch was also deeply associated with the unthinking animal. For any society to "achieve respectability," they had to rise above the "'animal' life of the body."[32] To associate a human animal with touch was to associate them with the lower life form of the "brute." The cultural understanding of the touch-centered, unthinking, unfeeling nonhuman animal was so powerful that it was used to reinforce oppression based on race, class, gender, and disability. And this hierarchy of the senses remains intact today, as these forces of oppression remain intact today.[33]

In our verbal, "higher"-sensing, neurotypically-human-designed world, how can we understand haptics? "Skin" does not have a term for passively receiving information; so, can we *listen* to haptics? Auditory information is vibratory, after all. To flip the question away from ability: what is the rhetorical nature of haptic sensation? By way of listening with our feet, I offer my grasp of four haptic modes: thermodynamics, texture-pressure, vibration, and pain. I find haptic rhetoricity to be a multivalent experience and haptics to be an infrastructure of circulation. By examining haptic modes of rhetoricity, we hold rhetorical theories accountable for what puts the *vulnus*, Latin for "wound," in "vulnerable." We also necessarily become accountable for habilitating the animalized—that degraded being for which we need a better word, that degradation that many of us experience—through resisting the hierarchy of sensation constantly implemented by human animals in power.

HAPTIC MODES: THERMODYNAMICS

When Lynn Worsham and Gary Olson interview autistic animal scientist Temple Grandin, they overlook the haptic mode of thermodynamics. In their introduction, they reflect on Grandin's discussion of emotion, language, and haptics. She reminds them: "physical touch may be our first language, one that is essentially an emotional language that ideally soothes and communicates reassurance and care and thereby reinforces social bonding." They provide moments in Grandin's research where she shows animals responding to the pressure of human touch. They reference the work of Robin Dunbar, who sees grooming as the precursor to language, citing Grandin's book *Animals in Translation* when they do.[34] The only way haptics is conceived of is as pressure intentionally applied, while the *warmth* of pressure goes undiscussed. I'm assuming their unintentionally limited perception of haptics causes them to miss Grandin's discussions of thermodynamics in *Animals in Translation*. Indeed, haptics are easy to over*look*. Haptic thermodynamics, or temperature registered and responded to through the skin or body, is sociality. Warmth orients us toward others, and cold orients us away from others. Grandin explains that thermodynamics are at the very root of signaling to an other, and Keller shows us how.

Keller's literacy education involves this nexus of relationality and thermodynamics. The first abstraction that Anne Sullivan tried to teach Keller was love. Keller didn't understand. She asked Sullivan if love meant warmth—and gestured to the sunlight.[35] Warmth seems to set Keller at ease. Keller would go out at night in her boat on the lake. She describes a feeling she gets when her surroundings change: "A luminous warmth seems to enfold me. Whether it comes from the trees which have been heated by the sun, or from the water, I can never discover. I have had the same strange sensation even in the heart of the city. I have felt it on cold, stormy days and at night. It is like the kiss of warm lips on my face."[36] Keller's sense of thermodynamics is finely tuned. Warmth seems to create spatial relationality, as it positions her with the water and the trees—making her feel a sense of sociality, of relation—the sensation becomes warm lips kissing her face. When she recounts the information she gets from handshakes, she connects the thermodynamics of the hand to an understanding of the person: "I have met people so empty of joy, that when I clasped their frosty finger-tips, it seemed as if I were shaking hands with a northeast storm." Her implication is discomfort. Some "hands have sunbeams in them...there is as much potential sunshine in it for me as there is a loving glance for others."[37] Keller illustrates the inseparability of sociality, relationality, and thermodynamics. Psychologists have been working to understand this phenomenon. Williams and Bargh concluded that "experiences of physical temperature per se affect one's impressions of and prosocial behavior toward other people"; warmth and trust go hand-in-hand.[38] This gives us a *sense* of the traces we choose to follow and those we turn from.

It makes social and physical sense to follow warmth, and Grandin explains why. One primary reason to call out is the vulnerability of thermodynamics. Grandin investigates what it means for a dog to make a separation distress call. Nonhuman animals (and she extends this to human animals) cry out for three reasons: if they are in pain, if their familiar space changes, or if their thermodynamics become unsafe. As psychologists Williams and Bargh show, Grandin ventures that our sociality is deeply related to thermodynamics, as human animals consider *people* to be warm or cold.[39] Living bodies touching each other generate warmth. A living body alone works to generate warmth. She says that lonely animals usually want to be touched, a response rooted in keeping close to parents for heat. She concludes: "Researchers believe that social warmth evolved out of the brain system that handles physical warmth"—we need both physical and emotional warmth to survive.[40] Parrish adds a second relational dimension here. To be cold can often mean to be exposed, to be vulnerable. To be so cold that you call out multiplies that vulnerability—not only is your body vulnerable to dangerous thermodynamics, but your location and level of vulnerability is now exposed to any hungry predator in range. Every call has a cost.[41] Every time you draw someone near, you simultaneously decrease and increase the intensity of your vulnerability. Nearness means both the asset of warmth and threat of violence of the ensuing tracer; distance means threat of cold and reduction of threat of violence of the ensuing tracer. As with the theoretical experience of the trace, haptic vulnerability is multivalent.

HAPTIC MODES: TEXTURE-PRESSURE

Texture-pressure is what I assume comes to mind when neurotypical human animals consider touch. A hand reaching out to make contact with a brick wall. Someone becoming suddenly aware of a scratchy sweater. Two human animals hugging after a social transgression. Texture here means the felt consistency of surfaces, but also any fine detail of touching, like when Sullivan spells with her fingers into the palm of Keller's hand. The meaning derived from an encounter with texture comes from a single moment of pressure or a sequence of single moments of pressure. By exploring texture-pressure, we deepen our understanding of how haptics is a driving force of tracing, how it complicates the threat-asset balance of vulnerability further, and how it is involved in navigating the individual relational traces that organize the huddle.

Generally, the pressure of (nonviolent) touch feels good; therefore we seek it. When animals we trust touch us, our brains release endorphins, natural painkillers.[42] Deep pressure is relaxing. Grandin discovered that it made her feel positive social feelings. She developed the "squeeze machine," an apparatus that relaxes human and nonhuman animals and helps autistic people manage their neurological systems to become more normatively social. When she started practicing with the squeeze machine, it first produced calming feelings, and then started to produce attitudes of "kindness and gentleness toward other people—social feelings."[43] She explains, "I had to have the nice feeling of being held to have nice thoughts about people."[44] Then she noticed that on days after she used the machine, she felt more anxious. She ventures that the squeeze machine was raising her endorphin levels, causing her to become "physiologically dependent" on it, the way people without autism rely on social interaction to manage their endorphin levels.[45] Following traces, getting closer to others, helps regulate thermodynamics and manage pain. The more distant others are from you, the less intense your vulnerability from threat, *and, the worse your health (your thermodynamics and your pain).* When considering texture-pressure, the proximity of the other, the trace, is more than just physical threat and thermodynamic asset—proximity maintains health.

How texture-pressure prepares the body for the vulnerability of sex—in any animal—is equal parts obvious, elusive, opaque, and enormous. It should be the focus of another time (especially because, following Parrish's notion of adaptive rhetoric, a major reason rhetoric evolved is to persuade others to mate).[46] Here, I describe a type of pig sexual experience because it elucidates a way the trace captures life potential. This process of artificial insemination provides an interesting perspective on sexuality, as often the human animal must simulate the touching ritual that would otherwise take place between pigs. A sow must be fully aroused to have a full litter, as the amount of semen her uterus will "pull" in is linked to her arousal.[47] A major factor in that arousal is pressure on her back. The breeder must not be involved in any fear-inducing relationships with her (like vaccinations), as she will reject a feared breeder. And if somehow the breeder and the pig(s) do create a litter in a fearful situation, the litter will be smaller and, research suggests, the piglets will be less healthy than piglets from a mother who was not frightened during insemination.[48] In this case, the threat of vulnerability outweighs the assets. This example suggests vulnerability-without-trust may *literally deteriorate health.*

We can also see texture-pressure employed to navigate individual relationships—to be actions of reconciliation. *Grooming, Gossip, and the Evolution of Language* presents Dunbar's research on a gelada monkey community. He observed reconciliatory grooming following acts of social transgression.[49] For example, the dominant male towered over a young female, grinding his teeth, scolding her. Her mother chose not to stand up to him. When he left, the mother made a "soft grunt" to beckon her daughter. When the daughter joined her mother, the mother began grooming her.[50] In addition to grooming, chimpanzees participate in reconciliatory scrotal rubbing (and many other gestures designed to secure goodwill).[51] Texture-pressure is used in nonhuman primate communities to navigate individual relationships. These haptic practices help a being navigate multivalent vulnerability with intention.

Texture-pressure is an organizing principle in the huddle. Dunbar argues human verbal languages developed out of grooming.[52] Grooming is an experience that is simultaneously "physical sensation and a social intercourse."[53] If you spend time grooming someone, you aren't spending time with someone else. Your choice to groom someone is recognized as an "investment" in them. It's your investment in giving pleasure to an other that communicates your social connection to them.[54] The only way to know if someone has a pattern of making bad returns on investments is to catch them mistreating someone else. Dunbar premises that the need to gossip, to share information about someone quickly, grows out of the need to form solid groups, balancing the benefits and disadvantages of larger numbers under the limits of time—and this need is what led to verbal language.[55] Grooming creates group "cohesion" and a "sense of belonging"—it organizes the social investments of group members.[56] Here, haptic forces negotiate vulnerability at the level of the huddle. Even though a group of beings generate power, what organizes and structures the group, what rubs the beings into being, are the texture-pressures (the traces) between individuals.

HAPTIC MODES: VIBRATION

Texture-pressure takes up a single contact. And while temperature doesn't require contact, when considering the thermodynamic presence of another body to a neurotypical human body, it does imply proximity. Vibratory haptics take us out of that proximity, allowing haptic information to travel distances, and, perhaps, time. Vibration carries two meanings: the making of immediate or historical meaning from pulses, and the haptic felt-sense of a room, house, or neighborhood. Both meanings carry the understanding that haptic information is traveling from the source through a medium. Keller says, "I perceive countless vibrations. I derive much knowledge of every-day matter from the jars and jolts which are to be felt everywhere in the house."[57] In the landscape of rhetoricity, vibration is connected to the distance perspective; it carries information about the movement of traces and tracers.

Animals large and small participate in communication through vibration. Caitlin O'Connell-Rodwell's research has shown that elephants use "*seismic communication*—making the ground rumble by stomping on it—to communicate with other elephants as far away as twenty miles."[58] Researchers watched as elephants leaned forward on their feet, or shuffled their weight around, paying attention to the ground. They learned elephants have special sensors in their feet and trunks to "detect vibrations," as do skunks and rabbits.[59] Funnel-web spiders have legs designed for creating and feeling drumming vibrations. In their preconflict ritual, instead of throwing themselves into a fight, they determine the stronger-bigger spider, allowing the smaller-weaker spider to retreat. When a spider invades the web of another, they drum. Both spiders judge each other's weight vibrationally: the "assessment is of weight, and that is made through vibrations of the web and not visually."[60] Vibration can take haptic rhetoric the distance. To study vibration is to study the *movement* of traces, and the selves marked in the rush.

Keller, as she says, "listened with her feet," understanding multitudes through vibration.[61] General haptic ambience, like milk pouring, flames, clocks are a part of her environmental awareness.[62] She reads gaits and can distinguish the performance of bodies *and* the relationship of that body to the environment, like how different waiters move through tables and respond to the music. She can tell the posture of her dog when he leaps up to the window. She knows when a carpenter is working down the street. She can also sense mood and disposition, particularly in those whom she knows well.[63] And, all that is with her feet on the ground. Bodies can detect vibrations without intentionally listening with hands and feet; feeling can, Keller says, "penetrate the skin, the nerves, the bones, like pain, heat and cold." She feels a snare drum in her chest and shoulders.[64] Her perception is so precise that when she enters the room of a man bent on discrediting her, she said she "found there persons who seemed to doubt me," and "felt that there was something hostile and menacing in the very atmosphere."[65] Here, vibration describes the landscape of vulnerability (who and what is near and far)—bringing an awareness of movement (who is tracing).

Animals can take immediate information from vibration, and Keller used this haptic mode to pick up historical feelings as well. She says she *feels* what has "marked man's progress" in "the great heart-throbs of humanity through the centuries; and if one does not feel [it] in these pulsations..., one must indeed be deaf to the harmonies of life."[66] Pulse connects Keller to what she imagines as a larger sweep of human experience. Keller appears to organize history linearly—not along a visual progression, but through the temporality of pulse. This temporality can be swiftly restarted at a distant present by hapticly connecting Keller to information preserved in texture. Thanks to her many privileges and fame, museum directors allowed her to touch sculptures. She remembers: "I should think the wonderful rhythmical flow of lines and curves could be more subtly felt than seen .... I can feel the heart-throbs of the ancient Greeks in their marble gods and goddesses."[67] She could mean this metaphorically, but as a fellow touch-sensitive, neurodivergent individual, I choose to take her a bit more literally. By touching sculpture, she can feel the texture of the labor of the artist—the rhythmic pulses of hammers and chisels, a haptic, three-dimensional telegraph. Those pulses, evidence of physical labor, imply a pumping heart. Keller's depiction of vibratory haptics brings the element of temporality to vulnerability, the rushing of the trace through time and space. The trace necessarily comes *before* the being. I feel in vibration the *before*-ness, the *already*-ness of the trace—the feeling of following, the progress and pulses Keller feels when touching statues.

But what of vibratory stimulation created immediately through direct contact? Texture-pressure feels good because pressure creates endorphins in the brain that help to manage pain.[68] Dunbar concurs: "the opiate system probably evolved in order to dampen the effects of pain caused by injury."[69] After responding to the danger that pain alerts me to, the pain still remains, and endorphins, slow acting and steady, help me manage the immediacy. Vibration, or "monotonous repetition of low-level stimulation," is what "the endorphin system responds best to."[70] Dunbar likens this to the repetitive pressures of jogging[71] and grooming.[72] We see the multivalence of vulnerability again: corporeality leaves us vulnerable to pain at the hands of an other, and at the same time, vulnerability, getting close to someone, mitigates suffering. This also opens questions about more conventional vibrations: what might this imply about the rhetoricity of personal vibrators? Of vibrating text alerts?

HAPTIC MODES: PAIN

And finally, what warmth, vibration, and texture-pressure all seem to work against: pain. The way we breed nonhuman animals often results in their pain and suffering. Breeding chickens for bigger breasts also causes their legs to deform under their unbalanced weight. Research has shown that these chickens will choose to eat food laced with painkillers over normal food, a reminder that endorphins can't always manage pain.[73] Pain occupies that more nebulous space shared by thermodynamics and pressure. It can be felt at all levels of the body; it doesn't necessitate contact with the skin. Often though, physical pain *is* a result of contact with the skin. And if haptics is a vocabulary for rhetoricity, I want to spend a moment considering the relational fracturing that is physical violence.

Nonhuman animals rape. Grandin's chapter called "Animal Feelings" opens with a scene of slaughtered hens, lying, their bodies broken open, on the pen's ground. The roosters at this farm raped-murdered hens so regularly that the farmer thought this was normal behavior.[74] Hens only choose to mate after they've seen roosters dance, and through the breeding process, the roosters lost the knowledge of their mating ritual.[75] The same is true when stallions are isolated in stalls. When a stallion wants to mate with a mare—when free to—he initiates a straightforward exchange of whinnies, allowing the mare to give consent. If she isn't willing, he goes elsewhere. Long hours of isolation in a stall causes a stallion to enter an aggressive state. In order for mating to occur, the owners must "tie up the mare so she can't get away, and then they hobble her feet so she can't kick the stallion." Then they release the stallion, who "rapes her."[76] Rape occurs in the wild also. Animal researcher Rachel Smolker has shown that male dolphins will form groups with the intent to rape, corner a female, then force themselves on her.[77] Haptics is the organizing force of the huddle, and all animals are socially implicated in the huddle, tracing and leaving traces. How does the *corrosion* (social, emotional, and physical) of rape fair in *irreducible* rhetoricity? Pain inflicted by the other can deteriorate corporality and life potential, and yet our vulnerability, our relational rhetoricity, remains fully intact. If a physical body is hobbled in place, the others pursuing her traces don't reduce their speed.

Nonhuman animals murder too. Grandin shares several examples, one of which is: a pod of orcas began to kill for sport. They separated a mother and calf, and spent six hours "crashing their bodies on top of" the calf repeatedly, a deadly and slow vibration, until it drowned. Then they ate only its tongue.[78] G. A. Bradshaw, a respected psychologist and ecologist, investigated a pattern of young male elephants (herbivores, remember) forming groups to rape and murder rhinoceroses.[79] While "wild" is a conflict-heavy "field," human animals of course influence the ecologies of all nonhuman animals. Bradshaw concludes that a young elephant's experience of the poaching of adult elephants, and therefore the lack of parenting and unbalanced regrouping of herds, allows for a diagnosis of post-traumatic stress disorder (PTSD).[80] She shares the example of (elephant) orphans "who exhibit wild-eyed panic and screaming in their early days" at the rehabilitation center.[81] We know a potential reason to vocalize is thermodynamic disruption, and we can assume a lack of texture-pressure and vibration hinders the management of pain. Perhaps this suggests that when irreducible rhetoricity, the constant tracing and leaving of traces, is coiled through pain and violence, it has the potential to beget further pain and violence, cycles of chasing the necessities of the haptic through violence.

If an understanding of rhetoric includes sensory rhetorics, particularly haptics, it can take up physical violence with discursive violence. I have expressed a commitment to understanding extralinguistic rhetorics; however, I don't intend to imply that we should exclude the influence of and presence of language in extralinguistic rhetorics. In *The Body in Pain*, Elaine Scarry draws a powerful connection between language and pain. She argues that the very indescribability of pain, the absence of language for pain, is what allows those in power to appropriate it, to cause it, to use it for their benefit.[82] By determining the nature of haptic rhetoricity, the intensive, originary, and corporeal feeling of vulnerability takes shape. Our vulnerability, that shared capacity to suffer and simultaneously to impress and be impressed upon by others, can become ruthless mortality. By taking up haptic rhetoricity, the physical reality of pain, violence, and death fall under the purview of rhetorical theorists. How we can and cannot describe these haptic modes of communication, this haptic state of rhetoricity, matters.

HAPTIC RHETORICITY

Davis builds her concept of rhetoricity from a foundation of Derridian anti-anthropocentism. When the rhetoricity of a being transitions into the ontology of a being, when a being becomes self-referential, they have felt a trace. The trace brings the multivalent experience of vulnerability: feeling sensation, self-recognition, leaving behind the mark of the self, following the marks of others. The nature of thermodynamics orients us toward the traces to follow and turn from. Together, the thermodynamics and texture-pressure modes struggle with the threat-asset balancing act of vulnerability and compound the assets (warmth and endorphins) of vulnerable proximity; and the mode of pain, of course, makes threat heavy. To be in the exposed, affectable state of vulnerability actually seems like a practical state—to be away from others is to be cold and in more pain. Yet, to be vulnerable is to be at the mercy of an other, to be followed, traced, hunted, potentially raped, potentially murdered by another tracer. The haptic modes of texture-pressure (used oppressively) and pain are agents of corporeal degradation, deterioration, and death. As a sensory landscape of the trace, haptic rhetoricity is experientially multivalent:

I am pressing, I am being pressed, there are pressures to follow, there are pressures following, and I feel this.

The relational quality of rhetoricity organizes Derrida's huddle, and haptic modes are an infrastructure allowing traces to circulate. Even though a group of beings generates power, what structures the group, what rubs and rushes the beings into being, are the texture-pressures (traces) passing between individuals. Texture-pressure practices help a being navigate the multivalence of vulnerability. While at any moment we are implicated in a version of this huddle, time, of course, means that traces are constantly circulating (through pressure), reorganizing the huddle, instituting who has the power. The vibratory mode of haptic rhetoricity emphasizes movement, distance, temporality, and history. Haptic rhetoricity is the circulation of the trace through the huddle in the conduits of texture-pressure and vibration.

Scientists see the thermodynamics, texture-pressure, and pain modes of haptic rhetoricity as origins of human animal sociality and language. After an exploration of these modes, it is obvious, though overwhelmingly indescribable, how haptics exercise the violence that is the foundation of discursive violence. Haptic rhetoricity necessarily takes up the corporeal degradation, deterioration, and death that remains behind the theoretical discussion of rhetoricity and the trace. The theory of rhetoricity is anti-anthropocentric and anti-ableist in its roots, and by presenting a theoretical exploration of the haptics of rhetoricity, I've hoped to present a radical take on rhetoricity.

Scarry argues that the indescribability of pain is indescribable a function of how it is wielded by those in power, and I extend that argument to haptics more generally (and perhaps even to the olfactory and gustatory, the "lower senses," the intimate senses, at large). Scarry's argument is an equally important second step. Though the totality and individuality of pain is indescribability, elements of pain *can* be laboriously worded into representation. And when that happens, either by those in pain or by those observing and advocating, *the appropriation of pain by those in power can be interrupted.*[83] Keeping animals (human and nonhuman) in pain (in the haptic realm, in the "lower senses" and out of the painless, seemingly body-less realms of sight, sound, logic, and independence) is one way those in power animalize and therefore justify the oppression and inhuman treatment of the animalizable. Understanding—and beginning to describe—haptic possibilities for persuasion and for being persuaded (the haptic modes of communication) and the felt dimensions of vulnerable relationality (haptic rhetoricity) are steps toward resistance that originate in the body and in what the body is put through. When studying haptic rhetoricity, we are not only accountable for understanding the relational workings of pain, violence, and death—we are also accountable for habilitating the animalized through resisting the hierarchy of sensation constantly implemented by the human animals in power.

NOTES

1. Because humans are animals, when I use the word "animal," I am referring to human and nonhuman animals. "Human" and "nonhuman" are used as adjectives. When I consider living beings beyond "animal," I use the word "being."

2. You might exchange "animalized" for "dehumanized." Both refer to a being whose life potential has been denigrated by the human animals in power. A better term would be "de-life-d" or "de-being-ed." For the sake of familiarity, and because "dehumanize" is filled with the complicated baggage of defining human, I'll use the term "animalized." When I do, I imply "de-being-ed."

3. Cary Wolfe, "Human, All Too Human: 'Animal Studies' and the Humanities," *PMLA* 124, no. 2 (2009): 567.

4. Jacques Derrida, *The Animal That Therefore I Am*, ed. Marie-Louise Mallet, trans. David Willis (New York: Fordham University Press, 2008), 160.

5. Debra Hawee's recent work takes up the rhetoric of sensation. See bibliography.

6. Diane Davis, "Creaturely Rhetorics," *Philosophy & Rhetoric* 44, no. 1 (2011): 89. Derrida and Davis frequently use italics for emphasis. For all quotations in this chapter, I have preserved the author's emphasis and have specified my own.

7. Alex Parrish, *Adaptive Rhetoric: Evolution, Culture, and the Art of Persuasion* (New York: Routledge, 2014).

8. Natasha Seegert, "Play of Sniffication: Coyotes Sing in the Margins," *Philosophy & Rhetoric* 47, no. 2 (2014): 158–178.

9. Cynthia Lewiecki-Wilson, "Rethinking Rhetoric Through Mental Disabilities," *Rhetoric Review* 22, no. 2 (2003): 157.

10. Parrish, *Adaptive Rhetoric*, 53.

11. Wolfe, "Human," 570.

12. Derrida, *The Animal*, 27.

13. Ibid.

14. Ibid., 28.

15. Diane Davis, "Autozoography: Notes Toward a Rhetoricity of the Living," *Philosophy & Rhetoric* 47, no. 4 (2014): 536–537.

16. Ibid., 548.

17. Davis, "Autozoography," 536; Davis, "Creaturely," 89.

18. Davis, "Creaturely," 89.

19. Ibid.

20. Davis and Derrida also consider how feeling a trace causes the self to split into a "me" and "I," a subject and an object that can mournfully never rejoin (see Davis "Autozoography," 543). In this way, the self becomes other. Due to the limits of this project, I will focus on the way the trace marks otherness, and this can refer to an other or the self-as-other.

21. Davis, "Autozoography," 548.

22. Ibid., 547.

23. Ibid. Davis is quoting Derrida.

24. Derrida, *The Animal*, 55.

25. Ibid., 56.

26. Ibid., 10.

27. This recording is presented on the investigative journalism podcast *Serial*, and I have quoted from the online transcript. See "Bergdahl" in the bibliography.

28. Shannon Walters, *Rhetorical Touch: Disability, Identification, Haptics* (Columbia, SC: University of South Carolina Press, 2014), 5–6.

29. Helen Keller, James Berger, Anne Sullivan, and John A. Macy, *The Story of My Life* (New York: Modern Library, 2003), 96.

30. Constance Classen, *The Deepest Sense: A Cultural History of Touch* (Urbana: University of Illinois Press, 2012), xii.

31. Ibid., 52, 71, 168.

32. Ibid., xii.

33. This is not to say that taking on gustatory, olfactory, and haptic rhetorics alone is the most radical step. Often, especially in the worlds of disability and neurodivergence, white and otherwise privileged voices are those recorded and circulated, hence my reliance on Grandin and Keller (both white, neurodivergent women). An important next step is to gather sources on the "lower" senses that don't begin in privileged positions.

34. Lynn Worsham and Gary Olson, "Temple Grandin, Translator: Sounding Autism, Seeing Animals, Making a Difference," *Journal of Advanced Composition*, 32, no. 1–2 (2012): 25.

35. Keller, *Story*, 26.

36. Ibid., 94.

37. Ibid., 105.

38. Lawrence Williams and John Bargh, "Experiencing Physical Warmth Promotes Interpersonal Warmth," *Science* 322, no. 5901 (2008): 607.

39. Temple Grandin and Catherine Johnson, *Animals in Translation: Using the Mysteries of Autism to Decode Animal Behavior* (New York: Scribner, 2005), 111–112.

40. Ibid., 112.

41. Parrish, *Adaptive Rhetoric*, 54–59.

42. Grandin, *Decode Animal Behavior*, 112.

43. Ibid., 114.

44. Ibid.

45. Ibid., 114–15.

46. Parrish, *Adaptive Rhetoric*, 3.

47. Grandin, *Decode Animal Behavior*, 104.

48. Ibid.

49. Robin Dunbar, *Grooming, Gossip, and the Evolution of Language* (Cambridge, MA: Harvard University Press, 1996), 26.

50. Ibid., 25–26.

51. Parrish, *Adaptive Rhetoric*, 46.

52. Dunbar, *Grooming*, 78–79.

53. Ibid., 1.

54. Ibid., 44.

55. Ibid., 78–79.

56. Ibid., 35.
57. Helen Keller, *The World I Live In* (New York: The Century Co., 1908), 43.
58. Grandin, *Decode Animal Behavior*, 60.
59. Ibid., 61.
60. John Maynard Smith and David Harper, *Animal Signals* (Oxford: Oxford University Press, 2003), 105.
61. Keller, *World*, 44.
62. Ibid., 45.
63. Ibid., 44.
64. Ibid., 49.
65. Keller, *Story*, 56.
66. Ibid., 83.
67. Ibid., 100.
68. Grandin, *Decode Animal Behavior*, 112.
69. Dunbar, *Grooming*, 36.
70. Ibid., 37.
71. In "On Speech and Public Release," rhetorical theorist Joshua Gunn also takes up jogging. Gunn explores how the human voice plays at the boundary between public and private spaces, and how uncontrolled speech (gasps, grunts, cries, yawps, etc). can help us understand late capitalism's merging of the public and private into intimate publics. He believes utterances of uncontrolled speech bring a private-seeming unruly body into a public space, exciting or unsettling listeners as they rupture the masculine, logical norms of public discourse. The upset comes from the sexual/sexualized (and therefore private) nature of uncontrolled speech being brought into public spaces. He argues that the discipline of rhetoric, as it seeks to repress the uncontrolled speech arising from the unruly body, is the disciplining of speech (192). He ties these private/public-threshold-crossing utterances directly to *the felt dimensions of the body*—sex, pleasure, pain, and the endorphin rush of running a marathon (in addition to the haptic, he also uses "feeling" to imply emotion). And when he does so, he also connects these felt dimensions to animality, disability, unfitness, misogyny: "The cry, the grunt, the scream, and the yawp index the sexual understood as *the body in feeling*. These also bespeak an absence of self-knowledge, a loss of control, even a tacit mindlessness associated with extreme affective states" (178). Gunn is tangibly close to haptic rhetorics here, as he is addressing the repression of the body through the "lower senses."Because Gunn argues that uncontrolled speech will help us to understand communication in this moment of intimate publics, and because uncontrolled speech is the vocal reaction to a *felt* experience, in future work, I hope to extend his questions through considering the haptic rhetoric that leads to uncontrolled speech. Indeed, the feline encounter that led Derrida to disrupt the anthropocentrism of continental philosophy happened along the private/public threshold. In the telling, Derrida emphasizes his nakedness, and particularly the exposure of his penis. And while the shame he describes seems to come from the cat *seeing* his penis, haptic rhetoricity is of course at work—not only in the vulnerability of exposing oneself to a predator ("although that threat remains on [the cat's] lips or on the tip of the tongue"), but also in the notion that his cat, with a set of whiskers and skin to sense the thermodynamics

of the shower and the vibrations of movement (let alone the olfactory possibilities), has already crossed the threshold, has already been encountering his privates far before this moment (Derrida, *The Animal*, 4).

72. Ibid., 36–37.

73. Grandin, *Decode Animal Behavior*, 71.

74. Ibid., 69.

75. Ibid., 70.

76. Ibid., 105.

77. Ibid., 151.

78. Ibid., 151–52.

79. G. A. Bradshaw, *Elephants on the Edge: What Animals Teach Us About Humanity* (New Haven: Yale University Press, 2009), 34–35, 78–80.

80. Bradshaw, *Elephants*, 80–87.

81. Ibid., 83.

82. Elaine Scarry, *The Body in Pain: The Making and Unmaking of the World* (Oxford: Oxford University Press, 1985), 14.

83. Ibid.

BIBLIOGRAPHY

Agamben, Giorgio. *Homo Sacer: Sovereign Power and Bare Life*. Translated by Daniel Heller-Roazen. Stanford, CA: Stanford University Press, 1998.

Bergdahl, Bowe. "DUSTWUN." Serial Podcast. Narrated by Sarah Koenig. Serial Podcast and WBEZ Chicago. Accessed February 1, 2017. https://serialpodcast.org/season-two/1/dustwun/transcript/.

Bradshaw, G. A. *Elephants on the Edge: What Animals Teach Us About Humanity*. New Haven: Yale University Press, 2009.

Classen, Constance. *The Deepest Sense: A Cultural History of Touch*. Urbana: University of Illinois Press, 2012.

Davis, Diane. "Autozoography: Notes Toward a Rhetoricity of the Living." *Philosophy & Rhetoric* 47, no. 4 (2014): 533–553.

———. "Creaturely Rhetorics." *Philosophy & Rhetoric* 44, no. 1 (2011): 88–94.

Derrida, Jacques. *The Animal That Therefore I Am*. Edited by Marie-Louise Mallet. Translated by David Willis. New York: Fordham University Press, 2008.

Dunbar, Robin. *Grooming, Gossip, and the Evolution of Language*. Cambridge, MA: Harvard University Press, 1996.

Grandin, Temple, and Catherine Johnson. *Animals in Translation: Using the Mysteries of Autism to Decode Animal Behavior*. New York: Scribner, 2005.

Gunn, Joshua. "On Speech and Public Release." *Rhetoric & Public Affairs* 13, no. 2 (2010): 175–215.

Hawhee, Debra. "Rhetoric's Sensorium." *Quarterly Journal of Speech* 101, no. 1 (2015): 2–17.

Keller, Helen. *The World I Live In*. New York: The Century Co., 1908.

Keller, Helen, James Berger, Anne Sullivan, and John A. Macy. *The Story of My Life*. New York: Modern Library, 2003.

Kennedy, George A. "A Hoot in the Dark: The Evolution of General Rhetoric." *Philosophy & Rhetoric* 25, no. 1 (1992): 1–21.

Lewiecki-Wilson, Cynthia. "Rethinking Rhetoric through Mental Disabilities." *Rhetoric Review* 22, no. 2 (2003): 156–67.

Maynard Smith, John, and David Harper. *Animal Signals*. Oxford: Oxford University Press, 2003.

Parrish, Alex. *Adaptive Rhetoric: Evolution, Culture, and the Art of Persuasion*. New York: Routledge, 2014.

Scarry, Elaine. *The Body in Pain: The Making and Unmaking of the World*. Oxford: Oxford University Press, 1985.

Seegert, Natasha. "Play of Sniffication: Coyotes Sing in the Margins." *Philosophy & Rhetoric* 47, no. 2 (2014): 158–178.

Walters, Shannon. *Rhetorical Touch: Disability, Identification, Haptics*. Columbia, SC: University of South Carolina Press, 2014.

Williams, Lawrence and John Bargh. "Experiencing Physical Warmth Promotes Interpersonal Warmth." *Science* 322, no. 5901 (2008): 606–607.

Wolfe, Cary. "Human, All Too Human: 'Animal Studies' and the Humanities." *PMLA* 124, no. 2 (2009): 564–575.

Worsham, Lynn and Gary A. Olson. "Temple Grandin, Translator: Sounding Autism, Seeing Animals, Making a Difference." *Journal of Advanced Composition* 32, no. 1–2 (2012): 11–56.

Chapter 4

Human Boundary Seepage, Bacterial Rhetorics

Jennifer Saltmarsh

> Who, out of the theory of the earth and of his or her body understands
> by subtle analogies all other theories,
> The theory of a city, a poem, and of the large politics of these States
>
> —Walt Whitman, "Kosmos"

The general public can only understand bacterial communication through scientific publications, but recent research suggests that we can feel with our bodies the effects of bacterial communication, and that we act upon these feelings to make the most basic decisions in our daily lives. This creates a bit of a paradox: we must choose whether we will believe what science tells us about our bodies' innate intelligence and decision-making processes, though we may not be able to consciously affirm these processes as they unfold inside our bodies. Our understanding of bacteria's reach into human emotion and cognition is blurring the distinction between the two. In this chapter, I will analyze research that may help us expand our definition of rhetoric to further include the nonrational and nonverbal, the body and sensation. I suggest not that we directly apply bacteriological theories to cultural phenomena, but that we investigate our commonalities in communication and the potential feedback between microscopic and macroscopic, biological and cultural phenomena.

The biological basics of persuasion might be found in bacterial *quorum sensing*: the intercellular chemical signaling process by which individual bacteria cells communicate to make decentralized, context-bound decisions, acting collectively as one multicellular organism. Bioluminescence spurred some of the earliest research into quorum sensing when scientists discovered that individual *Vibrio fischeri* bacteria, residing inside the light organs of marine creatures, would luminesce only upon sensing a quorum sufficient to collectively produce visible light when needed, thereby allowing host creatures to have startling physical effects upon their environment. An example of this phenomenon occurs in species of squid capable of counterillumination, an antipredation strategy that "enables the squid to avoid casting a shadow beneath it on bright clear nights when the light from the moon and stars penetrates the seawater."[1] An even more spectacular version of counterillumination occurs in the Hawaiian bobtail squid, which dons the brilliant speckles of a starry sky, a celestial camouflage from predators and prey beneath. The bobtail squid's bioluminescence demonstrates a nonhuman and environmental intelligence both within the bacteria collective, and between bacteria and squid, ocean, and starlight. These sophisticated responses to environment, occurring outside of spoken language and (perhaps) conscious intent, may show us the biological rudiments of decision-making. In quorum sensing, individual bacteria release diffusible molecules that bind to receptors on other bacteria, allowing them to sense their population density in a given environment. Many different species of bacteria may inhabit a particular area, but each species uses a different chemical signal to recognize one another and to respond collectively to rapidly varying environmental conditions. When bacteria sense a sufficient population density (a quorum), the individual bacterium act simultaneously in unison—for example, by regulating gene expression. Bacterial geneticist J.A. Shapiro says his decades-long research on *E. coli* bacteria pattern formations has convinced him that quorum sensing dismantles Cartesian dualism because bacteria display cognition by making decisions using only their bodies.[2] Later in this chapter, I will further address the question of cognition and intelligence, and whether these are merely metaphors. But regardless of how we define cognition, bacteria do make environmentally informed decisions, both as individuals and as groups. Bacterial quorum sensing may help us to reassess what and how human bodies decide, both prior to and in tandem with the brain. Bacteriology may allow us to better appreciate the decision-making capacities we have in common with other animals, and along with this the relationship between spoken language, body, and sensation.

George Kennedy initiated a conversation on animal rhetorics when he reframed rhetoric not as human art or science but as the emotional energy prior to speech and conscious intent, which humans and animals share as a "'deep' universal rhetoric." Kennedy therefore extends the potential for rhetoric to any life form that can give signals.[3] Diane Davis engaged with Kennedy's long-ignored argument by addressing what Freud called the Darwinian revelation—"that man shares a common ancestor with apes, which indicates that he is not inherently 'a being different from animals or superior to them.'" She argues that the "panicked deflection" of this revelation "continues to ground contemporary theories of rhetoric," manifesting as rhetoric's requisite "engagement with the symbolic.[4] While Davis challenges the (Burkean) notion of human as symbol-using animal, as well as contemporary rhetoricians who have been hesitant or unable to see engagement with the symbolic in nonhuman creatures, a few bacteriologists seem to be on a quest to find something like symbolic behavior in quorum-sensing bacteria, who have no brain and are not even categorized as animals. Combinatorial communication—a phenomenon believed to occur only among humans and select primates—was recently found in bacterial quorum sensing. I will argue that this research pushes us to reconsider our definitions of the symbolic, and with it, nonsymbolic motion/symbolic action. This research also implicitly confers with Davis's suggestion that animal studies offers rhetorical studies the chance "to reopen the question of the language relation, to reconsider not only what it involves but who or what engages it."[5] At the level of human-bacteria communication, the field of microbial endocrinology investigates the biological circuitry of the microbiota-gut-brain (MGB) axis—the vagus nerve route through which bidirectional signals travel between the gastrointestinal tract and the brain, allowing bacteria to affect digestion, emotion, cognition, sociality, gregariousness, visceral perception, and sexual behavior.[6] The holobiont and hologenome theories create a new microbiological framework that challenges the idea of animals as autonomous entities and instead posits the notion of animals as biomolecular networks, comprised of the host creature and its associated symbiotic microbes.[7] Additionally, work on bacterial pheromone production may encourage us to elaborate upon theories of the transmission of affect.

Taken together, this emerging research offers a new lens for questioning human exceptionalism, mind-body dualism, and the myth of the self-contained individual that makes autonomous, intentional, brain-based decisions. We do not classify bacteria as animals, but the complex bodies that we do classify as animals are not possible without the bacteria that inhabit them. Interested bacteriologists are beginning to ask rhetorically oriented questions and to invent a corresponding vocabulary. I hope that my reading of this literature will help rhetorical studies continue to push against the boundaries of human communication—the boundaries between bodies and their environs—while opening an interdisciplinary exchange that enlivens both rhetorical studies and bacteriology.

COMBINATORIAL COMMUNICATION IN BACTERIA: NONSYMBOLIC MOTION? SYMBOLIC ACTION?

One laboratory research team comprised of microbiologists, bioinformaticians, and cognitive and evolutionary anthropologists seems to have set out with the goal of unsettling human exceptionalism. In their publication "Combinatorial Communication in Bacteria: Implications for the Origins of Linguistic Generativity," Scott-Phillips *et al.* detail a famous study on primate communication and how they replicated it with bacterial quorum sensing. Combinatorial communication, ubiquitous in human language, exists when two signals combined create an effect different from the sum of component parts. Elsewhere, Thom Scott-Phillips, the cognitive and evolutionary anthropologist who took part in this research, explains the combinatorial power of spoken language by showing how we can extend any existing sentence simply by adding the expression "Pinker wrote that." This addition creates a new sentence, the meaning of which is not simply the combination of the two individual sentences, but rather, depends upon how they are combined. He offers the following example: if the existing sentence is "I didn't know that," then "the meaning of the new sentence would depend on the order in which this was combined with the additional expression 'Pinker wrote that.'"[8] The research team that Scott-Philips leads notes that this combinatorial quality "gives language its expressive power," and that outside of language, the most famous example of combinatorial communication is "the alarm call system of putty-nosed monkeys" which includes "a distinct call for each of two predators" (glossed as "pyow" for leopards—climb up; and "hack" for eagles—climb down); however, when the two distinct calls are produced together, the call sequence effect "is not simply the composite effect of the two component parts, but something different, namely group travel in non-alarm situations [e.g. to forage]."[9] When they replicated the study with bacteria, they encountered similar results.

The putty-nose monkey call system, considered alongside Burke's notion of human as *symbol-using animal*, prompts the question: what counts as a symbol? General, nondisciplinary definitions suggest that a symbol is a material object, sound, gesture, or idea that represents another idea, abstraction, action, or material entity. If putty-nose monkeys call "pyow" to signify leopards and climbing up, and "hack" for eagles and climbing down, then this behavior seems already symbolic because "pyow" means leopards but automatically translates into the command "climb up." Furthermore, if combining these two symbolic calls (pyow and hack) for opposing actions (*alarm* calls to climb up and climb down) results in an opposing meaning and response (*nonalarm* group travel to forage), then we can say that these symbols both produce abstraction and lead to action. Debra Hawhee says that essentially, "nonsymbolic motion names strictly physical movement, human and nonhuman, while symbolic action names the interpretive, communicative activity of language, the story-ing of motion."[10] According to these parameters, putty-nose monkey communication is symbolic action, because the monkeys tailor their calls to the environment; they create and respond to two distinct, combined signals as an altogether different signal—action that in humans would be deemed interpretive.

Prior to replicating the primate study with bacteria, it was believed that combinatorial communication occurred exclusively among humans and a few nonhuman primate species.[11] Accordingly, it was also previously assumed that "(i) these differences in the complexity of observed communication systems reflect cognitive differences between species; and (ii) that the combinations we see in nonhuman primates may be evolutionary precursors of human language." However, when they replicated the experiment with bacterial quorum sensing—pure chemical, material communication—researchers discovered the same pattern generally found in human communication: the "effect of the combined signal differs from the composite effect of the two individual signals."[12] In a related 2014 article, a team comprised of some of the same researchers further elaborates upon what the combinatorial communication experiment implies for bacteria sociality. They define quorum sensing as the process by which bacteria secrete and detect diffusible molecules to make frequent regulatory decisions based on "limited information about their external world"[13]; they explain that bacteria are able to make these decisions because they can respond to multiple signals (with different decay rates) using what the researchers call "heuristics" or "rules of thumb." They call this ability in bacteria the power to *infer* both *social* density (how many individual bacteria occupy a given space) and *asocial* "physical (mass-transfer) environment"[14](or, the rate at which the chemical signals diffuse or wash away in a given environment) (italics mine). So, according to the researchers, an individual bacterium responds to two distinct signals differently than if it were to receive each distinct signal separately, and it can also make what they call a robust estimate based on the rate at which the environment washes away those chemical signals—communicative behavior that the researchers term "inference." And the greater the variety of molecules a given bacteria species secretes, the more information they can cull, since the environment affects each kind of molecule differently.[15]

If we say that combinatorial communication in monkeys is symbolic action, then is combinatorial communication in bacteria also symbolic action? The researchers themselves are a bit ambiguous; they say their findings suggest that "advanced cognitive abilities and large brains do not necessarily explain why some species have combinatorial communication systems and others do not," and argue that it is "premature to conclude that the systems observed in nonhuman primates are evolutionarily related to language."[16] But as I will discuss later, the human neurosensory system, and thus human communication, is inextricable from bacterial communication. Even if combinatorial communication does not lead *directly* to language use (in the primate-to-human evolution sense), human language use nevertheless relies upon bacterial communication systems. Although the researchers do not extend advanced cognitive abilities to bacteria, they nevertheless do extend to bacteria the power to infer, estimate, and communicate to each other their social and environmental conditions—behaviors that we do associate with advanced cognitive abilities. However, they also undermine the long-held assumption that combinatorial communication is the special purview of linguistic or protolinguistic species. This confusion seems resolvable only by stating outright that advanced cognitive abilities either do or do not require brains. One might argue that extending these abilities to bacteria confuses our definitions of these abilities in humans, but I suggest that recognizing the commonalities between human and nonhuman behaviors can allow us to dig deeper into what exactly makes the human version of these abilities distinctly human. We may have it backwards. We see advanced cognitive abilities in humans and assume these abilities arose only within the past 50,000 years, along with modern human behavior (abstraction, symbolic behavior, exploitation of large game, etc.). However, it might be more productive to entertain the idea that the evolving human body adopted (and adapted) cognitive and communicative abilities from prior sources that arose during the previous 3.5 billion years of life on earth. Alex Parrish convincingly argues that "I must be careful to use honeybee rhetorics analogically, rather than homologically, as I might do with chimpanzee gestures and the physical suasion of humans. Because we share a recent ancestor with chimps, it is quite possible we are hardwired to perform similar behaviors in response to similar situations."[17] This assessment seems fair but difficult to apply when considering bacterial rhetorics. They are certainly our most remote evolutionary ancestors, but should we read them analogically or homologically when they reside within us and are impossible to disentangle from our own bodies? Rather than fear that we anthropomorphically impose a human framework upon nonhumans, we might instead imagine the human body as a distinct conglomeration of cognitive abilities that previously manifested in myriad other species. But, this could also mean that our cognition and communication patterns are less conscious and intentional than we have thus far assumed, or that our conscious and intentional communication relies upon an array of noncoconscious, nonintentional microdecisions and communications.

J. A. Shapiro argues that by reorienting our thinking, "we can find cognitive and computational phenomena" in "classic bacterial systems" that have recently been opened up to molecular study; additionally, he points out the long-time custom of comparing bacterial chemotaxis (movement in response to a chemical signal) with the operation of a neurosensory system.[18] Bacterial chemotaxis developed billions of years prior to neurosensory systems. Either scientists are using convenient operational metaphors for bacterial behavior, or they are finding in bacteria the rudiments of neurosensory systems; or perhaps they are doing some admixture of the two, muddying the issue of anthropomorphism and metaphor use. Pamela Lyon argues that although unicellular organisms are less complex than animals, their behavior is not necessarily so. The ability to "navigate, become familiar with, value, learn from and solve critical existential problems within its world of experience, including coordinating action with conspecifics," Lyon says, may explain why microbial researchers have "helped themselves to cognitive terminology (i.e., 'decide,' 'talk,' 'cheat,' 'eavesdrop,' 'lure,' 'vote,') to describe complex bacterial behavior, often without caveat." Lyon argues that this language use may not be entirely metaphorical, that "there is something going on at the microscopic level that doesn't just 'look' cognitive, it *is* cognitive" and is typically considered so "when studied in animals more like us."[19] Our first instinct as rhetoricians may be to question bacteriologists' metaphor choices, but given the fully entangled bacteria-human symbiosis, we might give pause to the critical impulse until we have given enough consideration to claims like Shapiro's and Lyon's. Extending cognition to bacteria need not diminish the unique ways that human cognition manifests. To make better sense of human communication patterns, we might entertain bacteriologists' imaginative leaps and parenthetically treat them as *more than* imaginative.

For example, we could see differently human- and language-centered distinctions between nonsymbolic motion and symbolic action. Hawhee says critics obfuscate the nonsymbolic motion/symbolic action pairing when they rigidly dichotomize the two terms and ally Burke with symbolic action" (the interpretation, communication, languaging, story-ing of motion).[20] If bacteriologists now extend to bacteria the power to make group inferences about social-plus-environmental conditions, then what bottom-line requirement makes communication symbolic? Bacterial chemotaxis is the first kind of chemotaxis, the first kind of motion and decision-making performed by life on earth. But can we call it symbolic action if brains are not involved? For Hawhee, it is important that Burke neither advocates liberation from bodily matters, nor believes that motion is 'senseless,' even bestial," but rather, that motion is sense-full: "nonsymbolic motion names ... sensory perception" and "sheer physical movement" which hovers at the edge of language, or symbolic action, neither juxtaposed nor mutually exclusive to it."[21] Bacterial quorum sensing (the decision) is indistinguishable from bacterial chemotaxis (the motion), because it is based on bacterial inference or interpretation of social-plus-environmental conditions; in this instance, nonsymbolic motion (strictly physical chemotactic movement) is indistinguishable from symbolic action. Bacteria make collective decisions using material, chemical secretions; this chemical sensing is symbolic action that results in group response. A chemical response at the group level is no less a response or decision than is a verbal response.

Our current definitions of the symbol, and therefore of nonsymbolic motion/symbolic action, break down when we analyze monkeys and bacteria. What makes human language use unique, and what nonsymbolic human actions might we place in the realm of the symbolic? This distinction may allow us to participate in what we call rational or reasonable discourse, but the distinction may also prevent us from more fully understanding certain features of rationality. To classify quorum sensing as symbolic action may upend the sensory/symbolic dichotomy and open a floodgate of related questions: If quorum sensing never leads to or "hovers at the edge of" spoken language, then what of human sensing and sensation? If bacteria comprise 90 percent of the cells in the human body and communicate extensively inside of and with it, then what material combinatorial communications do our bodies practice outside of spoken language and conscious intent? How does this in turn shape our conscious attempts at rational deliberation? I propose that it might not be possible for the human body to enact *sheer physical movement* in an inference-free, story-less vacuum. If bacteria "converse" with each other and their environs to make decisions, then we should inquire into the wordless conversations between human bodies and their environs. The nonhuman physical movements occurring inside the human body may contribute to the story-ing of human movements.

BACTERIAL COMMUNICATION AND THE HUMAN BODY: MICROBIAL ENDOCRINOLOGY AND THE HOLOBIONT

In his introduction to *Microbial Endocrinology: The Microbiota-Gut-Brain-Axis in Health and Disease*, editor Mark Lyte defines microbial endocrinology as the intersection of microbiology and neurobiology, or the study of microorganisms' ability "to both produce and recognize neurochemicals that originate either within the microorganisms themselves or within the host they inhabit."[22] Microbial endocrinology blurs the distinction between bacteria and host, and asks how bacteria affect hosts, complicating a host's ability for conscious intent and decision-making. Lyte says that although it is still unclear why bacteria produce neuroendocrine hormones, it is clear that many of the "*neurohormones produced by bacteria also function in mammals as part of the neurophysiological system*" and "*can impact the neurophysiological aspects of the host including cognition.*"[23] Chemical happenings inside the body do not easily translate into engagement with language and symbols (in the traditional sense), and may remain in the realm of unspoken and ineffable sensations. These sensations would be neither entirely human nor entirely nonhuman but a synthesis of the two. Human cognition, as Lyte suggests, may not be as unique as we have believed it to be, but might be better understood as a distinct assemblage of nonhuman cognition: "The ubiquitous presence of neuroendocrine hormones in non-mammalian systems means that the presence of the very same neuroendocrine hormones in mammalian systems has a long evolutionary shared history."[24] In other words, as Shapiro and Lyon noted above, Lyte attributes to bacteria cognitive capacities, but he goes even further, suggesting that human cognitive capacities originated in, and rely upon, bacterial cognition. Lyte also proposes that neuroactive chemicals produced by the microbiota may not only interact with the host but also act as a signaling mechanism between members of the microbiota, and would thus make up a kind of "primitive nervous system" reminiscent of "an organ—namely, that the cellular elements which comprise the organ can be influenced, and in turn influence, the host" or a kind of "microbial organ."[25] This bidirectional influence, like a conversation between host and microbiota, seems to dissolve any rigid distinction between brain and gut, human and bacteria. A newfound microbial nervous system would mean that we have been giving our mind-brains too much credit for cognitive functions that are distributed across our bodies and their symbionts. Microbial endocrinology can open conversations about nonconscious bodily impulses toward belief and action that may inform deliberate and rational decision-making.

These concerns are not altogether new. Bodily cognition has concerned humans for quite some time. Nonscientific, nonacademic conversations can offer a different perspective on these matters. For example, Alan Watts uses metaphors of light—spotlight consciousness and floodlight consciousness—to explain bodily cognition:

> When the teacher in class says, "Pay attention!" everybody stares, and looks right at the teacher. That is spotlight consciousness; fixing your mind on one thing at a time … flip, flip, flip, flip, flip. However, we also have floodlight consciousness. For example, you can drive your car for several miles with a friend sitting next to you, and your spotlight consciousness may be completely absorbed in talking to your friend. Nevertheless, your floodlight consciousness will manage the driving of the car, will notice all the stoplights ... without even thinking about it. ... However, our culture has taught us to specialize in spotlight consciousness, and to identify ourselves with that form of consciousness alone. ... Although we very largely ignore it, the floodlight consciousness is working all the time, and every nerve ending that we have is its instrument … the floodlight consciousness is undervalued, we have the sensation of ourselves as being just the spotlight, just the ego that looks and attends to this and that and the other. So we ignore and are unaware of the vast, vast extent of our being. People, who by various methods become fully aware of their floodlight consciousness, have what is called 'a mystical experience,' or what the Buddhists call *bodhi*, an awakening. The Hindus call it *moksha,* or liberation, because they discover that the real deep, deep self, that which you really are fundamentally

> and forever, is the whole of being—all that there is, the works, that is you. Only that universal self that is you has a capacity to focus itself at ever so many different here-and-nows ... we forget and we do not know who we are. We are brought up in a special way so that we are unaware of the connection, and unaware that each one of us is the works.[26]

Floodlight consciousness, based in the body and the nerves, allows us to navigate, process, and adapt to our environs without thinking—a consciousness without intent. This mode of being and knowing might be considered the antithesis of rational deliberation. The body has been opposed to the intellectual as a mode of engaging with the world, partly due to its association with mysticism. For example, Freud disparages what he calls the "oceanic feeling" (a visceral sense of oneness with the world), associating it with the "primitive ego feeling" of infancy, an inability to differentiate between (subjective) self and (objective) world.[27] The opposition between body/sensation and mind/intellect draws attention to the academic discomfort with bodily sensation. However, if we are to open up new avenues of inquiry into rhetorical theory, we might gravitate toward modes of being and knowing that have been repressed, or disparaged as unknowable. Though it may be challenging to introduce into scholarly conversations concepts such as floodlight consciousness, we need not reject them as mystical claptrap. Massumi is not too far off from the idea of floodlight consciousness when he points out that the "vast majority of the world's sensations are certainly nonconscious."[28] And Diane Davis approaches Watts' notion of *the works* when she declares, "the body is not a discrete phenomenological entity. It's unclear where it begins and ends."[29] Taken together, these concepts may help us to reflect on how the body factors into the way humans develop a deep sense of belonging to something greater than themselves, like a religious community, a political group, or the natural environs. In their own way, each of these theorists moves toward the unknown contours of bodily sensation and the limits of knowing and reason. We can neither pin down sensations nor explicitly understand our bodily enmeshment in the world, but this does not mean that we should ignore the messy ways that bodily phenomena factor into our deliberative and rhetorical capacities. Perhaps the unknowns of bacteria-human symbiosis can allow us to theorize such rhetorical concepts without slipping into biological reductionism.

The holobiont theory revises organismic boundaries and gestures toward unity between organisms, but it is still in its formative stages. Bordenstein and Theis clarify this emerging concept, and attempt to make it widely understood by linking it to current theories in ecology and environmental biology. They say plants and animals are no longer understood as "autonomous entities, but rather as 'holobionts,' composed of the host plus all of its symbiotic microbes." But a holobiont is still an inherently unknowable kind of entity, given the countless symbiotic microbes inhabiting a particular host, and accordingly, they acknowledge the importance of selecting the appropriate terminology to build this new conceptual framework: "the term [holobiont] importantly fills the gap in what to call such assemblages."[30] Not only do they acknowledge that there is a language gap when referring to these material entities (which are perhaps unknowable in each given situation) but they also use the term assemblage, thus navigating terrain familiar to rhetoricians retheorizing the subject-object dichotomy. Regardless of whether terms like holobiont accurately represent material truth, they seem especially pertinent given the current human condition. Bordenstein and Theis say that symbiotic microbes are fundamental to "nearly every aspect of host form, function, and fitness," and to "traits that once seemed intangible to microbiology: behavior, sociality, and the origin of species."[31] The holobiont theory aims to make tangible the invisible human-bacteria processes and relationships, but, as mentioned, it is still in the formative stages. What we might take from it at this point is a shift in perspective. If we view ourselves not as biologically autonomous but rather as enmeshed and shifting assemblages, then this material condition redistributes human desire, agency, and actions across human and nonhuman bodies and populations. We are perhaps too much in the midst of things to draw conclusions about the cultural factors informing terms like *holobiont*; however, a desire to unify disparate biological parts would make sense, given the uncertainty about our place in a now fragile ecology, and fraught global and domestic politics. It remains to be seen whether and how re-imagining the *biological* assemblage of human communities can also help us to re-imagine the *cultural* assemblage of human communities, or, how our immediate cultural circumstances influence how scientists *see* our biological circumstances. Understanding ourselves as a species-cooperative at the micro level could influence our ideas about cooperation at the macro level. But what of agonism?

To better understand how competition works within the holobiont, we must look at the closely related hologenome theory. The most current effort at distilling the hologenome and holobiont theories comes from an interdisciplinary team of fifteen researchers and theorists who say that "the host and microbial genomes of a holobiont are collectively defined as its hologenome."[32] The hologenome theory looks at the holobiont in evolutionary terms, so this is where we need to tread lightly with bacteriological research. Given the criticism that the holobiont and hologenome theories have received from the scientific community, the researchers clarify that cooperation is not the only driving force in the coevolution of the hologenome. Rather, they say that "conflicts of interest resulting from the nature of the host transmission of microbes could select for microbes that manipulate hosts to improve their own transmission."[33] The idea that resident bacteria can manipulate host behaviors may cause discomfort for those of us concerned with biological determinism (not to mention human exceptionalism). However, cultural factors needn't be dismissed nor explained away simply because we factor in new information and theories about human-bacteria symbiosis. The researchers add that holobionts and hologenomes are not the *primary* unit of selection but rather "*a* unit of selection in evolution" (italics mine).[34] The way we speak to each other has influenced the way we have evolved, but so have nonspoken forms of rhetoric. Human bodies and bacteria have evolved both cooperatively and agonistically, much like larger human societies have evolved. We should beware of determinism and reductionism, but we cannot continue to act as though culture does not emanate from—and create feedback with—a collective of bodies within bodies, each emanating its own distinct rhetorical energy.

Humans generally believe that rational discourse exists and that human language produces it. A rational statement cannot be made, articulated, and evaluated as such outside of human language or symbol systems. Rationality, as we understand, it, is a *closed* system; as such, it is difficult to flesh out its relationship with bodies as *open* systems. Joshua Gunn points out that rhetorical studies was "founded on the object of human speech" and that its "demotion of feeling over the last century has become particularly problematic."[35] He argues that uncontrolled speech, gendered as female, "represents a dimension of speech that we tend to repress: the bodily and affective. This dimension is repressed because it threatens masculine norms of control and order … the discipline of rhetoric concerns the disciplining of speech."[36] It's easy to lose sight of the simple fact that rhetoric attempts to manage sensation, affect, and bodies. Rationality is determined by an ever-shifting quorum of bodies that desire, hunger, envy, and nurture offspring, regardless of how well–educated said group may be. Our decisions may depend upon environmental factors more than we have been able to understand or acknowledge (which seems especially pertinent during a time of environmental crises and population upheavals). Bodily sensation may be primary, and language may be secondary; deliberation may involve verbal confirmation of group decisions that our bodies have already created and/or sensed. However, accepting this notion would threaten our belief in rational discourse as we currently understand it. We hope to be persuaded along the lines of reason, or at least a well-reasoned pathos. Gunn says that "although we acknowledge the centrality of affect and suasive ascent, we no longer seem particularly invested in explaining just exactly what influence or persuasion actually is."[37] Citing Diane Davis (2008), he says that "persuasion may be nothing more than suggestion, a form of hypnosis."[38] Focusing on the ur-decision-makers could help us put bodily matters into perspective. If bacteria make decisions using only their bodies and material surroundings, then we should be asking how brains and language factor into our own bodily decision-making processes. With a shift in perspective, we can see speech as an intangible bodily secretion and understand rational discourse as a black box between brains and bodies. This prospect need not alarm us if it helps us to reassess what humans are up to in their language-based attempts at rationality.

AFFECTIVE TRANSMISSION ACROSS BODIES: THE MICROBIAL INFLUENCE ON GROUP COHESION

Instead of questing after what makes humans different, we could think of ourselves as an assemblage of preexisting nonhuman communication systems, to get at what makes human communication distinctly human. Our speech manifested from myriad biological communication systems, and may be more like these systems than we've been able to see or acknowledge. E. P. Greenberg says that, by and large, "the eyes of microbiology" only began to see bacteria sociality in the 1990s, as quorum sensing and biofilm research rapidly developed. He says it has now become "apparent that the same selective pressures that led to the evolution of sociality in animals are forces for evolution of sociality in bacteria."[39] Preexisting values about humans' place in an ecological hierarchy may have once prevented bacteriologists from seeing bacteria as social creatures; likewise, preexisting values may now prevent us from seeing our own sociality and communication (especially linguistic) as anything other than human. But if bacteria help regulate nonconscious, intrabodily communication in humans, then we face the challenge of discussing where bacterial communication ends and human communication begins, which is both difficult and disconcerting. If we wish to track microbial influence upon human sociality, there are several channels we can pursue.

Perhaps the simplest and least controversial route for bacterial transmission across bodies is via physical touch. In their study of wild baboons (excluding for kinship, shared diet, and environment) Jenny Tung *et al.* found that "an individual's contacts in a grooming-based social network, as well as its membership in a given social group [highly predict] its gut microbiome composition," thus implicating "direct, affiliative physical contact" as a determinant of gut microbiome … with potentially important consequences for the evolution of sociality."[40] The researchers highlight the importance of frequent hand-to-mouth contact, especially in the grooming that adult males direct to estrous females, which concentrates "heavily on the ano-genital region, increasing the probability of fecal-oral transfer." They add that physical contact in humans (hugging, kissing, holding hands) "may provide a similar route through which close social partners transmit gut bacteria."[41] We might be encouraged by the idea that we are only microbially influenced by those with whom we share physical proximity, but given the microbiota's role in producing human neuroendocrine hormones, *affective* transmission is possible without physical touch. Archie and Theis (2011) report that many mammalian odors, which animals use to recognize members of the same species, are produced by symbiotic bacteria. They say "some of the best evidence that bacterial communities signal information about their hosts comes from human armpits," with different bacteria metabolizing different organic compounds (i.e. testosterone, sebum, sweat) to produce different odors, which humans also use as recognition cues, with each person likely having an individual "microbial signature" that reflects "genotype and/or kin relationships."[42] The wild baboon study, coupled with evidence about microbial signatures, indicates what looks like a microbiota-animal behavior feedback loop, in which animal social behavior both influences microbiota composition and microbiota composition in turn influences animal social behavior. If we recognize relationships (or the lack thereof) through nonconscious odor cues, we may be inclined to communicate and interact differently toward different people, which could possibly interfere with our desires for and conscious efforts at inclusion and tolerance.

Bacteriologists are beginning to inquire into how microbiota composition and human cultural behaviors interact, which creates more space for rhetorical theorists to join in their conversations. For example, Panchin *et al.* note that "[s]ome microorganisms would gain an evolutionary advantage by encouraging human hosts to perform certain rituals that favor microbial transmission"; they hypothesize that "certain aspects of religious behavior … could be influenced by microbial host control and that the transmission of some religious rituals could be regarded as the simultaneous transmission of both ideas (memes) and parasitic organisms."[43] This hypothesis is still untestable because it would require advances in genome sequencing, but it presents a perhaps exaggerated imagining of how microbiota might influence human beliefs and values. While it is fascinating to consider the implications of this hypothesis, it nonetheless has the potential to degrade many cultural behaviors—not just religious ones—as irrational, and to explain away the behaviors of other groups as mindless. However, it is based upon evidence about bacterial influence upon human and animal sociality. Where do we draw the line when turning to scientific research in hopes of learning more about human behavior patterns? How do we rely on scientific explanations for group behavior while avoiding ethical landmines? Ignoring biological influence upon group behavior may be as unethical as resorting to biological determinism. We need to strike a balance between critiquing bacteriological discourse and using it to gain new insights into human emotion and communication. Teresa Brennan argues that the turn away from affect in group psychology research reflects "the turn toward the self-contained individual"; she links the belief in the "self-contained state" with the trend in social movement literature, which despite good intentions shows "the same escalating prejudice ... caught up in the very process it should be analyzing ... the idea of self-containment is tied to the belief that cognition, more than emotion, determines agency ... as [self-containment] comes to dominate in the history of ideas, so does [cognition]."[44] If the idea of individual cognition can be linked to national self-containment, then bacteriology's theories of distributed cognition and affect might be able to counteract ideas of self that underpin the self-contained state.

Understanding bacterial communication, both inside and outside of humans, may help us to better understand human group behavior in the realms of crowd psychology, military training, and education practices. However, we would need to find a way to discuss bacterial suasion without directly and reductively applying it in the hopes of explaining away the unknowns of human behavior, for this seems to be where we may encounter ethical pitfalls. Encouraging further research into human pheromones, Archie and Theis cite a study finding that bacteria "play a causal role in invertebrate social behavior ... by producing pheromones that promote cohesion in locust swarms," though "similar research on vertebrates lags far behind."[45] While we cannot expect the same results among humans, neither should we be surprised to find similarities. Brennan argues that "research on entrainment by olfactory and other sensory means accounts for situations where people act as of one mind [and] points to a reevaluation of evidence for the postnatal social shaping of factors we think of as biologically given: human hormones."[46] What Brennan calls postnatal social shaping could be complimented by microbial endocrinology's research into human hormone production. She nominates human hormones as "the leading candidates for how affects are transmitted, as they involve the deployment of all the senses."[47] Hormones do not act reductively, in isolation, but are integrated into the body-brain processing of information. Interested in biological factors in group cohesion, Brennan asks, "[i]f olfactory communication turns a hormone into a pheromone and changes another's affects, does it also change their hormones in a way that (temporarily) changes their habitual affective disposition? Are such changes, in turn, communicated by additional pheromones?" She predicts that if "such cycles can be shown to hold in groups, then the contagion of affects has been explained."[48] Though Brennan is correct in suggesting the value of inquiring into how hormones become pheromones and what this means for group psychology, we must also be wary of using biology to fully explain something like the contagion of affects. Bacteriology can help us interrogate the myth of the self-contained individual and to blur the distinction between nonconscious persuadability and conscious rhetorical practice, but our biology exists in feedback with cultural factors. Microbial influence on the transmission of affect brings us to the nether-reaches of understanding nonconscious bodily communication. While we should not be surprised to learn that our co-evolutionary symbionts have afforded us some of their own social capacities, they are not the only influence upon us.

CONCLUDING THOUGHTS

Recall Diane Davis's concern with the "panicked deflection" of the Darwinian revelation, which manifests as rhetoric's requisite "engagement with the symbolic." If microbes do promote human group cohesion via pheromone production, then we must address the postsensation function of language accordingly, as well as the cultural systems—ideas, philosophies, ethics, religions—that bind (and repel) human societies. Where Gunn calls for attention to instances of uncontrolled speech to parse cultural values around self-control, I argue that we look to the hologenome to parse cultural values of self-containment. In 1966, Lynn Margulis put forth her endosymbiotic theory of evolution (or symbiogenesis): that eukaryotic (plant and animal) cells developed when one bacteria ingested, but was unable to digest, a smaller bacteria—which then evolved into eukaryotic organelles like mitochondria and chloroplasts. In the 1980s, DNA sequencing allowed her theory to gain wide acceptance. In light of this agonistic cellular synthesis, we might hazard to say that we are bacteria, all the way down, uniquely arranged. After a decades-long feud with Margulis, *The Selfish Gene* author Richard Dawkins conceded that cooperation and altruism may be vital components of evolution,[49] though he primarily holds fast to his original notion of the individual selfish gene that drives evolution. However, attention to creatures' entire genomes plus environment is beginning to leverage more explanatory power than does attention to the goals of isolated genes, a hopeful prospect given the cultural sway that science holds forth. David Dobbs culls research on gene expression and plasticity to explain the genetic (non)difference between grasshoppers and locusts.

> [E]very locust was, and technically still is, a grasshopper—not a different species, but a sort of hopper gone mad. If faced with clues that food might be scarce, such as hunger or crowding, certain grasshopper species can transform within days or even hours from their solitudinous hopper states to become part of a maniacally social locust scourge … these phase changes … occur when crowding spurs a temporary spike in serotonin levels, which causes changes in gene expression so widespread and powerful they alter not just the hopper's behaviour but its appearance and form. Legs and wings shrink. Subtle camo colouring turns conspicuously garish. The brain grows to manage the animal's newly complicated social world, which includes the fact that, if a locust moves

> too slowly amid its million cousins, the cousins directly behind might eat it. ... Does something happen to their genes? Yes, but … they don't mutate or in any way alter the genetic sequence or DNA. Nothing gets rewritten. Instead, this bug's DNA—the genetic book with millions of letters that form the instructions for building and operating a grasshopper—gets reread so that the very same book becomes the instructions for operating a locust... . Same genome, same individual, but … quite a different beast … Perhaps better then to speak not of genes but the genome—all your genes together. And not the genome as a unitary actor, but the genome in conversation with itself, with other genomes, and with the outside environment … it's not a selfish gene or a solitary genome. It's a social genome.[50]

Nonhumans offer rhetorical studies the opportunity to reassess assumptions about individualism and agonism. A hologenomic re-orientation might help lay culture to retrace the contours of the social, of rhetoric, and rational deliberation. Entertaining the idea that human behavior evolved so that microbes could be more easily transmitted across our bodies may humble us, but reconceiving ourselves as biological networks could open new avenues for thinking the body and culture. This could also afford a rich material perspective on what Davis calls "an always prior rhetoricity, an affectability or persuadabilty," based upon "the *exposedness* of corporeal existence," making rhetoric not "an essence or property '*in* the speaker' ... but an underivable obligation to respond that issues from an irreducible relationality." [51] This irreducible relationality may be uncomfortable, as is the difficulty of addressing "communal sensation without presuming *sameness*," as Hawhee puts it.[52] Humans strive toward immaterial, rational ideals to elide material, bodily constraints. Understanding our collective efforts at rationality as attempts to transcend bodily impulses, we might come to see deliberative rhetoric, both our own and other groups', as less alien and more contingent, revisable, and akin to our own.

NOTES

1. Melissa B. Miller and Bonnie L. Bassler, "Quorum Sensing in Bacteria," *Annual Review of Microbiology* 55 (2001): 167.
2. James A. Shapiro, "Bacteria are Small but not Stupid: Cognition, Natural Genetic Engineering, and Socio-bacteriology," *Studies in the History and Philosophy of Biological and Biomedical Sciences* 38 (2007): 807.
3. George A. Kennedy, "A Hoot in the Dark: The Evolution of General Rhetoric," *Philosophy & Rhetoric* 25 (1992).
4. Diane Davis, "Creaturely Rhetorics," *Philosophy and Rhetoric* 44 (2011): 88.
5. Ibid. 92.

6. Augusto J. Montiel-Castro *et al.*, "The Microbiota–Gut–Brain Axis: Neurobehavioral Correlates, Health and Sociality," *Beyond the Borders: The Gates and Fences of Neuroimmune Interaction* 7 (2013).

7. Seth R. Bordenstein and Kevin R. Theis, "Host Biology in Light of the Microbiome: Ten Principles of Holobionts and Hologenomes," *PLoS Biol* 13 (2015).

8. Thom Scott-Phillips, *Speaking Our Minds: Why Human Communication is Different, and How Language Evolved to Make it Special* (London: Palgrave MacMillan, 2014), 44.

9. Thom Scott-Phillips *et al.*, "Combinatorial Communication in Bacteria: Implications for the Origins of Linguistic Generativity," *PloS One* 9.4 (2014): 1–2.

10. Debra Hawhee, *Moving Bodies: Kenneth Burke at the Edges of Language* (Columbia: University of South Carolina Press, 2009), 156.

11. Scott-Phillips *et al.*, "Combinatorial Communication," 1–2.

12. Ibid. 1–2.

13. Daniel M. Cornforth *et al.*, "Combinatorial Quorum Sensing Allows Bacteria to Resolve their Social and Physical Environment," *Proceedings of the National Academy of Sciences* 111.11 (2014).

14. Ibid. 4280.

15. Ibid. 4283.

16. Scott-Phillips *et al.*, "Combinatorial Communication," 1.

17. Alex C. Parrish, *Adaptive Rhetoric: Evolution, Culture, and the Art of Persuasion.* (Florence: Taylor and Francis, 2013), 47.

18. Shapiro, "Bacteria are Small but not Stupid," 812.

19. Pamela Lyon, "The Cognitive Cell: Bacterial Behavior Reconsidered," *Frontiers in Microbiology* 6 (2015).

20. Hawhee, *Moving Bodies,* 157.

21. Ibid. 157.

22. Mark Lyte, "Microbial Endocrinology and the Microbiota-Gut-Brain Axis." In *Microbial Endocrinology: The Microbiota-Gut-Brain Axis in Health and Disease*, ed. Mark Lyte (New York: Springer, 2014), 3.

23. Ibid. 5.

24. Ibid.

25. Ibid. 6.

26. Alan Watts, *The Tao of Philosophy* (Vermont: Tuttle Publishing, 1995), 11–12.

27. Sigmund Freud, *Civilization and its Discontents,* trans. James Strachey (New York: Cape and Smith, 1930), 4.

28. Brian Massumi, *Parables for the Virtual* (Durham: University of North Carolina, 2002), 14.

29. Davis, "Creaturely Rhetorics," 89.

30. Bordenstein and Theis, "Host Biology in Light of the Microbiome."

31. Ibid.

32. Kevin R. Theis *et al.*, "Getting the Hologenome Concept Right: An Eco-Evolutionary Framework for Hosts and their Microbiomes," *Msystems* 1 (2016): 1.

33. Ibid. 4.

34. Ibid.

35. Joshua Gunn, "On Speech and Public Release," *Rhetoric and Public Affairs,* 13.2 (2010): 191.

36. Ibid. 192.

37. Ibid. 191.

38. Ibid. 211.

39. E. P Greenberg, "The New Science of Sociomicrobiology and the Realm of Synthetic and Systems Ecology." (2011): 213.

40. Jenny Tung *et al.*, "Social Networks Predict Gut Microbiome Composition in Wild Baboons" *eLife* 4 (2015): 8.

41. Ibid.

42. Elizabeth A. Archie and Kevin R. Theis., "Animal Behaviour Meets Microbial Ecology," *Animal Behaviour* 82.3 (2011): 427–28.

43. Panchin *et al.*, "Midichlorians—the Biomeme Hypothesis: Is there a Microbial Component to Religious Rituals?," *Biol Direct* 9.1 (2014): 4.

44. Teresa Brennan, *The Transmission of Affect* (Ithaca: Cornell University Press, 2004), 62–63.

45. Archie and Theis, "Animal Behaviour Meets Microbial Ecology," 431.

46. Brennan, *The Transmission of Affect*, 52.

47. Ibid. 53.

48. Ibid. 72.

49. Richard Dawkins, cited by Lynn Margulis in "Gaia is a Tough Bitch." In *The Third Culture: Beyond the Scientific Revolution,* ed. John Brockman. Simon and Schuster, 1995.

50. David Dobbs, "Die, Selfish Gene, Die," *Aeon Magazine*, 12/03/13, accessed 07/23/17, https://aeon.co/essays/the-selfish-gene-is-a-great-meme-too-bad-it-s-so-wrong

51. Davis, "Creaturely Rhetorics," 89.

52. Debra Hawhee, "Rhetoric's Sensorium," *Quarterly Journal of Speech* 101.1 (2015): 13.

BIBLIOGRAPHY

Archie, Elizabeth A., and Kevin R. Theis. "Animal Behaviour Meets Microbial Ecology." *Animal Behaviour* 82.3 (2011): 425–436.

Bordenstein, Seth R., and Kevin R. Theis. "Host Biology in Light of the Microbiome: Ten Principles of Holobionts and Hologenomes." *PLoS Biol* 13 (2015): e1002226.

Brennan, Teresa. *The Transmission of Affect*. Ithaca: Cornell University Press, 2004.

Cornforth, Daniel M. *et al.* "Combinatorial Quorum Sensing Allows Bacteria to Resolve their Social and Physical Environment." *Proceedings of the National Academy of Sciences* 111.11 (2014): 4280–4284.

Davis, Diane. "Creaturely Rhetorics." *Philosophy and Rhetoric* 44 (2011): 8E4.

Dobbs, David. "Die, Selfish Gene, Die." *Aeon Magazine*. December 3, 2013. Accessed July 23, 2017. https://aeon.co/essays/the-selfish-gene-is-a-great-meme-too-bad-it-s-so-wrong

Freud, Sigmund. *Civilization and its Discontents.* Translated by James Strachey. New York: Cape and Smith, 1930.

Greenberg, E. P. "The New Science of Sociomicrobiology and the Realm of Synthetic and Systems Ecology." In: Institute of Medicine (US) Forum on Microbial Threats. The Science and Applications of Synthetic and Systems Biology: Workshop Summary. Washington, DC: National Academies Press (U.S.) (2011). A7

Gunn, Joshua. "On Speech and Public Release." *Rhetoric and Public Affairs* 13.2 (2010): 175–215.

Hawhee, Debra. "Rhetoric's Sensorium." *Quarterly Journal of Speech* 101.1 (2015): 2–17.

———. *Moving Bodies: Kenneth Burke at the Edges of Language*. Columbia, SC: University of South Carolina Press, 2009.

Kennedy, George A. "A Hoot in the Dark: The Evolution of General Rhetoric." *Philosophy & Rhetoric* 25 (1992): 1–21.

Lyon, Pamela. "The Cognitive Cell: Bacterial Behavior Reconsidered." *Frontiers in Microbiology* 6 (2015). https://doi.org/10.3389/fmicb.2015.00264

Lyte, Mark. "Microbial Endocrinology and the Microbiota-Gut-Brain Axis." In *Microbial Endocrinology: The Microbiota-Gut-Brain Axis in Health and Disease*, edited by Mark Lyte, 3–24. New York: Springer, 2014.

Margulis, Lynn. "Gaia is a Tough Bitch." In *The Third Culture: Beyond the Scientific Revolution,* ed. John Brockman. Simon and Schuster, 1995.

Massumi, Brian. *Parables for the Virtual.* Durham: Duke University Press, 2002.

Miller, Melissa B., and Bonnie L. Bassler. "Quorum Sensing in Bacteria." *Annual Reviews in Microbiology* 55 (2001): 165–199.

Montiel-Castro, Augusto J., *et al.* "The Microbiota–Gut–Brain Axis: Neurobehavioral Correlates, Health and Sociality." *Beyond the Borders: The Gates and Fences of Neuroimmune Interaction* 7 (2013): 63–78.

Panchin, Alexander Y., Alexander I. Tuzhikov, and Yuri V. Panchin. "Midichlorians—The Biomeme Hypothesis: Is there a Microbial Component to Religious Rituals?" *Biol Direct* 9.1 (2014): 1–14.

Parrish, Alex C. *Adaptive Rhetoric: Evolution, Culture, and the Art of Persuasion.* Florence: Taylor and Francis, 2013.

Scott-Phillips, Thomas C., *et al.* "Combinatorial Communication in Bacteria: Implications for the Origins of Linguistic Generativity." *PloS One* 9.4 (2014): e95929.

Scott-Phillips, Thomas. *Speaking Our Minds: Why Human Communication is Different, and How Language Evolved to Make it Special.* London: Palgrave MacMillan, 2014.

Shapiro, James A. "Bacteria are Small but not Stupid: Cognition, Natural Genetic Engineering, and Sociobacteriology." *Studies in the History and Philosophy of Biological and Biomedical Sciences* 38 (2007): 807–19.

Theis, Kevin R., *et al.* "Getting the Hologenome Concept Right: An Eco-Evolutionary Framework for Hosts and their Microbiomes." *Msystems* 1 (2016): 1–6. e00028–16.

Tung, Jenny, *et al.* "Social Networks Predict Gut Microbiome Composition in Wild Baboons." *eLife* 4 (2015): e05224.

Watts, Alan. *The Tao of Philosophy.* Vermont: Tuttle Publishing, 1995.

Part II

EXPANDING BOUNDARIES—EXTERNALLY

Chapter 5

The Biotic Turn in Rhetoric

Ethical Internatural Communication as Suasory Peacebuilding

Ellen W. Gorsevski

> Many years [later] ... . I find myself back at the same point, seeking a communication discipline that is inclusive of all animals—indeed of all life—and views the theoretical resources of our discipline as starting points for a greater understanding of how best to live together.
>
> —Emily Plec[1]

Nonviolent and non- or less-coercive forms of human-to-nonhuman animal communication have long flourished but have received less attention than prevalent violent modes, which have tended to receive the lion's share of attention in research. Following in the steps and spirit of Frans de Waal's[2] work emphasizing the nonviolence of the bonobo chimpanzees, and Alison Suen's work "reconceptualizing language and reconsidering what it means to be a 'speaking' animal,"[3] I offer this chapter as a means to contribute to resources moving Communication studies in general and critical rhetorical and discourse analyses in particular to swing toward a biotic sensibility. This chapter offers two productive means to make this shift. First, I offer transitional definitions of key terms that will facilitate the telos of discussing other sentient beings communicating with receptive humans in nonviolent ways that may be considered to be rhetorical or suasory. Second, I present, as evidence of this transition in Communication and rhetoric, representative scientific knowledge of experiential learning and communicative interactions between humans and nonhuman animals; these exemplars demonstrate that greater levels of fair, just, and equitable interactions among humans and other sentient beings exist as nonviolent means of communication. I suggest that this trajectory of mutual learning and communication between and among sentient beings could yield productive insights for ethical rhetorical theory and for the conduct of rhetorical criticism that is less human-centric and more attuned to the biotic world.

TRANSITIONAL TERMS TOWARD A BIOTIC TURN

While established scholars of rhetoric like George Kennedy and Debra Hawhee have pondered the place of what have commonly been referred to as "nonhuman animals" in rhetorical theory, they did so in "special issue" formats of mainstream Communication journals that publish research on rhetoric.[4] Further afield, in select anthologies such as this one, and likewise with Emily Plec's edited volume *Perspectives on Human-Animal Communication: Internatural Communication*, scholars of rhetoric engage with nagging existential questions about relationships and communicative interconnections among people and creatures on earth, all while peering from the outside in to the heart of Communication. Before substantive, topical, and rhetorical issues in communication may circle into the field's vibrant center via its journals, questions will continue to be debated surrounding the artifice and artificiality of the bifurcation between human communication and so-called "nonhuman animal" communication. The discussion that follows here stems from important contributions to date, in hopes that this addition may add to the process of conjoining what have thus far been referred to clunkily as "nonhuman animal communication" with mainstream literature about communication that has long been implicitly, explicitly, and solely focused on that of people as human animals. Paying homage to its theory-based origins in critical animal studies (CAS), while extending both theoretical and practical concerns to the field of Communication, this chapter, then, arises from the merging of human and nonhuman animals as a *humanimal* turn in rhetoric. This is a transitional phase that is moving the field in a continuous shift toward a fully *biotic* turn.

This chapter will draw on and advance key interconnections between CAS and Communication studies, linking rhetorical and critical discourse analysis (CDA) with exciting innovations in multidisciplinary inquiry. Theoretical and practical connections of human and nonhuman animal interrelationships facilitated by internatural communication are instantiating a reconceptualizing of rhetoric. In this chapter, I will offer as evidence some examples of internatural communication to illustrate how and why *humanimal* persuasive interaction operates in ways that can be similar to and different from person-to-person communication. In offering pivotal examples of nonviolent suasory interspecies interactions, this chapter is designed to move the needle in internatural communication away from its presently rather apocryphal status and toward a more mainstreamed acceptance of humans' existential, moral, and ethical responsibilities to animals and how we view and interact with them.

As I will demonstrate, one implication of greater appreciation for internatural communication is that some of its forms that are nonviolent also share implications for human communication.

Viewing and interacting biotically in the communicative sense encompasses rhetorical epistemology and ontology. People both characterize and interact with nonhuman animals discursively, symbolically, and in embodied and sensory ways. This chapter invites possibilities for humanimal orthopraxy and orthodoxy, which can help to transition from predominantly violent modes of humanimal communication and interactions to more nonviolent ones. In brief, this chapter highlights as instructive some less well known theoretical inquiries and practices of interaction and communication among species, including humans and other animals, ranging from aquatic life, to large and small mammals, to insects. For space limitations, I will leave out covering microorganisms here, as that is ample fodder for a whole separate study unto itself.

Rhetorical studies have in recent decades expanded into queering, peacebuilding, and environmental areas of the field of Communication. Violence, particularly in forms of communication that normalize and instantiate racism, sexism, classism, homophobia, and environmental destruction, has been decried by foremost scholars in our field. Considering this justice-based arc of inquiry, so, too must rhetorical theory grapple with detecting and rejecting speciesism (prioritization of humans' needs over nonhuman animals' needs). Recognizing *humanimal* communication is the first step to shifting terms, symbols, and resulting interactions toward a more holistic sensibility of *biotic* communication relationships that form an integral part of striving for peace with justice. From Dr. Martin Luther King, Jr. to peace and conflict studies (PACS) scholars/practitioners such as Johan Galtung, *peace with justice* is referred to (in PACS literature) as "positive peace," meaning legal, cultural, and other structural supports have been created and maintained and are being actively upheld to ensure equality and fairness in the treatment of all human beings, regards of race, class, age, sexual orientation, nationality, and other determinants of existential being.[5] Then, by including nonhuman animals into the fold, this chapter explores the implications of communicative *humanimal hybridity*, with a shift toward a *biotic* turn within rhetorical studies.

Humanimal hybridity refers to interdependency between and among humans and other nonhuman animals. Take, for instance, "seeing-eye dogs" that are trained, and work for a living to the end of their lifespans, to assist visually unsighted persons. Dogs are being trained to sniff out illnesses such as certain kinds of cancer, or low blood-sugar levels, in humans. There are also bomb-sniffing dogs ubiquitous in airports these days. In these instances, the trained dogs' skills are being used by people, so it is a hierarchical arrangement with no 'choice' in it for the dogs. That said, the dogs also benefit from and depend on people to be well fed and have their veterinary health maintained. In cases such as these, people's lives very much depend on the dogs, whether it is at a busy crosswalk in town or at an airport, or at home in the middle of the night when a diabetic might fall into insulin shock without a dog to wake the person and prompt her to take insulin. While most dog lovers would recognize that a dog wagging its tail is communicating something akin to the dog equivalent of joy, only recently has science confirmed that human nonverbal communications of affection in the form of hugging one's dog typically causes discomfort or suffering to many dogs. Internatural communication that is anthropocentric, then, can be harmful. *Humanimal hybridity* connotes communication that recognizes such harms and seeks to avoid them and better yet, to promote nonviolent, peaceful interactions.

Another mammal example of symbiotic communicative interdependency that is the hallmark of *humanimal hybridity* is that many cities worldwide still employ horse-mounted police for traffic, crowd, and riot control. In prior generations of training horses, violent methods of "bomb-proofing" horses to remain calm in parades or other crowded settings included terrorizing horses repeatedly to attempt to create desensitization to loud noises. Today, better understanding of horses' behavioral and cognitive learning has led to far less invasive, more humane training and applications. For example, job or task selection of equine coworkers may be based on factors such as horse breed and personality, plus safety gear such as horse earplugs and blinders. Such *humanimal hybridity* is a function of real, pragmatic applications by humans with nonhuman animals that have yet to have been supplanted by electronic gadgets or robots. Cars and other motorized vehicles have largely supplanted the use of horses, donkeys, and mules for everything from basic farming (with exceptions such as American Amish communities) and transportation to the conduct of warfare in most developed parts of the world. From technological advancement, species preservation, and other science-driven reasons, there has been a biotic turn toward attempting to restore some balance to humans' treatment of and interactions with domesticated animals like dogs and horses and other endangered animals such as gorillas, elephants, or tigers. Worldwide, there is recognition that wild animals are under threat from human population explosions with concurrent cityscape incursions into formerly wilderness areas. Meanwhile, in developed nations, events ranging from Seaworld shows to circuses, which were culturally acceptable until around twenty years ago, today are facing pressures to change or close and are increasingly phasing out or ending many uses of wild animals for human entertainment.

On the other hand, for domesticated creatures, horse-riding event competitions and horse racing remain as popular as ever, as do dog racing and dog shows. Horse and dog racing have not been completely replaced by car racing or other kinds of motorized vehicle "sports" as popular forms of entertainment. At least some of the obsolete, brutal training methods for animals like horses and dogs have become frowned upon, and in many places in developed nations they are banned outright so as to ensure the safety and well-being of both humans and nonhuman animals. Such developments socially, culturally, and scientifically are part of the *biotic turn*. The *biotic turn*, which will be further elaborated here, refers to an *attempt to recuperate balance of ecological, environmental conditions that more closely approximate respect for all sentient beings, not just for privileged humans*.

Why are terms like "human animal" versus "nonhuman animal" problematic, and why am I deploying *humanimal* as a transitory discursive marker toward *biotic*? The problem of the appellation of the majority of biotic species on the planet via terms such as "nonhuman" is at least twofold when it comes to discipline specific analyses as well as multidisciplinary avenues of research that entail communication studies and especially rhetorical studies, the latter two of which are concerned with exploring contacts expressed audibly and nonverbally that influence interactions among interlocutors. First, the "non" in "nonhuman" implies—as ethically suspect—that any species that is *not* human is of a lesser significance or status in the natural world that we call the environment; *nonhuman* evinces a rhetoric of anthropocentrism. Second, segregating according to species as a mere thematic rather than integrating known interactants in felt and sensed forms of communication, such as audible, sensory, nonverbal forms of communication, favors humans over animals. Frans de Waal indicates that a corollary problematic to hierarchical segregation of species with humans at the top and animals below is as old as Aristotle's ancient but persistent schema.[6] DeWaal notes that in an effort to avoid anthropomorphizing, scientists instead have resorted to using excessively distancing discourse that has been persuasive in uncoupling humans from other animals, with dire ethical and material consequences. He suggests this discursive maneuvering has been "detrimental to cognitive science, as it has kept us from developing a truly evolutionary view. In our haste to argue that animals are not people, we have forgotten that people are animals, too."[7] Perhaps counterintuitively, discourse about animals that stringently attempts to avoid anthropomorphizing has often proven to be detrimental to animals' well-being.

Some qualified anthropomorphizing, such as human communicative concepts of "speaking" or "listening" are gaining support in CAS literature. For instance, Alison Suen decries masculinist philosophical treatments of human-animal communication that privilege reason over emotion and finds fault with animals' extradiscursive forms of communication. Suen posits that rather than saddling animals with the "burden" of proving their communicative merits based inequitably on human measures of "speaking," it is "our [human] responsibility to listen carefully" to animals' diverse means and modes of communication.[8] With the present *humanimal hybridity* phase that reaches toward a *biotic turn*, then, there is a recognition of the need to develop and apply a more balanced view, in theory and praxis, of the continuum of cognition and of communicative articulation among all creatures.

Similarly, until recently, much research in Communication in general and rhetoric in particular has fallen short of the mark in honoring a more ethical, critical orientation toward a *biotic sensibility of the interplay and interconnections in communication between and among all animals*, including humans as one among many kinds of animals. From cave paintings or stick figures inscribed as indigenous peoples' carvings into ancient rocks, comprising among the earliest recorded human artistic communicative conveyances of their life worlds or environments, animals such as deer and horses figure prominently and attain parity of perceived biotic status in visual representations that symbolize value and reverence.[9]

The recognition of speaking, writing animals like we humans engenders a downplaying of researchers' need to grasp what ephemerally here may be called *humanimal hybridity* as a transitional term that conveys the interdependence of people with other sentient beings. *Humanimal hybridity* is an evanescent moment that opens the way for productive discussions of *biotic* interplay in communication writ large and in rhetoric specifically. Anyone who has ever related to a pet, whether a beloved dog, cat, fish, or turtle, understands intuitively this nonhuman animal communication beyond and with the human community, and vice versa. The innovative work of Gala Argent makes way for researchers, animal lovers, and animal practitioners such as equine therapy coaches to lighten the load of some aspects of mainstream approaches, which Argent perceptively notes have overemphasized domination over animals.[10] Argent has explored how, increasingly, attempts at attaining nonviolent human/horse training and unobtrusive, subtle equestrian techniques have evolved to the point where a well-trained horse, when teamed with an equally well versed rider, can be ridden bareback, with no saddle, bridle, whip, nor any other accoutrements; both horse and rider communicate with soft touches, called "aids," moving in flow, in perfect synchrony. Following Argent's wise example, this discussion suggests the rhetorical choices that scholars make and reify in conceptualizing the *humanimal* to *biotic turn in communication* should purposefully advocate for a relationship that fosters a more holistic, less anthropocentric state of mutual respect and of reverence for all animal life forms (even—controversially—life forms that humans may end up using, eating, or processing for meat and hide). The research goal of reaching a state of greater equilibrium among humans and nonhuman animals teleologically and theoretically displaces an overemphasis of human domination over animals with a more equal emphasis on ways people can coexist with nonhuman animals in a state of *humanimal hybridity. Humanimality*, as a step toward the biotic turn, fosters a spirit of participation in mutually beneficial forms of verbal communication as well as extraverbal, tactile, sensory, and visual nonverbal communication.[11] Regarding assessing moral worth from language, Suen also suggests that humans' frequently botched attempts at understanding animal silences are often oppressive and can serve violence, so for Suen, it is both the communicative and noncommunicative *being* of animals as sentient beings that must be respected.[12]

THEORISTS CREATING SPACE FOR HUMANIMALITY

Humanimal is a heuristic explanatory working term to begin with, which makes sense from the outset for the purposes of broaching this discussion. Deeply informed by the CAS canon, *humanimal* refers to and suggests the drawing down of humans from their present pedestal of primacy among complex species, as a discursive and symbolic means to start to reject speciesism in rhetorical inquiry, and to reinsert humans back into being reconsidered as one among the broader category of complex biotic life forms. This discussion springs from key Western thinkers in the CAS canon, beginning with Peter Singer's utilitarian views, while encompassing gendered perspectives such as Catharine MacKinnon's and Carol Adams' feminist critiques of the politics, images, discourses, and practices normalizing violent oppression and treatment of both women and animals.[13] Since 1975, Singer has advocated to "Open the Cages!" on behalf of animals, lately showcasing their plight within food production, linking consumption of meat to climate change.[14] In the past two decades, CAS scholars have been productively expanding lines of inquiry, with Tom Regan's moral ideal of "empty cages" sharing Singer's utilitarianism, while being pragmatic and accommodating to "muddlers" as people who are conscious of animal suffering but are not sure what or how to do something about it.[15] Other CAS researchers influencing my thinking in this chapter include Mark Rowlands, who offers a contractualist perspective that sociopolitical treatment of sentient beings should be consistent whether one is a human or nonhuman.[16] Notably, James Rachels, like de Waal, uses the context of evolutionary theory to confer agency to nonhuman animals.[17] Gary Francione's work is helpful from a rhetorical perspective, since it bases an ethical stance on the *humanimal* continuum by pointing out inconsistencies in animal welfare discourses and uses humans' commonsense attitudes toward animals and other humans to advocate for "abolitionism," the ending of humans' using other animals the way that humans have historically enslaved other humans.[18] While Francione's absolutism is vital and laudatory, there is insufficient space in this chapter to address obviating internatural communication altogether; Francione's views put the cart before the horse, destroy the cart, and release the horse to run free.[19] I suggest that while humanity is capable of imagining and may praise such a scenario, for the present time being, my concern in rhetorical studies with ethical implications is that not only is the horse is still pulling the cart, literally and figuratively, but also the horse is often needlessly mistreated along the way. Thus in articulating *humanimal hybridity*, I mainly invoke CAS perspectives that are helpfully converged upon and advanced by an obscure (for most Americans) French philosopher, François Laruelle, whose ideas are gaining traction now that translations of his works have been recently made available in English. Laruelle uses traditional Western thought to open up discussions, proffering subversive ideas on humans' existential sharing of suffering with animals and humans' moral obligations to express in rhetoric equality of value and preciousness of both humans and nonhuman animals alike. This shared experience of life in his ethical philosophy enables conceptualizing what I call *humanimality* in the discussion that follows.

Humanimality leans toward the *biotic turn* in Communication studies; *humanimality* is enriched by and depends on the well-supported theoretical and practical scaffolding established through CAS scholarship, among other areas of study, such as philosophy and theology harnessed by François Laruelle. Sharing kinship with de Waal's critique of Aristotle's vertical segregation of god-man-animal, Laruelle contends that man-animal-plant (MAP) is a schema that has been rhetorically appealing but is morally bankrupt and outmoded. In his *General Theory of Victims*, Laruelle argues that there is a possibility for nonviolent rapport among people and other sentient beings.[20] Laruelle proposes that like humans, so, too, may animals be respected as actual and potential victims of suffering and violence. Laruelle invokes Christianity as one possibly suasory means of enhancing moral *humanimal* relations. He argues that just as Christ was crucified, that victimization today serves as a symbol for all Christians to feel kinship with the persecuted. The implication here is that it is morally objectionable and avoidable that humans persecute other humans and that humans persecute nonhuman animals, too. In comparative religious terms, the moral equivalent could be, for example, Buddhism's Noble Truth that the suffering of birth, life, and death are parts of every sentient being's existence. Thus, to aid another sentient being by preventing or alleviating suffering is a moral imperative. Laruelle suggests there is both agency and nonviolent potential in transforming shared suffering: "the victim is no longer exclusively defined by suffering, but has the capacity to rise up against the world's persecution. Based on this … a new ethical role [arises] for … intellectual in which s/he does not merely represent the victim, but imitates or clones …[her], thereby assisting the victim's uprising within thought."[21] This viewpoint synergizes with Suen's views that humans must imaginatively and actively *listen to and for* animals, not as a way to "give voice over" but rather to create a means for "listening and helping" animals as others to "articulate" their biotic messages. Laruelle, like Suen, finds transcendence of humans over violence (whether human to human or human to animal) is philosophically and materially possible and is, moreover, "a benedictory definition for the glory of God."[22] John Maoilearca offers a way to apply Laruelle's stance, taking us from the *humanimal* to a *biotic* perspective. Maoilearca affirms that

> Laruelle's work can be seen (or re-visioned) as performing something other than an inflation or deflation of either side of one fixed philosophical dyad (human/non-human); rather, it can be seen as unilateralising the couple, that is, expanding (amplifying, mutating, superposing) the meaning of the two alongside the One Real that resists any lone philosophical version's attempt to commandeer that meaning (and so the essence of what is human and what is nonhuman). It is the performative mutation that counts.[23]

In short, Maoilearca believes that Laruelle is offering an ethical stance via a bio-/theo-logical confluence, one that is suffused with practical potential, and one that hinges on communicative intent and a "performative" rhetorical action of "mutation." In other words, Laruelle articulates the moral, ethical, and communicative imperative of *humanimal hybridity*. This "performative mutation" means a *biotic turn* in communication is actionable. For Laruelle, humans' compassion for one another must "extend even to [animals]," so in all life forms from birth to death, there is a "universal life" that is shared, which implies an equality and equilibrium for seeking and applying standards of ethical conceptualizing and treatment of nonhuman animals and of biotic life forms that match standards understood for ethical representation and handling of humans.

In tandem with Laruelle's thinking, Dennis Soron offers a similarly poignant and nontheological contribution to theorizing *humanimality* through a Marxist study of the rhetoric of roadkill.[24] Soron argues convincingly that discursive, visual, and commodified products rhetorically portraying roadkill in a humorous light constitute structural violence. Soron offers ample examples to demonstrate the visual and product jokes about roadkill, ranging from art works to gummi candies, all combine unethically to function as a devaluing, violent mode of persuasion. Soron demonstrates the ways that jokes making fun of dead, run-over animals constitute powerful rhetorical arguments, which are directed in commodified form as candies that essentially teach children not to care about the needless suffering of nonhuman animals. Soron further suggests that roadkill jokes and products normalize car culture, which destroys domestic and wild animals, their habitats, and, via vehicle exhaust, is contributing to perilous climate change on the planet. In combining the logic of Laruelle's theologically grounded views calling for ethical relationships between humans and animals as existential partners, and Soron's decidedly Marxist approach, which applies CDA and critical rhetorical methods to the study of debased international representations, there is a strong basis for grounding theory and practice in *humanimality*.

EXAMPLES OF *HUMANIMALITY* AS SENSORY "MUTATION" RHETORIC

Humanimality is the ephemeral or, in Maoilearca's terms, the "mutation" phase of hitching theory to practice, which is heavily dependent upon philosophical formulations of interdependence in the natural world, which are then expressed through sensory and performative modes of communication. *Biotic* "universal life" refers to living organisms that depend on one or more ecosystems and that range in complexity from highly intricate sentient beings, like people and their pets, to simple microorganisms that one can see in a microscope. This is not to say, however, that people or pets should be equated with or treated like microorganisms in a petri dish. At the same time, this *humanimal* to *biotic* shift shares a measure of literal and symbolic substance with de Waal's critique of how Charles Darwin's evolutionary ideas have been oversimplified and misleadingly reduced to separating humans from chimps. Thus, while abjuring the *reductio ad absurdum* of equating humans with single-celled microorganisms in a petri dish, the *humanimal* move first takes a discursive, rhetorical, and symbolic step to conjoin human with animal. In equestrian sciences, for example, horseback-riding coach Sally Swift calls this state of being, a condition of one-ness in peacefully and nonviolently riding one's horse, the "horse-person."[25] Swift is the precursor to Argent's reconceptualizing of the objective ideal in equestrian activity as striving toward its potential for a greater measure of nonviolent *humanimal hybridity*. A rider could harshly jerk on the reins to move a horse, causing pain to the horse. In Swift's teachings, such inexpert, fear and pain-based treatment of horses and callousness in riders is to be avoided. Instead, using many internaturally communicative sensory as well as human visualization techniques, Swift conceptualizes and explains that the "horse-person" communicates mutually; the rider is attuned to, anticipates, and responds to silent, non-, and extraverbal communicative expressions of the horse and vice versa. Optimally, the desired sensory and extraverbal, symbiotic communication between rider and horse obviates the need or reversion to violent, pain-based methods of equine pedagogy, benefiting both horse and rider. Thus enacting performatively, the "horse-person" constitutes a nonviolent communication of *humanimality*. To attain a state of being of "horse-person" is equal parts moral, physical, and rhetorical maneuver.

Stephen Newmyer, like de Waal, contends that as a means to omit animals from moral consideration and treatment, "no human capacity has been so jealously defended as unique to humans as has language possession."[26] Fortunately, the field of Communication, with its emphases on intercultural and extraverbal forms of "language," has demonstrated effectively that written or discursive language is no longer the sole marker or measure of study; rather, communication in all its rich diversity is open to all manner of exploration to enhance understanding, reduce, or prevent conflict and to make the world a better place. Communication is rhetorical, and rhetorical inquiry within this broader context expands the possibilities of rhetoric to effect meaningful, ethical changes to enhance internatural interactions among all sentient beings.

Rights and protections are highly rhetorical modes of communication expressing internatural relationships, which set the tone for the kinds of language and *humanimal* interactions that occur within those auspices. Rights or protections are established for animals that are dependent on humans for complicated reasons including civilization, culture, livelihood, and so forth; some discourses of ownership may be viable forms of human representation of nonhuman animals within the biosphere. For instance, at the horse barns where riders board their horses, each horse is legally "owned" by the rider, whose legal "ownership" for that horse sets in place an economic and legal protection for this domesticated animal so that said horse can be fed, provided with medicines, and tended and cared for. For many people with horses, the blunt characterization of "ownership" of a horse falls short, since it confers superiority in multiple ways, such as anthropomorphically or legally. Instead, "horse people" tend to use parental or familial terms, such as being the horse's "mom." When people ask, horse people may use conventional discourse, such as "here's my horse." This particular phrase "my horse" may also confer familial sense of kinship, similar to "here's my sister." For the time being, given legal and rhetorical constraints, with regard to engaging in and supporting a *humanimal* turn, many horse people like me are works in progress. For although on both a personal level and on a theoretical level, I reject the notion of overly deterministic, simplistic, and bifurcatory human/animal relationships of "ownership," on a practical and methodological level, it is grudgingly conceded that horse people must both maintain and communicatively express legal "ownership" of a horse to support a healthy, stable environment for the horse. Insofar as nonverbal expressions of horse mirth that horse playfulness indicates, optimal horse-keeping conditions foster "owners'" awareness of horse signals denoting emotion, such as of being content. How does a horse evince or communicate a "happy" state? Horses are understood by humans to communicate with each other and with humans, a measure of what for lack of better words may be called "horse happiness," is expressed auditorily by sounds that we can hear, such as a "nicker," and which may be expressed nonverbally by a horse when she runs up to greet a human, such as a regular rider or feed-bringer—who may constitute something like a peripheral or non-horse-herd member—when the horse notices the human is there, or if she comes trotting over in greeting when the human calls the horse by name. Argent's work reconceptualizing horse-human communication opens the door for a biotic turn in rhetoric as a

> mode of interaction [that] is at least as potent as human verbal "language," … [since] horses are more highly attuned at nonverbally reading intentionality than are humans. ... [A]n intra- and interspecific model of assessing nonverbal communication and embodied synchronous movement, when coupled with an allowance for the different abilities and types of intelligences afforded by these modes of interacting, can open doors for new appreciations of animal minds.[27]

As with Argent's important recognitions of nondominant and nondominating relationships between humans and horses,[28] there are nonviolent relationships and forms of interspecies communication possible among diverse contact partners, including auditory, tactile, visual, and nonverbal interactions. The bridle, which is a system of leather head-straps, is an example of riding tack that can run the gamut from extremes such as being cruel and painful in some kinds of bit bridles, to being a device for nonviolent horse-person communication, as in nonbitted hackamores or bitless bridles. For instance, "the earliest artistic evidence of use of some form of bitless bridle [is] … in illustrations of Syrian horsemen, dated approximately 1400 B.C."[29] The newest, contemporary version of the bitless bridle is gaining popularity today in the feminist, pain-free equine movement.[30] From an ecofeminist perspective, it is possible that the bitless bridle is becoming more popular now that women, more so than men, are participating in equine sports and gaining experience with international communication between women and horses. Earlier, patriarchal methods of horse "breaking," and primary usage of horses for transportation and warfare, were often violent and painful. Today, horses are seldom used for most warfare. Considering the prevalence of women in training and equine sports, using pain, fear, and violence-based methods for the training or upkeep of horses is increasingly being questioned, eschewed, or banned. Following World War I, American and European armies decommissioned horses and replaced them with motorized vehicles, with a few exceptions such as small numbers of horses used ceremonially for military parades, or occasionally for police usage in big-city traffic or crowd control. The move away from military horsemanship toward mostly sports or recreation oriented "horsewomanship" also created more gender balance because no longer were horses used mainly by men for men's pleasure, profit, or goals. In the ensuing decades, women have become numerically and economically influential in transforming horse/human interactions. For instance, in the US Equestrian federation, which is the equestrian equivalent of or akin to the NBA for basketball or the NFL for football, a whopping 85 percent of members are women.[31] In short, ecofeminist, internatural interactions between women and horses are having a major impact in developed nations on horses and humans alike through women's pressing for nonviolent changes in equine training, in which internatural communication is key.

Similarly, for millions of pet owners who live and often sleep with their animal companions, such as dogs or cats, there is also an intuitive understanding of this experience of "animal-person-ness." Cesar Millan, the popular so-called "dog whisperer," refers to this relaxed, mutually empowered state as one that is an expression and sharing of "calm, assertive energy."[32] Such experts in observing animal behaviors and encouraging appropriate human nonverbal communication to foster healthy, nonviolent, functioning relationships between people and animals are making some headway in gaining a hearing from skeptics. It is important to emphasize the distinction between mere animal trainers, and those, like Swift or Millan, who operate under a premise of *humanimal hybridity*. As Millan puts it, "I'm a dog behaviorist, which means that I rehabilitate dogs, and I train people. It's an important distinction, and one that can be hard for people to understand."[33] Such behavioral expertise entails seeking, reading, and learning animals' communicative cues. What Swift, Millan, and Argent are conveying is part of their own substantive, actionable, and rhetorical shifts in their chosen terminologies to express forms of *humanimal hybridity*: these burgeoning communicative utterances constitute rhetoric that is supported by ethical moral philosophy, which is made actionable in the real world, and which moves both theory and practice closer to a *biotic turn*.

Thus far in this discussion, much of the familiarity with pets is what theorists and practitioners are using to facilitate the biotic turn, but what are we to make of nondomesticated wild animals and their relationships to and communications with people? Granted, many human-to-pet communicative discourses and rhetorical characterizations have justifiably been dismissed as merely anthropomorphism. However, in wilderness studies and exploration as well as in ecofeminist scholarship and activism, there has been an increasing recognition that the rhetorical configurations from the domination model of "man versus animal" are outdated. For instance, in wilderness and conservation education, there has been a gradual shift in understanding and reconfiguring discursive and symbolic portrayals of predators in wilderness contexts, whether it is in national park settings or in the world's oceans. Long-time experienced sea divers such as "shark whisperer" Andre Hartman and sea conservation writer Joseph Dovala emphasize, for example, that outdated misconceptions and symbolizations of sharks have contributed to sharks' rapid decline in numbers worldwide, with many shark species on a path toward extinction.[34] A veteran of free dives (without any protective gear or cages), Dovala asserts,

> The week I spent in the company of someone who knows these animals from over 30 years of diving with them helped me reach a new awareness. Day after day I could get into a rhythm with whatever sharks came around. It didn't take long to see patterns emerge. Some liked to come in close while others kept their distance. Some paced around the reef taking the same path every time and some never repeated the same track twice. You could "feel" some of the bolder animals as they swam by, their dark eyes seeing, trying to understand what the noisy, clumsy, vertical beast was in their domain ... . I quickly got the hang of it. I made 23 more dives swimming with [a variety of shark species] tigers, lemons, nurse, black tip, grey, and hammerheads.[35]

Dovala, Hartman, and many other wilderness interactionists confirm there is sensory heightening process of non- and extraverbal communication. Dovala says one can "feel" the communicative interaction of *humanimality* between shark and diver. Like dog whisperer Millan, Dovala and Hartman articulate *humanimality* in international communication as energy-level modes of interaction and interbeing. Sea divers express sensory experiences of communicating with sharks as animals formerly thought to be single-minded killers, finding instead these animals to be highly intelligent, sensitive, and, importantly for this discussion, in large part, nonviolent, beings. Indeed, such kinetic conversations harken to George Kennedy's claims about rhetorical energy preceding communication in all animals. A corollary of energy-attuned communication in person-to-person health communication, for instance, is autism counselor Cress Spicer's experienced observation that "autistic children are very sensitive to the energy of others"; Spicer increases awareness of this "energy" as a message-laden nonverbal exchange to help attune educators and family members as a means to improving communication and overall interactions with autistic children.[36] Whether with animals or with dis- or alternately-abled humans, these kinds of articulations and purposeful practices of internatural communicative interaction constitute enabling, rhetorical spaces that hone a sensory communicative "feel" that is based on looking, listening for, reading, and responding to one's non-locutionary interlocutor's exertions of "energy," whether that nonverbal other is human or not.

As with human child behaviors, animal behavioral trainers such as Toledo, Ohio's Lara Joseph uses applied behavior analysis (ABA) from the field of psychology to rehabilitate animals.[37] Joseph, who has a Master's in ABA, has practiced this method for ten years as a means to prevent animals with behavioral issues from being euthanized for unwanted behaviors, and to enable the animals to be adopted by responsible people. Joseph believes that ABA is a significant interspecies communication tool that removes old fashioned circus styled "coercion," such as the use of violent forms of "force" meted out by things like "shock or prongs" from the interactional equation, replacing the harmful communicative contacts with positive reinforcement through the use of items like toys or food treats. Joseph has found that dogs with disabilities such as deafness, and parrots, which can live as long as humans, have responded particularly well to ABA as a nonviolent interspecies communication method. Joseph explains, "If an animal can see, hear or smell you—you're training it whether you realize it or not. Training is a form of communication, both verbal and nonverbal."[38] Recognized nationwide as a specialist and educator in animal behavior, Joseph offers a weekly live broadcast as well as an annual, three-day workshop event to train veterinarians, animal technicians, and staffers from various conservation centers or zoos.[39] While work such as Joseph's is far from ideal in terms of the views of animal rights advocates like Tom Regan, it is part of the *humanimal* turn toward a holistic biotic understanding of interspecies communication. This phase still privileges humans in terms of who is directing the interaction, yet the phase's uses of "energy" attunement and positive reinforcement in *humanimal* communication is greatly lessening or preventing the suffering of different others in the communicative transaction.

Arising from *humanimal* communication, there is the vital symbolic role in advocacy of human experts in championing the protection of wild species. As Suen puts it, presently "there is no such thing as a free-floating, autonomous animal voice free from human representation … [humans] cannot abnegate our obligation to" do so in a way that reduces harm or benefits nonhuman animals.[40] Rhetorical advocacy is an example of Suen's ideal of humans listening to animals. Suen reiterates that, rather than "speaking for the animal uncritically ... an animal solution requires us to listen to the animals as active participants in their liberation .... it requires us to problematize the linguistic divide."[41] The concept of *humanimality* participates in this problematizing by actively narrowing the gap. For instance, beyond the reciprocity of humanimal communication, "lion whisperer" Kevin Richardson also recognizes the rhetorical role that his own celebrity brings to symbolize and support awareness, fundraising, and conservation efforts for great African lions, which continue to be hunted at rates of about 1,000 a year in "canned hunts," despite the alarming decline in wild lion numbers related to causes such as loss of habitat and poaching.[42] Barbara Willard has critiqued place-based rhetorics that are similar to Richardson's, dismissing them as "bioregionalism," which she believes is insufficient in rhetorical potential to foster identification with broad audiences.[43] Yet *humanimalistic* suasory invention, from anthropomorphizing to celebrity or place-based marketing, does contribute to Laruelle's aim of including nonhuman animals into humans' existential fold by having prominent thinkers publically avow inclusivity. Thus, when enriched with sophisticated, politically voiced biological knowledge of animal physiology and psychology, increasingly scholars, researchers, daredevil conservationists, and hikers or divers into the world's wilderness areas are expressing *humanimality*, thereby contributing to the biotic turn.

Suen's work on linguistic and discursive refinements in conceptualizing *humanimality*, in concert with Laruelle's ethical, moral reconfigurements, form links that are pulling many fields, including Communication, toward theory and practices that are dissolving human and animal into all that is *biotic*. From a feminist, cultural studies perspective, Karin Weil underscores, "We may not always understand …[animals], but it is imperative that we acknowledge that they may have things to say."[44] In short, the *biotic* turn incorporates animals, both human animals and nonhuman animals. The conceptual merging of all biotic life forms in theory and practice encapsulates multidisciplinary, scientific and layperson attempts at achieving a literal and substantive shift in communicative interactions among beings. A *biotic* turn in rhetoric necessitates a humbling and diminution of the implicit rationale for humans to have their way with both biotic and abiotic forms on planet earth (presently, and possibly, given space travel, to do so on other habitable planets). An EBSCO search in 2017 using as key terms "biotic communication" returns about 4122 peer-reviewed research articles; add to the initial two key terms "rhetoric," and the articles yielded drops dramatically from thousands to a mere six articles. Many scholars, even the most animal-revering, animal-rights respecting researchers in all fields, including Communication and rhetorical studies, have yet to have made the shift of productively recasting and reconceptualizing human influence, including discursive power, regarding purposeful interactions of biotic life in our biosphere.

Multidisciplinary Convergence: Implications for Rhetorical Theory and Criticism

Rhetoric has long been, and continues to be, crucial in pressing for the remediation of human suffering worldwide. Interdisciplinarity with complementary areas of study, such as cultural studies, legal studies, economics, or human sexuality studies, has been part of rhetoric's vital role in advocating civil and human rights for those harmed by communicative practices that reify variables such as race, class, nationality, or sex and gender as lesser statuses. Communication scholars can equally embrace knowledge across disciplines such as biology, veterinary studies, conservation, and environmental studies to press for animal respect and rights. Hybridity in *humanimal* communication exists but presently lacks institutional legitimacy. From preservationists fighting poaching to ecofeminists pressing for creation and enforcement of nonhuman animal rights and recognition, to acknowledging that humans are animals too, theorists and practitioners are, in effect, enabling a recontextualizing of rhetorical theory within *humanimal hybridity*. I suggest that this active process, which theoretically and practically is altering metacommunication, international communication, and intranatural communication, can be applied equally well to enriching rhetorical theory in ways that contribute to peacebuilding and fostering justice for all sentient beings.

In creating rhetorical grounds for animal liberation, giving voice to animal suffering is a creative endeavor. Willard has critiqued the rhetoric of "bioregionalism," or place-situated activism, as a predominant faction and rhetoric within what she has categorized as "radical ecologists."[45] Willard has argued that poetic nature writing such as poet Gary Snyder's, or what she has referred to as "disclosive" rhetoric, ought to be considered as being ethically and pragmatically preferable to biotic rhetorics that convey human and animals' senses of place in the natural world through "performative" and "carnivalistic" discourses that she has questioned as being too alienating and sensationalistic, as well as for being tainted by anthropomorphism.[46] Willard's point is well taken regarding the inability of local people in public performances, such as protests, commemorations, or marches, to use rhetoric to persuade apathetic and increasingly ever urbanized larger publics to care about the natural world. On the other hand, Willard seems to have overstated the existence of understandings of fluidity and interconnectedness of biotic beings worldwide as part of her attempt to downplay the importance of the rhetoric of valuing a special place on the ground, a georeferenced and physical spot on the globe. Yes, carnivalistic rhetorics may alienate mainstream or apathetic audiences regarding the need to safeguard biotic community members in the natural world, but so too do dispassionate scientific discourses frequently appear off-putting and boring to those same audiences. Worse than being merely dry, de Waal's work suggests that scientific discourses have more often dulled humans' ability to imagine and to understand how other sentient beings communicate and exist in alternative yet equally worthy and viable ways apart from humans. Reimagining communication and suasory processes is a moral and practical necessity of an ethical evolution of philosophy and theory of communication; a rhetorical turn toward the biotic entails appreciating communication that is purposeful and goal oriented, and which occurs between and among the species sharing our earth.

A rhetorical theory and practice of rhetorical criticism that is oriented toward the natural world and the other nonhuman sentient beings within it can invite *humanimal hybridity* as a part of international communication to join the established canon. Somewhere in the middle ground of environmental discourses may be found representational symbols that capture and evince respect for place and for the biotic beings that exist in that place. Regarding both architecture and biotic species, there are, for example, World Heritage sites that logically commingle reverence for naturally built and human-built additions to the environment. A commitment to human-built environments that are in harmony with natural materials, such as rocks, and environmental processes, such as a dearth or excess of water, can provide this middle ground of rhetorical tropes of persuasion and identification in ways that formal or scientific arguments and informal or carnivalesque performances may not.

Anthropocentrism is a totalizing focus on human beings, which facilitates anthropomorphism, or the overemphasizing or misinterpreting of nonhuman animals' appearance, actions, or communications as being human-like, even when they are not. Insightfully, de Waal's work critiques overly scientistic discourses that enable callous disregard for and mistreatment of nonhuman animals; de Waal's research suggests that some levels of comparison or anthropomorphizing may actually gesture discursively and politically toward animal liberation. Anthropocentrism is implicit in emotionally distancing scientific lingo, which, while removing the appearance of goals, nonetheless presumes emphasizing humans' interests over that of nonhuman animals. Regarding the problem of anthropomorphism, Emily Plec concedes in her edited volume that, "the anthropocentric impulse holds fast" and that "some authors use the language of ownership to describe human relationships with companion animals while others make rhetorical choices that seek to challenge our ways of understanding interaction with other animals."[47] If some forms or expressions of anthropocentrism may at times or in certain contexts serve as a means to promote understanding and appreciation for communication as an interdependent channel to foster biotic equity and respect for the natural world as a fragile environment, then that kind of anthropomorphizing may not always be a negative aspect of humans' communicative portrayals of the biotic sphere.

In an ideal or utopian world, the current and longstanding conflicting and conflicted status of *humanimal* relations would be resolved by enacting Tom Regan's global goal of "empty cages" for all captive animals on the planet.[48] As a movement toward that laudable, albeit perhaps impossible, goal, this chapter has *linked rhetoric to participatory action of community within and for the biosphere.* This work acknowledges the reality of the difficulty—if not outright impossibility at least in the short term—of getting the majority of humans to cease dysfunctionally dominating the planet and its nonhuman animals. Abstaining from harming sentient beings by swapping reliance on cars for public transit, and by safeguarding wilderness areas globally, constitute two practical means among many that could help to lessen climate change. There are benefits from aspiring to Regan's emancipatory goals of freeing all animals while reaching attainable markers along the way. Present discourses expressing *humanimal* communication as a phase address Regan's uncompromising positionality, but reach Suen's practical conclusion that in contexts ranging from the everyday to the political, we humans must listen to, sense the energy of, and speak for nonhuman animals.

Insofar as studying persuasion is concerned, the trend of "animal whisperers" and proponents of ABA is interesting and relevant because it entails not only modifying the animal's behavior, but also, importantly, modifying the human's behavior too. Whether it is becoming "calm and assertive" to act as a pack leader to influence one's dog, or introducing a toy to elicit a desired behavior from a pig or parrot, these communicative modes are a palpable shift from outmoded, coercive, and often brutal techniques of *humanimal* interactions that have long but unnecessarily overshadowed interactions between people and their biotic counterparts. Especially in the past several hundred years, the lopsided coercive persuasive relationships between humans and their biotic counterparts in ever-disappearing wilderness areas have escalated to the present-day species extinction clock, fast ticking in the twentieth and twenty-first centuries. Today, the pressing need for a biotic turn that promotes nonviolent communicative exchanges and relationships among all biotic life forms is heightened by climate change.

Scholars Maneesha Deckha and James Stanescu have applied Judith Butler's research on ways that the concept of disabilities as human frailties and vulnerabilities can serve as a unifying trope.[49] In an exciting move toward the biotic turn in rhetoric, beyond unifying gender and sexuality in mainly human beings as Butler has done, Deckha and Stanescu extend Butler's notion of human "precariousness" and "vulnerability" to other animals as a means for understanding and respecting difference within and across species. Such acknowledgement and sharing of differences has implications for persuasive communication and for nonviolent conditions of existence that foster justice and peace. Deckha has studied legal discourses and implications and found that "Vulnerability theory offers a discourse that can avoid sameness arguments and thus better respect animals' differences from humans and from each other."[50] Elaborating into politics from the juridical realm, Stanescu adds to Deckha's recasting of relationships by implicating militarism into the artificial and destructive forces of violence that he argues need not be the measure or biopolitical rule for humans and other animal life forms. Stanescu maintains

> systems of immunity against our vulnerability are our biggest threat. Thus, we have wars in Iraq and Afghanistan to protect us; we have nuclear weapons to protect us; we have the Patriot Act to protect us; we lock up millions of people in the United States to protect us... . Immunity is always a function

> of disavowal, of a refusal to connect our shared embodied finitude. This is why Butler insists upon the bios of the animal, and insists that the human animal is always an *animal* ... . It is only by avowing our animality that we can avow our precariousness. To do this shifts everything. It changes ... who gets to be the bearer of rights.[51]

In short, as a tool for interspecies communication and persuasion, understanding difference and disabilities unites many forms of biotic, sensate beings' predictable and shared life experiences of vulnerability to suffering and ultimately to death. Laruelle's work takes these notions of shared suffering ever further into the theoretical territory that offers room for nonviolent transcendence in the form of action, led by ethical intellectuals whom he hopes inspire and enact change.

CONCLUDING DIRECTIONS FOR BIO-TURNING

In this chapter, I have emphasized some of the linguistic challenges to effecting changes in theory and practice in international communication generally and rhetorical inquiry specifically, and in offering examples of changes ongoing in multidisciplinary theory and international practices, I have pointed out that advances in international understanding bode well for conceptualizing, theorizing, creating, and enacting nonviolent rhetorics when humans and other sentient beings can merge into *humanimal hybridity*. The transitory notion of *humanimality* can foster movement toward a fully biotic turn in rhetorical studies. Interesting developments in multidisciplinary biotic studies, from biology to CAS to psychology and zoology, hold implications for enabling a reconceptualizing of rhetoric, exploring moral and pragmatic avenues into how and why persuasion works, particularly in interspecies interactions. This discussion has emphasized the exciting possibilities of shifting from predominantly violent modes of *humanimal* communication and interactions to the less well known but just as persistent inquiries and practices of interaction and communication among species including humans and other animals, such as fish, large and small mammals, and even insects.

Nonviolent and noncoercive forms of *humanimal* communication have long existed, although they may have been lost, forgotten, or have been misportrayed as less persuasive than the violent modes of energy as communicative exchange with which most people are, unfortunately, already familiarized. A shift toward a biotic sensibility presses for marrying scientific knowledge to common sense. This shift is exemplified by reaching for and sometimes attaining greater levels of just and equitable interactions among all sentient beings. Merging theory and practice regarding other-oriented communicating and caring is an ethically grounded (and ideally brief) phase of *humanimality*. *Humanimality* as a productive "mutation" promises to open up spaces for taking the "animal turn" along an evolution toward a unifying *biotic* turn.[52] This process suggests nonviolent means of communication are possible and have long been underway: it is up to all of us to enact, live, and share this paradigm advancement with every creature with whom we come in contact.

NOTES

1. Emily Plec, "Preface." in *Perspectives on Human-Animal Communication: Internatural Communication* (New York: Routledge, 2013): *xiii*.
2. Frans de Waal, "What I Learned from Tickling Apes." *The New York Times Sunday Review.* Accessed July 6, 2016, http://www.nytimes.com/2016/04/10/opinion/sunday/what-i-learned-from-tickling-apes.html.
3. Alison Suen, *The Speaking Animal: Ethics, Language, and the Human–Animal Divide* (Lanham, MD: Rowman & Littlefield, 2015).
4. Debra Hawhee, "Toward a Bestial Rhetoric." *Philosophy & Rhetoric* 44, (2011): 81–87; George Kennedy, "A Hoot in the Dark: The Evolution of General Rhetoric." *Philosophy & Rhetoric* 25 (1992): 1–21.
5. In contrast, "negative peace" is theorized as a state of being for humans where no active, direct conflict (such as war) is ongoing that results in easily visible or discernible violence. In states of negative peace, while there is no warfare going on, there is still rampant inequality and injustice. Negative peace has been observed in postcolonial societies like the United States or South Africa, where persistent cultural and structural forms of racism contribute directly or indirectly to the physical, social, psychological, economic, and other forms of harm done to members of minority, ethnic, racial, and/or other underrepresented populations. In short, negative peace is responsible for both direct and indirect forms of violence being perpetrated on individuals and afflicted groups, often at large scales. See: Johan Galtung, "Positive and Negative Peace." In *Peace and Conflict Studies: A Reader*, eds. Charles Webel and Jorgen Johansen (New York: Routledge, 2012), 75–79.
6. Frans de Waal, "What I Learned from Tickling Apes." *The New York Times Sunday Review,* April 10, 2016. Accessed July 6, 2016, http://www.nytimes.com/2016/04/10/opinion/sunday/what-i-learned-from-tickling-apes.html.
7. De Waal, "What I Learned," 2016.
8. Suen, *The Speaking Animal,* p. 17.
9. See: Steven J. Mithen. *The Prehistory of the Mind: A Search for the Origins of Art, Religion, and Science* (London, Thames and Hudson, 1996); Jean Clottes. *What is Paleolithic Art? Cave Paintings and the Dawn of Human Creativity*. Translated by Oliver Y. Martin and Robert D. Martin (Chicago, IL: University of Chicago Press, 2016).
10. Gala Argent, "Toward a Privileging of the Nonverbal: Communication, Corporaeal Synchrony, and Transcendences in Humans and Horses." In *Experiencing*

Animal Minds: An Anthology of Animal–Human Encounters, eds., Julie A. Smith and Robert W. Mitchell (New York: Columbia Univ. Press, 2012), 112–128.

11. Argent, "Toward a Privileging of the Nonverbal," 113.

12. Suen, pp. 14–17.

13. See: Carol Adams, *The Sexual Politics of Meat*: *A Feminist Vegetarian Critical Theory, 20th Anniversary Edition* (Bloomsbury Academic, 2010); Catharine MacKinnon, "Of Mice and Men: A Feminist Fragment on Animal Rights." In *Animal Rights: Current Debates and New Directions*, eds., Cass R. Sunstein and Martha C. Nussbaum. (Oxford: Oxford University Press, 2004), 263–276.

14. Peter Singer, *Animal Liberation: The Definitive Classic of the Animal Movement* (New York: Harper Perennial, 2009); Peter Singer, "Open the Cages!" *The New York Review of Books* 63 (2016): 22–26.

15. Tom Regan, *Empty Cages: Facing the Challenge of Animal Rights* (Oxford, UK: Rowman and Littlefield Publishers, 2004).

16. Mark Rowlands, *Animal Rights: Moral, Theory and Practice*, 2nd ed. (New York: Palgrave Macmillan, 2009).

17. See: James Rachels, *Created from Animals: The Moral Implications of Darwinism* (New York: Oxford University Press, 1999).

18. Gary L. Francione, *Animals as Persons: Essays on the Abolition of Animal Exploitation (Critical Perspectives on Animals)* (New York: Columbia University Press, 2008).

19. Gary L. Francione on Animal Abolitionism. Podcast. Institute of Philosophy, School of Advanced Study, University of London, 2012. Accessed July 6, 2016, http://hwcdn.libsyn.com/p/8/5/c/85cd9a0d58eb12e6/Gary_Francione_on_Animal_Abolitionism.mp3?c_id=5041718&expiration=1469564941&hwt=a8e7132e367f783e784c7f07c42213ac

20. Francois Laruelle, *General Theory of Victims* (Cambridge, UK: Polity Press, 2015).

21. Laruelle, *General Theory of Victims,* 117.

22. Ibid.

23. John Ó Maoilearca, "The Animal Line: On the Possibility of a 'Laruellean' Non-human Philosophy." *Angelaki: Journal of the Theoretical Humanities* 19 (2014): 113–129. DOI:10.1080/0969725X.2014.950867.

24. Dennis Soron, "Road Kill: Commodity Fetishism and Structural Violence." *Topia 18* (2008): 107–125.

25. Sally Swift, *Centered Riding 2: Further Exploration* (VT: Trafalgar Square Books, 2014).

26. Stephen Newmyer, "Speaking of Beasts: The Stoics and Plutarch on Animal Reason and the Modern Case against Animals." *Urban Quarterly of Classic Culture [Quaderni urbinati di cultura classica]* 63 (1999): 99–110.

27. Argent, "Toward a Privileging of the Nonverbal," 124.

28. Ibid. 124.

29. Anonymous. Ancient Peoples. [blog]. Accessed July 6, 2016, http://ancientpeoples.tumblr.com/post/29549387655/the-history-of-the-horse-tack-in-ancient-terms

30. Robert Cook, "FAQ: The Bitless Bridle." Accessed July 6, 2016, http://www.bitlessbridle.com/index.php?main_page=page&id=1

31. Anonymous. "Horse Industry Statistics." *The Equestrian Channel.com*, 2002. Accessed July 6, 2016, http://www.theequestrianchannel.com/sitebuildercontent/sitebuilderfiles/2002USEFdemographics.pdf

32. Cesar Millan, "How To Be Calm and Assertive." Cesar's Way: How to Achieve Balance and Harmony, 2016. Accessed July 6, 2016, https://www.cesarsway.com/dog-psychology/pack-leader/how-to-be-calm-and-assertive

33. Cesar Millan, "Well-trained but not well behaved." Cesar's Way: Achieving Balance and Harmony, 2016. Accessed July 6, 2016, https://www.cesarsway.com/cesar-millan/cesars-blog/well-trained-but-not-well-behaved

34. Lauren Said-Moorhouse and Marc Hoeferlin, "What I Learned from Freediving with Great White Sharks." CNN. 2014. Accessed July 6, 2016, http://www.cnn.com/2014/12/24/world/africa/andre-hartman-freediving-with-great-whites/ See also: Meera Dolasia, "Meet Andre Hartman, South Africa's Shark Whisperer!" DogonNews.com. April 25, 2012. Accessed July 6, 2016, http://www.dogonews.com/2012/4/25/meet-andre-hartman-south-africas-shark-whisperer. Dolasia notes that "finning" is a practice that has led to the alarming decline in shark numbers, for products like China's delicacy "shark fin soup" selling for $150 a bowl; global overfishing has also drastically reduced sharks' food sources.

35. Joseph C. Dovala, "Swimming With Sharks: Always Be Looking Down Current!" *California Diver*, July 7, 2015. Accessed July 6, 2016, http://californiadiver.com/intelligence-and-personalities-sharks-are-our-friendsnot-food/

36. Cress Spicer, "Autistic Children are Very Sensitive to the Energy of Others." Blog. June 24, 2015. LinkedIn.com. Accessed July 6, 2016, https://www.linkedin.com/pulse/autistic-children-very-sensitive-energy-others-cress-spicer

37. Marisa Rubin, "Vultures, Parrots and Mini-Pigs! Oh My!" *Toledo City Paper*, March 9–22, 2016, 16–17.

38. Joseph as cited in Rubin, "Vultures, Parrots and Mini-Pigs!" 17.

39. TheAnimalBehaviorCenter.com

40. Suen, *The Speaking Animal*, 75.

41. Ibid.17.

42. Susan Orlean, "What Makes the 'Lion Whisperer' Roar?" *Smithsonian Magazine*. June 2015. Accessed July 6, 2016, http://www.smithsonianmag.com/science-nature/what-makes-lion-whisperer-roar-180955290/?no-ist

43. Barbara Willard, "Performing Nature: The Bioregional Performing of the Biotic Community." *Journal of the Northwest Communication Association* 25 (1997): 23–41.

44. Karin Weil, "A Report on the Animal Turn." *Differences: A Journal of Feminist Cultural Studies* 21 (2010): 1–23.

45. Willard, "Performing Nature," 23–41.

46. Ibid. 35–38.

47. Emily Plec, "Introduction." In *Perspectives on Human–Animal Communication*, 2.

48. Tom Regan, *Empty Cages*.

49. Judith Butler, *Precarious Life: The Powers of Mourning and Violence* (London, UK: Verso, 2004); Maneesha Deckha, "Vulnerability, Equality, and Animals." *Canadian Journal of Women and the Law* 27 (2015), 47–73; James Stanscu, "Species Trouble: Judith Butler, Mourning, and the Precarious Lives of Animals." *Hypatia* 27 (2012): 567–582.

50. Deckha, "Vulnerability, Equality, and Animals," 70.

51. Stanescu, "Species Trouble: Butler, Mourning, and the Precarious Lives of Animals," 576.

52. For theoretical merging or mutating, see: Maoilearca, "The Animal Line." For a clear summary of challenges to the "animal turn" see: Weil, "A Report on the Animal Turn."

BIBLIOGRAPHY

Adams, Carol. *The Sexual Politics of Meat: A Feminist Vegetarian Critical Theory 20th Anniversary Edition.* London, UK: Bloomsbury Academic, 2010.

Anonymous. Ancient Peoples. [blog]. Accessed July 6, 2016: http://ancientpeoples.tumblr.com/post/29549387655/the-history-of-the-horse-tack-in-ancient-terms

Anonymous. "Horse Industry Statistics." *The Equestrian Channel.com*, 2002. Accessed July 6, 2016: http://www.theequestrianchannel.com/sitebuildercontent/sitebuilderfiles/2002USEFdemographics.pdf

Argent, Gala. "Toward a Privileging of the Nonverbal: Communication, Corporaeal Synchrony, And Transcendences in Humans and Horses." In *Experiencing Animal Minds: An Anthology of Animal–Human Encounters*, edited by Julie A. Smith and Robert W. Mitchell, 112–128. New York: Columbia University Press, 2012.

Butler, Judith. *Precarious Life: The Powers of Mourning and Violence*. London, UK: Verso, 2004.

Clottes, Jean. *What is Paleolithic Art? Cave Paintings and the Dawn of Human Creativity*. trans., Oliver Y. Martin and Robert D. Martin. Chicago, IL: University of Chicago Press, 2016.

Cook, Robert. "FAQ: The Bitless Bridle." Accessed July 6, 2016: http://www.bitless-bridle.com/index.php?main_page=page&id=1

Deckha, Maneesha. "Vulnerability, Equality, and Animals." *Canadian Journal of Women and the Law* 27 (2015): 47–73.

De Waal, Frans. "What I Learned from Tickling Apes." *The New York Times Sunday Review*. Accessed July 6, 2016: http://www.nytimes.com/2016/04/10/opinion/sunday/what-i-learned-from-tickling-apes.html?_r=0

Dolasia, Meera. "Meet Andre Hartman, South Africa's Shark Whisperer!" DogonNews.com. April 25, 2012. Accessed July 6, 2016: http://www.dogonews.com/2012/4/25/meet-andre-hartman-south-africas-shark-whisperer

Dovala, Joseph C. "Swimming With Sharks: Always Be Looking Down Current!" California Diver, July 7, 2015. Accessed July 6, 2016: http://californiadiver.com/intelligence-and-personalities-sharks-are-our-friendsnot-food/

Francione, Gary L. *Animals as Persons: Essays on the Abolition of Animal Exploitation (Critical Perspectives on Animals).* New York: Columbia University Press, 2008.
———. Francione on Animal Abolitionism. (2012). Podcast. Institute of Philosophy, School of Advanced Study, University of London. Accessed July 6, 2016: http://hwcdn.libsyn.com/p/8/5/c/85cd9a0d58eb12e6/Gary_Francione_on_Animal_Abolitionism.mp3?c_id=5041718&expiration=1469564941&hwt=a8e7132e367f783e784c7f07c42213ac
Galtung, Johan. "Positive and Negative Peace." In *Peace and Conflict Studies: A Reader*, edited by Charles Webel and Jorgen Johansen, 75–79. New York: Routledge, 2012.
Hawhee, Debra. "Toward a Bestial Rhetoric." *Philosophy & Rhetoric* 44 (2011): 81–87.
Kennedy, George. "A Hoot in the Dark: The Evolution of General Rhetoric." *Philosophy & Rhetoric* 25 (1992): 1–21.
Laruelle, Francois. *General Theory of Victims*. Cambridge, UK: Polity Press, 2015.
Maoilearca, John Ó. "The Animal line: On the Possibility of a 'Laruellean' Non-human Philosophy." *Angelaki: Journal of the Theoretical Humanities* 19 (2014): 113–129.
MacKinnon, Catharine. "Of Mice and Men: A Feminist Fragment on Animal Rights." In *Animal Rights: Current Debates and New Directions*, edited by Cass R. Sunstein and Martha C. Nussbaum, 263–276. Oxford, UK: Oxford University Press, 2004.
Millan, Cesar. "How To Be Calm and Assertive." Cesar's Way: How to Achieve Balance and Harmony. Accessed July 6, 2016: https://www.cesarsway.com/dog-psychology/pack-leader/how-to-be-calm-and-assertive
Millan, Cesar. "Well-trained but not Well Behaved." Cesar's Way: Achieving Balance and Harmony, 2016. Accessed July 6, 2016: https://www.cesarsway.com/cesar-millan/cesars-blog/well-trained-but-not-well-behaved
Mithen, Steven J. *The Prehistory of the Mind: A Search for the Origins of Art, Religion, and Science*. London: Thames and Hudson, 1996.
Newmyer, Stephen. "Speaking of Beasts: The Stoics and Plutarch on Animal Reason and the Modern Case against Animals." *Urban Quarterly of Classic Culture [Quaderni urbinati di cultura classica],* 63 (1999): 99–110.
Orlean, Susan. "What Makes the 'Lion Whisperer' Roar?" *Smithsonian Magazine*. June 2015. Accessed July 6, 2016: http://www.smithsonianmag.com/science-nature/what-makes-lion-whisperer-roar-180955290/?no-ist
Plec, Emily. "Preface." In *Perspectives on Human–Animal Communication: Internatural Communication*, edited by Emily Plec, xiii–xiv. New York: Routledge, 2013.
Rachels, James. *Created from Animals: The Moral Implications of Darwinism*. New York: Oxford University Press, 1999.
Regan, Tom. *Empty Cages: Facing the Challenge of Animal Rights*. Oxford, UK: Rowman and Littlefield Publishers, 2004.
Rowlands, Mark. *Animal Rights: Moral Theory and Practice*, 2nd ed. New York: Palgrave Macmillan, 2009.
Singer, Peter. *Animal Liberation: The Definitive Classic of the Animal Movement*. New York: Harper Perennial, 2009.

———. "Open the Cages!" *The New York Review of Books*, 63 (2016): 22–26.

Soron, Dennis. "Road Kill: Commodity Fetishism and Structural Violence." *Topia* 18 (2008): 107–125.

Swift, Sally. *Centered Riding 2: Further Exploration*. North Pomfret, VT: Trafalgar Square Books, 2014.

Spicer, Cress. "Autistic Children are Very Sensitive to the Energy of Others." Blog. June 24, 2015. LinkedIn.com. Accessed July 6, 2016: https://www.linkedin.com/pulse/autistic-children-very-sensitive-energy-others-cress-spicer

Rubin, Marisa. "Vultures, Parrots and Mini-Pigs! Oh My!" *Toledo City Paper*. March, 2016, 16–17. TheAnimalBehaviorCenter.com

Said-Moorhouse, Lauren and Hoeferlin, Marc. "What I learned from Freediving with Great White Sharks." *CNN*. December 24, 2014. Accessed July 6, 2016: http://www.cnn.com/2014/12/24/world/africa/andre-hartman-freediving-with-great-whites/

Stanscu, James. "Species Trouble: Judith Butler, Mourning, and the Precarious Lives of Animals." *Hypatia* 27 (2012): 567–582.

Suen, Alison. *The Speaking Animal: Ethics, Language, and the Human–Animal Divide*. Lanham, MD: Rowman & Littlefield, 2015.

Weil, Karin. "A Report on the Animal Turn." *Differences: A Journal of Feminist Cultural Studies* 21 (2010): 1 –23.

Willard, Barbara. "Performing Nature: The Bioregional Performing of the Biotic Community." *Journal of the Northwest Communication Association* 25 (1997): 23–41.

Chapter 6

Toward a Rhetorical Ethology

Dustin A. Greenwalt

In an edition of *PMLA* partially dedicated to the burgeoning field of animal studies, Cary Wolfe points out that one "of the challenges involved in writing about animal studies ... is a daunting interdisciplinarity that is inseparable from its very genesis." More than a field that traffics in textual criticism or ponders the nature of particular objects such as books, cinema, or speeches, animal studies must grapple with its relationships to living bodies sorted by the sciences and culture into the category of the nonhuman. Wolfe argues, moreover, that the study of these bodies within the humanities owes much to the ethological sciences: "Animal studies ... would probably not exist, at least not in its current form, without the work done in field ecology and cognitive ethology over the past twenty to thirty years."[1] Engagements with the question of the animal within the humanities, then, have often unrecognized allies within the ethological sciences, or the study of animal behavior that often emphasizes its relationships to biology, which this chapter utilizes in theorizing interspecies rhetorics.

Following Wolfe's overtures toward interdisciplinarity, my goal in this chapter is to engage discussions of interspecies encounters by prominent ethologists and philosophers as inventional resources for reconsidering rhetorical studies' conceptualizations of contexts and critical methodologies. Specifically, I claim that interspecies rhetorics might best be understood by attending to the affective force of encounters with, and between, animals *in situ* as well as in circulating texts. Affects, according to Brian Massumi, are prepersonal energies that exert force on audiences' sensory inputs and become absorbed as intensities.[2] Intensities are then selectively filtered through bodies' neurological systems to inflect conscious experience, including reasoning and emotions, while the remaining energies remain virtual, or within the realm of potentialities, and propel our bodies in often unthought-of or taken-for-granted ways.[3] The power of these encounters, furthermore, partially issues from their interactions with the rhetorical ecologies that compose their contexts. Rhetorical ecologies, as I later argue, are complex systems composed of affects and symbols that circulate through, and are modulated by, media as well as *in situ* architectures and environments. This perspective subsequently expands on rhetorical theories that posit the entwined nature of affect, embodiment, and circulation to account for the persuasive power of interspecies encounters.[4]

More than simply expanding the theories of affective ecologies, this chapter inflects this perspective with an ethological sensibility that renders the attribution of rhetoric's persuasive power to a single stylistic form or substantive argument insufficient for engaging the force of interspecies encounters. Understood as a method of observation, ethology calls for pluralistic accounts of animal behaviors in their environments and distances itself from what Marc Bekoff calls "mere reductionistic description"[5] that sees simple linear relationships between animal behaviors and single external stimuli. Rather than assuming that animal rhetorics can be apprehended by studying a single aspect of the encounter, such as behaviors, sounds, smells, or tactile sensations, in other words, an ethological sensibility calls attention to the entire constellation of factors that affect audiences. This sensibility subsequently invites the development of another method for describing the complex interactions that contribute to animal rhetorics' affective power.

In response to these ecological and ethological dispositions, this chapter calls upon scholars to construct typologies of the affects produced by interspecies encounters as well as circulating texts featuring animals. Issuing from ethologists' construction of "ethograms" that schematize interspecies encounters, as well as Gilles Deleuze's readings of Baruch Spinoza,[6] these typologies would be comprised of complex lists and descriptions of how animals' communicative behaviors affect and are affected by critics, other audiences, and their broader contexts. This rhetorical ethology would compile depictions of nonhuman animals' rhetorical capacities in order to challenge their widespread devaluation because they lack the *logos*, or complex symbolic languages and the attendant faculty of reason. By calling attention to the capacity of different species to affect audiences, moreover, this mode of scholarship also harkens toward the invention of rhetorics that may expand public compassion for nonhuman animals as a part of what Jacques Derrida calls "a war (whose inequality could one day be reversed)"[7] over their suffering in factory farms, slaughterhouses, laboratories, and any other space in which they are denigrated for human use.

I begin the map of an ethological rhetoric by engaging George A. Kennedy's argument for the affective nature of interspecies rhetorics. Despite the value of Kennedy's work, he maintains an essentialist characterization of participants within encounters as stable subjects or beings that forecloses an attention to rhetoric's power to modulate their bodies' compositions, dispositions, and communicative capacities. As such, I turn to Donna Haraway's observation that interspecies affects propel *metaplasmic* becomings, or interactions between bodies that alter their mutual constitution, such as domestic cats' evolution of new vocalizations in order to elicit specific human responses.[8] In the following section, I argue that this affective force issues from the interactions between the multiple sensory elements produced by communicative behaviors as well as the ecological contexts of the encounter. I draw on the works of Bekoff, Massumi, and Jenny Edbauer to explicate the ecological nature of both rhetorical effects and sense perception, a move that complicates conceptualizations of experience as issuing from any particular facet of an interspecies encounter. The third section visits with ethological field methods, Aristotle's theory of *topoi*, and Deleuze's work on Spinoza in order to develop a methodology for capturing the complexities of nonhuman rhetorics. It calls for the construction of typologies that describe interspecies encounters as confluences of sense experiences, communicative behaviors, and their contexts. I conclude, then, by speculating about this ethological rhetoric's potential for challenging both the suffering of animals in contemporary society and the anthropocentric privileging of the *logos* within rhetorical studies as well as contemporary culture writ large.

ETHOLOGICAL RHETORICS

As a field that draws its justification from the ubiquity of persuasion in Western cultures at least since ancient Greece, rhetorical studies pairs easily with ethology, or the study of human and nonhuman behavior. Such, at least, is George A. Kennedy's argument in his now famous essay "A Hoot in the Dark: The Evolution of General Rhetoric."[9] For Kennedy, rhetoric is a cross-cultural and cross-linguistic process issuing from "the energy inherent in communication."[10] This rhetorical energy, according to Kennedy, is a quality of the communication itself, as well as the product of the cognitive labor on the part of rhetors and their audiences. Akin to Massumi, then, Kennedy posits that communication's force constitutes is most important feature. Yet, unlike Massumi's conception of affect, Kennedy's account carries an essentialist notion of the speakers and audiences as stable subjects or beings, whereby rhetoric merely moves them toward particular ends rather than altering their constitutions, a shortcoming that might be remedied by turning toward Donna Haraway's claim that interspecies encounters enact *metaplasmic* modulations in the participants' bodies. Before performing this maneuver at the end of this section, I further explicate Kennedy's account of rhetoric as affective and observable across a broad range of species. In so doing, I argue that scholars must attend to the affective force of communicative behaviors in order to understand animals' capacities for moving audiences.

Importantly, Kennedy's perspective broadens the possible range of communicative behaviors that might be considered rhetorical beyond the use of symbolic language. For him, the power of rhetoric can be found "in physical actions, facial expressions, gestures, and signs generally."[11] This view subsequently resonates with Debra Hawhee's claim that language's materiality renders rhetorical practice inseparable from bodily attitude and movement.[12] Pointing to Sir Richard Paget's theory that systems of language developed from gestures as an important intellectual forbearer of Kenneth Burke's concept of attitude as the embodied movements from which words derive their poetic power, Hawhee argues that "*sensuous* action,"[13] or emotional and bodily force, underlies all discourse. While this claim was criticized by Kennedy's contemporaries for blurring the distinctions between rhetoric and signs more generally,[14] Hawhee's resuscitation of Burke's emphasis on attitude, as well as recent turns toward theories of communicative labor and affect within rhetorical studies,[15] confirm that communication's persuasive force can be measured by what it does to bodies in addition to the meanings it carries.

This emphasis on the rhetorical nature of bodily gestures situates nonhuman animals as possible rhetors. Take, for example, how Kennedy's discussion of epideictic rhetoric among a murder of crows emphasizes the role played by the distinct tenor of their cries in affirming identities:

> Zoologists have identified among crows what is known as an "assembly call," which consists of a succession of long raucous cries. This brings other crows together. Another kind of utterance is the "contact call," which consists of a series of short caws spaced into groups of two. This is probably a way of establishing a relationship to others in the flock. The assembly and the vocalization thus perform something like the reaffirmation of group identity in ceremonial oratory on public occasions.[16]

While these communicative behaviors might be motivated by some intended meanings, Kennedy calls attention to how they move audiences rather than anchoring rhetoric to the use of complex symbols. Commenting on Kennedy's essay, Hawhee confirms that insofar as "rhetorical energy resides in places where human animals may not even tread,"[17] we might begin to "devote energy to rhetoric's nonhuman forebears."[18] Kennedy's engagement with the affective force of communicative behaviors thus provides grounds for an ethological rhetoric that observes the communications of humans and nonhumans alike.

Despite the importance of Kennedy's broadening of the range of potential rhetors, his theory maintains an essentialist understanding of the bodies within interspecies encounters and subsequently limits rhetoric's potential power. Notably, Kennedy's claim that rhetorical concepts might work across cultures has been criticized as a problematic form of cultural anti-essentialism (which ostensibly issues from an essentialist view of human biology),[19] a tendency that is mitigated by Alex C. Parrish's reframing of cross-species and cross-cultural rhetorics as an adaptation emerging from complex interactions of biology as well as cultural contexts.[20] My argument, however, emphasizes Kennedy's failure to acknowledge the power of rhetoric to modulate the constitution of communicating bodies despite their propensity to be moved toward action. Notice how, in his description of the crows, their calls bring the murder together and "perform something like the reaffirmation of group identity."[21] Rather than leaving open the possibility that the calls alter the bodies, or contribute to the formation of new identities, this description implies the existence of preformed, and somewhat immovable, bodies within encounters. As Barbara A. Biesecker points out, this essentialist conception of identity limits the potential for rhetoric to play a role in the constitution of subjects and bodies.[22] Indeed, if we stop here and accept the formulation of rhetorical energy as a survival tactic, as Kennedy would have it, we are likely to ignore the ways in which rhetoric enacts *metaplasmic* alterations in bodies' capabilities.

Inquiring about the rhetorical qualities of the energies within interspecies encounters provides one way of altering the essentialist conception of subjects that persists within Kennedy's theory. Notably, Donna Haraway has described interspecies encounters as taking the form of "*metaplasmós*, meaning remodeling or remoulding."[23] While *metaplasmós*, or *metaplasm*, is often denoted as the alteration of a word that changed its verbalization, such as contractions, Haraway provides a materialist spin on this trope by refiguring it as the alteration of the participants' bodies in interspecies encounters. Through these encounters, she notes, "Flesh and signifier, bodies and words, stories and worlds: these are joined in naturecultures."[24] Indeed, she describes how genetic variations have helped Scottish Border Collies better herd sheep and also enable their adaptation to the communicative situations of competitive agility games. In this manner, interspecies encounters have affected material changes in the biological composition of bodies and even the trajectory of entire species.

While single encounters might not affect such drastic changes in species' compositions, each rhetorical engagement nevertheless has the potential to modulate bodies' constitution. In his work on Louis Althusser's aleatory materialism, which itself draws upon the metaphysics of Lucretius, Donovan Conley contends that rhetorical encounters involve the "infinitesimal motions"[25] of atoms colliding in often unpredictable ways that modulate bodies, modify their trajectories, and thus alter history. Take, for instance, how interspecies encounters in zoos and aquariums amplify visitors' openness to being affected by conservation messages.[26] These encounters, even in domesticated settings, produce dispositions toward empathizing with and protecting the animals. The infinitesimal movements and metaplasmic modulations produced by interspecies encounters, which propel bodies along new paths and produce new possibilities for engaging the world, thus provide the object of study for an ethological rhetoric. This perspective builds upon widely accepted definitions of rhetoric as affective force to call attention to how calling birds, barking dogs, meowing cats, yipping foxes, and roaring lions, as well as human utterances and bodily indications, move audiences, modulate their bodies, and open the possibility for the development of new perspectives on the world.

ETHOLOGY AND RHETORICAL ECOLOGIES

Despite incorporating nonhumans into an understanding of persuasion that takes seriously its power to affect, and thus modulate, bodies, my refiguration of inquiry has heretofore assumed a relatively conventional understanding of rhetoric as issuing from a set of specific and identifiable behaviors with the relevant audience limited to a specific space. While providing for an easy, and sometimes revelatory, form of critical engagement, these strictures cannot account for the delimited and contextual nature of affect itself, nor can they sufficiently engage the role that ecological interactions of communicative behaviors play in producing rhetorical energies. As such, I argue that an ethological rhetoric must grapple with the interactions of communicative behaviors and their contexts as well as those *between* different facets of the behaviors themselves. These two facets of the rhetorical ecologies, importantly, call attention to how the contextually specific rhetorical energies circulating in encounters move participants and point to the necessity of the ethological method outlined in the third section.

ECOLOGICAL CONTEXTS

The contexts of interspecies encounters, including the immediate situation as well as the circulating affects that modulate bodies in their everyday lives, likely transverse and modulate the force of communicative behaviors. Rather than bracketing the rhetorical situation to the discrete relationship between a sender and a receiver, in other words, it might be more accurate to conceptualize communicative behaviors and the rhetorical energies they produce as occurrences within a field of always fluctuating relationships. Notably, Jenny Edbauer figures rhetorical ecologies as "effects, enactments, and events"[27] circulating through particular spaces. Attunement to affective ecologies, she argues, emphasizes processes over products, engagements with "amalgamations and transformations"[28] rather than stable texts fashioned by rhetors prior to speaking engagements, and the "*mutuality* of material practice"[29] as opposed to the linear relationships implied by speaker-audience models of rhetoric. Rhetorical energies are never stable in this field of relationships and effectivities. Instead, they shift and are modulated in their encounters with bodies, signs, and other energies. We are subsequently left to trace, encounter, and feel rhetoric's affective power when we encounter communicative behaviors.

These ecologies necessitate becoming attuned to our emotions and other more unexplainable affective inclinations within encounters, rather than immediately resorting to the formalisms and critiques of ideology that tend to dominate rhetorical studies. This is, in part, the aim of Rice's theory of rhetorical ecologies. Indeed, she calls upon us to "*encounter rhetoric*"[30] and to enact a "*thinking/doing*—with a razor thin slash mark barely keeping the two terms from bleeding into each other."[31] This call to encounter, to reflect on, and to interact with rhetoric resonates with Marc Bekoff's description of his ethological method. For Bekoff, the situatedness of the observer necessitates a *feeling with* animals at the same time one reflects on their behavior: "When I study coyotes I am coyote; when I study birds I am bird. Often when I stare at a tree, I am tree. There is a strong sense of oneness. We are all part of the same deeply interconnected and interdependent community in which I, the seer, am the seen, woven into a seamless tapestry of unity with interconnecting and reciprocal bonds."[32] Here, Bekoff points to how interspecies encounters encourage reflections on the ways in which scholars themselves are affected. The mutuality of seeing and being seen, as well as the feelings of connection and unity, are the types of intermingling necessary for the enactment of shifts in trajectories and perspectives characteristic of *metaplasmic* motions. These feelings, alongside the force of any audible communication, gestures, and other communicative behaviors, provide important material for descriptions of interspecies encounters. Following Edbauer, moreover, scholars might attend to how the specific context, as well as the observers' broader affective ecologies, contributed to these feelings of mutuality. Descriptions of how communicative behaviors interact with the specific location and its atmosphere to potentially alter observers' bodily states and trajectories, as well as how it might resonate with their previous experiences, constitute important material for an ethological rhetoric. An account of the force of communicative behaviors within an encounter can therefore only be understood by attending to how they are modulated by the ecologies of the specific situation as well as their broader contexts.

ECOLOGICAL COMMUNICATION

An ethological rhetoric also emphasizes how the particular characteristics of interspecies engagements, and the ecological nature of sense perception more generally, call attention to the interactions of an observers' senses during encounters. In so doing, it renders inadequate contemporary rhetorical methods for accounting for the force of interspecies encounters. Importantly, all rhetorical encounters, including those with texts as well as bodies *in situ*, are necessarily multisensory. Living bodies move, smell, and make noise, while computers encourage haptic, visual, and auditory engagements, and even books are rife with aromas and textures.[33] Yet, attending to a broad range of sensory inputs is particularly important for understanding interspecies encounters. Writing about the validity of ethological observations as evidence for animals' emotional lives, Bekoff writes that: "If you doubt that non-human animals have emotions, just look at them, listen to them and inhale the odours that pour out when they interact with friend or foe."[34] In another essay he also points out that "How researchers choose to summarise what they see, hear, or smell when studying animals depends on the questions in which they're interested."[35] Despite the impossibility of replicating or capturing an encounter in totality, in other words, ethological practice emphasizes attunement to multiple sensory inputs. Such attunements consequently resonate with ethology's *êthos* of complexity and pluralism outlined in the introduction. Observers might, then, construct multiple textured descriptions of a broad range of sensations they experience within encounters.

The ecological nature of sensory inputs and experiences further complicate critics' capacities to describe encounters by mitigating the assumption of a linear relationship between any one facet of communicative behaviors and the recognition of particular feelings. In his theorization of the virtual, or the plane of potentiality where prepersonal intensities reside, Massumi argues that consciousness's emergence is characterized by nonlinear movements and interactions between sensations: "Intensity would seem to be associated with nonlinear processes; resonation and feedback that momentarily suspend the linear progress of the narrative present from past to future."[36] The prior force of our everyday contexts and encounters, as well as previous thoughts and memories that have fed back from consciousness, compose the virtual and modulate our cognitive experiences. Massumi thus argues that the linearity of sensory experience is disrupted by the interaction of two "entirely different rules of formation."[37] While the virtual enfolds affects and receives feedback from memories and linguistic thoughts, consciousness emerges from the virtual and is modulated by the remaining intensities. The complex movements between cognition and the virtual may subsequently disrupt the linearity often associated with measuring rhetorical effect (i.e., this form produces a particular feeling or has some meaning). We are left, instead, to depict the relationships between different facets of interspecies encounters without making explicit cause-effect judgments or assigning them linguistic meanings. Such depictions would contribute to a catalog of nonhuman animals' rhetorical capacities from which rhetorical scholars might develop interventions for expanding public empathy for their position as objects within our contemporary socioeconomy.

RHETORICAL ETHOLOGY AS A TYPOLOGICAL ENTERPRISE

A turn toward constructing depictions of interspecies rhetorics constitutes one means of pursing a rhetorical ethology. This approach positions scholars as catalogers of *topoi*, or the common affects produced by rhetorical ecologies, and draws inspiration from Aristotle's *On Rhetoric*.[38] Yet, unlike Aristotle's handbook of the general forms and topics of persuasion, a rhetorical ethology would compile detailed descriptions of animals' rhetorical capacities as the possible contents of messages that accomplish the specific task of expanding public empathy for their suffering. To wit, I argue that Aristotle's work on typologies might provide grounds for the genesis of an ethological investigation of interspecies rhetorics in which one, as Deleuze puts it, makes "a list of affects."[39] Rather than being comprised of pithy statements akin to a shopping list, however, these typologies would be similar to ethograms, or complex descriptions and drawings of encounters, used by ethological scientists. A rhetorical ethology might, then, construct maps of the intensities produced through encounters with other species from which we might begin to better understand and publicize the rhetorical capacities of humans and animals alike.

TOPOI, TYPOLOGIES, AND ARISTOTLE'S ANIMALS

Aristotle's theory of *topoi*, as the common grounds for most public persuasion, provides rhetorical scholars with a program for theorizing the nature of public values and reasoning. While my method shifts its emphasis from Aristotle's attention to common preferences, reasons, and arguments to be deployed in other speaking situations toward describing contextually specific communicative capacities and affects, explicating the function and importance of *topoi* might show how this ethology resonates with the deep-rooted significance of typologies for rhetorical practice. Primarily, Aristotle's theory of *topoi* stems from his observation that most rhetoric is enthymematic insofar as it relies on the audience to fill in a set of unsaid assumptions or premises. Enthymemes are important, according to Aristotle, because they allow one to argue while avoiding lengthy demonstrations or "stating the obvious."[40] "Thus," he writes, "one should not speak on the basis of all opinions but of those held by a defined group … and the fact that what is said seems true should be clear to all or most people."[41] Therefore, there is a *doxa* composed of common opinions and traditions at work in given audiences that allows rhetors to judge the potential effectiveness of arguments and the extent to which they need to explain their reasoning. Although there is some debate over whether the audiences' assumptions, feelings, and dispositions are real or imagined, within the rhetorical enterprise, as Parrish points out, "inferences about the contents of another (social) mind" to are used to invent persuasive content that we will deliver in a form suitable to our opinions of what those minds want to hear."[42]

Aristotle's discussion of the *topoi* of happiness provides one example of how his concern for developing typologies of audiences' affective states as they relate to their broader contexts might resonate with a rhetorical ethology. Here, he provides a short map of the different mental, physical, and social conditions necessary for the achievement of happiness: "Let happiness be [defined as] success [*eupraxia*] combined with virtue, or as self-sufficiency [*autarkeia*] in life, or as the pleasantest life accompanied with security, or as abundance of possessions and bodies, with the ability to defend and use these things; for all people agree that happiness is pretty much one or more of these."[43] Such a conception of *topoi* as emergent from interactions between audiences' mental states and material conditions, whereby one is unlikely to have a pleasant life without the feeling of security or other bodies with which to interact, gives important purchase to my concern with animals' rhetorical capacities. Indeed, Parrish demonstrates how other species' evolved capacities to receive, recognize, or respond to communicative behaviors is crucial to considerations of their efficacy.[44] He points to how, for instance, the bright coloration of poisonous fish and insects both warns off predators with an evolved tendency to evade their flashy looks and potentially invites these species' extinction from the presence of other species that have yet to adapt to their communications. Listing *topoi* consequently invites attention to the roles of audiences' social contexts, as well as the biological capabilities of all bodies in an encounter, in depicting animals' rhetorical capacities.

Another example, this time from Aristotle's discussions of the use of animal fables in *On Rhetoric*'s second book, might demonstrate the suitability of his theory for grounding a rhetorical ethology that attends to both *in situ* encounters and as well as those in circulating texts. Take, for instance, how Aesop's use of the fable of the hedgehog and fox ostensibly resonates with common assumptions about humans' self-interested and power-hungry nature, a common *topoi* in works about state-craft (e.g., Hobbes, Machiavelli):

> Aesop, when speaking on behalf of a demagogue who was on trial for his life in Samos, told how a fox, while crossing a river, was carried into a hole in the bank. Not being able to get out, she was in misery for some time and many dog-ticks attacked her. A hedgehog came wandering along and, when he saw her, took pity and asked if he could remove the ticks. She would not let him and, when asked why, [said,] "These are already full of me and draw little blood, but if you remove these, other hungry ones will come and drink what blood I have left." "In your case too, O Samians," said [Aesop], "this man will no longer harm you; for he is rich. But if you kill him, other poor ones will come who will steal and spend your public funds."[45]

Here, Aesop's personification of the fox and the hedgehog serve as an example of how the removal of a demagogue (the ticks) invites worse rulers to take power. Despite functioning as representational analogs of the subjects of the speech, and therefore meant to convey meanings about them, the animal figures may affect the audience. At the very least, they may resonate with their fear that other people will always take power for their own financial gain, a possibility that points to how images and descriptions of animals in stories, films, and other media might affect audiences.[46] Aristotle's animals, as well as his theory of *topoi* with its attention to audiences, bodies, and capabilities, consequently contribute a rhetorical grounding for the development of typologies of the affects produced by animals' communicative behaviors.

COUNTING ANIMAL AFFECTS

Aristotle's proclivity for typologies, as well as my reformulation of rhetoric as circulating affects that issue from particular communicative behaviors but can only be encountered as the product of a series of complex ecologies, gesture toward an ethology in which we attend to bodies' potentialities "for affecting and being affected."[47] This ethological practice, which involves constructing lists of affects produced by rhetorical practices within, and across, species, seeks to avoid the domestication of animals through unnecessary anthropomorphism and encourages *metaplasmic* participations in interspecies encounters.

Rhetorical ethology's attentiveness to animals' communicative behaviors in relation to their contexts partially issues from Deleuze's pairing of

Spinoza's affect theory with Jakob von Uexküll's biological behavioralism. According to Deleuze, both the philosopher and biologist direct attention to how bodies affect, and are affected by, their environments:

> Long after Spinoza, biologists and naturalists will try to describe animal worlds defined by affects and capacities for affecting and being affected. For example, J. von Uexküll will do this for the tick, an animal that sucks the blood of mammals. He will define this animal by three affects: the first has to do with light (climb to the top of the branch); the second is olfactive (let yourself fall onto the mammal that passes beneath the branch); and the third is thermal (seek the area without fur, the warmest spot).[48]

"Such studies as this," continues Deleuze, "which define bodies, animals or humans by the affects they are capable of, founded what today is called *ethology*."[49] Notably, von Uexküll developed nearly exhaustive depictions of animals' dwelling-worlds, or *umwelten*. An animal's *umwelt* is defined by a series of tones, marks, or signs, such as mating calls or colors, that it has the capability to perceive, interact with, and potentially change.[50] Within their *umwelten*, animals can deploy different forms of communication to encourage actions from a variety of species, including humans. For instance, von Uexküll cites the case of a Jackdaw named Tschock that adopted fellow ethologist Konrad Lorenz as its mother because no other members of its species were available. Tschock called to him to be fed, attempted to persuade him to fly, and even performed mating dances in front of his domestic servant because, as a human, she carried the signs of a potential mate that transposed from Lorenz's position as the mother figure. Akin to von Uexküll's fieldwork, a rhetorical ethology would, as Deleuze and Felix Guattari put it, "list the affects"[51] by taking note of these rhetorical behaviors and attempting to describe them in fine detail.

While capturing the communicative behaviors of nonhuman animals within their shifting rhetorical ecologies is a difficult enterprise, we might take inspiration from the *ethogram*, the mode of cataloging animal behaviors deployed by ethological scientists. "An ethogram," writes Philip N. Lehner, "is a set of comprehensive descriptions of the characteristic behavior patterns of a species."[52] Similar to spreadsheets, ethograms are composed of a list of behaviors accompanied by an adjacent column containing detailed descriptions of them, as well as the notable parts of the contexts and their affective force.[53] Ethograms, furthermore, "should be clear and concise, yet complete,"[54] and observers can use "drawings to supplement the written descriptions."[55] This mode of cataloging, which emphasizes detailed descriptions in an organized form, might offer critics sufficient flexibility to account for the complexities characteristic of interspecies encounters. Indeed, ethograms need not be repeatable in every detail in order for their descriptions to serve as confirmation of animals' rhetorical capabilities or as inventional resources for expanding public empathy toward nonhumans.

While interspecies comparisons might serve as useful tools in challenging the subjugation of nonhumans, and while the feelings that propel the all-too-common positioning of animals as stand-ins for human qualities might point to affects worth interrogating, ethological rhetorics should seek to avoid reducing animals to anthropomorphic projections. Indeed, the processes of description and invention risks sliding into what Frans de Waal calls "anthropocentric anthropomorphism,"[56] or the attribution of "human feelings and thoughts to animals based on insufficient information or wishful thinking."[57] Rather than pointing us to animals' observable behaviors, anthropocentric anthropomorphisms position nonhuman bodies as characters in morality plays, comedies, or dramas that offer lessons about human society. Giving into these common forms of anthropomorphism, or even conflating animals' and humans' reasoning processes without substantial evidence, participates in the anthropocentric logic whereby "it is always one version or another of *the human* that falsely occupies the space of the universal and that functions to exclude what is considered non-human ... from ethical and political consideration."[58] A rhetorical ethology, in other words, must avoid situating animals as anthropomorphic projections, as doing so substitutes cultural productions for the communicative behaviors that constitute interspecies encounters and subsequently perpetuate the systems of violence against nonhumans.

In contradistinction to these anthropomorphic practices, scholars should attend to the feelings produced by encounters as indicators of how they might affect *metaplasmic* movements which "uproot one from humanity, if only for an instant"[59] and potentially alter how we understand animals' rhetorical capacities. Because interspecies encounters intermingle bodies and their affective forces, in other words, refusing these anthropomorphic tendencies opens possibilities for moving beyond our limited human perspectives. This refusal also invites attention to what Deleuze and Guattari call "pack or affect animals that form a multiplicity"[60] irreducible to human meanings. While humans and animals share common capacities worth identifying, only by attending to other species' irreducibility to the category of the human might we begin to understand what they are really capable of. David Abram provides a useful example of this practice: "I startled neighbors by chattering with squirrels who swiftly climbed down the trunks of their trees and across lawns to banter with me, or by gazing for hours on end at a heron fishing in a nearby estuary, or at gulls opening clams by dropping them from a height onto the rocks along the beach."[61] For Abram, participation with animals involves abandoning the preconceptions about the nature of human communication and becoming attuned to the feelings these encounters produce.

While we would add an attention to the ecologies that establish the ground for encounters and their affects, his conversations demonstrate the potential for feeling animals' communications, allowing them to momentarily change us, and then describing those feelings. A rhetorical ethology would subsequently construct typologies of affects produced by interspecies encounters, a task made possible by avoiding pernicious anthropomorphism and becoming attuned to how our bodies change in relationship to the animals and our contexts.

CONCLUSION

This chapter has mapped a rhetorical ethology that constructs typologies of animals' affective capacities. These capacities, it argues, only emerge through their interactions with the complex, interlocking, and shifting ecologies that simultaneously compose their contexts and blur the distinctions between particular facets of communicative behaviors. By calling further attention to *metaplasmic* movements effectuated by the encounter, moreover, it seeks to disrupt our settled human perspectives. As we near the edge of our map, then, I hope to address what Derrida so elegantly called "the *question* of the animal"[62] as a means of showing how a rhetorical ethology might intervene in the subjugation and exploitation of nonhumans within contemporary society. Here, I return to my claim from the introduction that this method calls for the invention and circulation of rhetorics meant to expand public empathy for animals and blur the distinctions between humans and animals.

Rather than relying on arguments and logical persuasion, a rhetorical ethology emphasizes the importance of affect and emotion, or *pathos*, as the grounds for encouraging audiences to become concerned with the status of animals. In a passage that addresses Jeremy Bentham's famous footnote that positions beings' ability to suffer as a precondition for their moral standing, Derrida situates widespread empathy as a necessary precondition for challenging the contemporary objectification of animals insofar as "No one can deny the suffering, fear, or panic, the terror or fright that can seize certain animals and that we humans can witness."[63] Challenging animals' exploitation on the grounds of *pathos*, in other words, might instill audiences with a certainty about their responsibility for ameliorating the suffering. One only needs to look at the history of public outrage over the killing of zoo animals,[64] as well as the hunting of big game in Africa,[65] to confirm Derrida's hypothesis that interventions into the status of nonhumans might work best by leveraging humans' affective and emotional connections to our sentient relatives.

Scholars might, therefore, invent new ways of publicizing both animals' capacities for suffering and their range of communicative behaviors in order to complicate the anthropocentric hierarchies that privilege humans because of their association with the *logos*. The success of media campaigns and documentaries about animal suffering points to the potential efficacy of such a strategy. Take, for instance, the success of the documentary *Blackfish* in creating public backlash against SeaWorld's practices of Orca captivity and breeding. The film, which depicted numerous instances of Orca suffering, spurred widespread online activism that contributed to a decline in SeaWorld's profits as well as changes in their treatment of their captive sea life.[66] While *Blackfish* may not fully exploit the capacities for media to affect audiences as it often emphasizes the scientific problems with Orca activities rather than foregrounding their communicative behaviors, its efficacy in spurring change demonstrates the potential of this strategy of invention and circulation to serve as a weapon in the war over the state of animals.

Scholars should also invent new ways to render visible animals' whole range of rhetorical capabilities, in addition to their abilities to communicate suffering, in the service of troubling the boundaries between humans and other sentient bodies. These texts might encourage the types of encounters that move participants toward new perspectives whereby, as Moira Gatens puts it, "distinctions between one thing and the next are reckoned in kinetic and dynamic terms."[67] Encouraging encounters with pack animals, in other words, will draw attention to the affective power of bodies rather than the biological qualities and comparisons of faculties so often used to justify the uniqueness of *homo sapiens*. In so doing, an ethological rhetoric might disrupt the popular tendency to police interspecies boundaries in the name of the *logos*.

NOTES

1. Ibid., 565.
2. Brian Massumi, *Parables for the Virtual: Movement, Affect, Sensation* (Durham: Duke University Press, 2002).
3. It's worth noting here that I follow Massumi in simplifying the relationships between affect and intensity for the purposes of clarity. For a more complex treatment, see: Gilles Deleuze, *Difference and Repetition,* trans. Paul Patton (New York: Columbia University Press, 1994).
4. C.f. Catherine Chaput, "Rhetorical Circulation in Late Capitalism: Neoliberalism and the Overdetermination of Affective Energy," *Philosophy and Rhetoric* 43 (2010): 1–25.
5. Marc Bekoff, "Description and Explanation: A Plea for Plurality," *Behavioral and Brain Sciences* 15.2 (1992): 270.
6. Gilles Deleuze, *Spinoza: Practical Philosophy*, trans. Robert Hurley (San Francisco: City Lights Books, 1988).

7. Jacques Derrida, *The Animal That Therefore I Am*, trans. David Willis (New York: Fordham University Press, 2008): 28.

8. Nicholas Nicastro, "Perceptual and Acoustic Evidence for Species-Level Differences in Meow Vocalizations by Domestic Cats (*Felis catus*) and African Wild Cats (*Felis silvestris lybica*)," *Journal of Comparative Psychology* 118 (2004): 287–96.

9. George A. Kennedy, "A Hoot in the Dark: The Evolution of General Rhetoric," *Philosophy and Rhetoric* 25 (1992): 1–21.

10. Ibid., 2.

11. Ibid., 4.

12. Debra Hawhee, "Language as Sensuous Action: Sir Richard Paget, Kenneth Burke, and Gesture-Speech Theory," *Quarterly Journal of Speech* 92 (2006): 331–54.

13. Ibid., 348.

14. C.f. Jo Liska, "The Role of Rhetoric in Semiogenesis: A Response to Professor Kennedy," *Philosophy & Rhetoric* 26 (1993): 31–38.

15. C.f. Chaput, "Rhetorical Circulation," 1–25.

16. Kennedy, "Hoot in the Dark," 5.

17. Debra Hawhee, "Toward a Bestial Rhetoric," *Philosophy and Rhetoric* 44 (2011): 85.

18. Ibid., 85.

19. Mary Garrett, "Short Reviews: Review of George Kennedy, *Comparative Rhetoric: An Historical and Crosscultural Introduction* (New York: Oxford University Press, 1998);" *Rhetorica: A Journal of the History of Rhetoric* 16 (1998): 431–33.

20. Alex C. Parrish, *Adaptive Rhetoric: Evolution, Culture, and the Art of Persuasion* (New York: Routledge, 2014).

21. Kennedy, "Hoot in the Dark," 5.

22. Barbara A. Biesecker, "Rethinking the Rhetorical Situation through *Différance*," *Philosophy and Rhetoric* 22 (1989): 110–30.

23. Donna Haraway, *The Companion Species Manifesto* (Chicago: Prickly Paradigm Press, 2003): 20.

24. Ibid., 20.

25. Donovan Conley, "Grid and Swerve," *Critical Studies in Media Communication* 27 (2010): 36.

26. Roy Ballantyne et al., "Conservation Learning in Wildlife Tourism Settings: Lessons from Research in Zoos and Aquariums," *Environmental Education Research* 13 (2007): 367–83.

27. Jenny Edbauer, "Unframing Models of Public Distribution: From Rhetorical Situation to Rhetorical Ecologies," *Rhetoric Society Quarterly* 35 (2005): 9.

28. Ibid., 17.

29. Ibid., 21.

30. Italics in the original. Ibid., 23.

31. Ibid.

32. Marc Bekoff, "The Evolution of Animal Play, Emotions, and Social Morality: On Science, Theology, Spirituality, Personhood, and Love," *Zygon* 36 (2001): 649.

33. For more on multisensory rhetorics, see: Dustin A. Greenwalt, "User Generated Videos of Urban Exploration and the Production of Affective Space," *Explorations in Media Ecology* 14.1&2 (2015): 125–39.

34. Marc Bekoff, "Are You Feeling What I'm Feeling?" *New Scientist* 194 (2007): 42.

35. Marc Bekoff, "The Emotional and Moral Lives of Animals: What Darwin Would Have Said," *Clio Medicia* 93 (2013): 320.

36. Massumi, *Parables for the Virtual,* 26.

37. Ibid., 26.

38. Aristotle, *On Rhetoric: A Theory of Civic Discourse Second Edition* Trans. George A. Kennedy (Oxford: Oxford University Press, 2007).

39. Deleuze, *Spinoza: Practical Philosophy*, 124.

40. Aristotle, *On Rhetoric*, 169.

41. Ibid., 169.

42. Parrish, *Adaptive Rhetoric*, 117.

43. Aristotle, *On Rhetoric*, 57.

44. Parrish, *Adaptive Rhetoric*, 117.

45. Aristotle, *On Rhetoric*, 163.

46. For more on the affective power of textual animals, see: Akira Lippit, *Electric Animal: Toward a Rhetoric of Wildlife* (Minneapolis: University of Minnesota Press, 2000).

47. Deleuze, *Spinoza*, 123.

48. Ibid., 124.

49. Ibid., 125.

50. Jakob von Uexküll, *A Foray into the Worlds of Animals and Humans with A Theory of Meaning,* trans. Joseph D. O'Neill (Minneapolis: University of Minnesota Press, 2010).

51. Gilles Deleuze and Felix Guattari, *A Thousand Plateaus: Capitalism and Schizophrenia Volume 2,* trans. Brian Massumi (Minneapolis: University of Minnesota Press, 1987): 257.

52. Phillip N. Lehner, *Handbook of Ethological Methods* (Cambridge: Cambridge University Press, 1998): 90.

53. For some examples of ethograms, see: *Ethosearch*, http://www.ethosearch.org.

54. Lehner, *Handbook of Ethological Methods*, 92.

55. Ibid., 92.

56. Frans de Waal, *The Ape and the Sushi Master: Cultural Reflections by a Primatologist* (New York: Basic Books, 2001).

57. Ibid., 72.

58. Matthew Calarco, *Zoographies: The Question of the Animal from Heidegger to Derrida* (New York: Columbia University Press, 2008): 10.

59. Deleuze and Guattari, *A Thousand Plateaus*, 240.

60. Ibid., 241.

61. David Abram, *The Spell of the Sensuous: Perception and Language in a More-Than-Human World* (New York: Vintage Books, 1996): 24.

62. Italics in the original. Derrida, *The Animal, 8.*

63. Derrida, *The Animal,* 28.

64. Marc Bekoff, "Why Was the Gorilla Harambe Killed at the Cincinnati Zoo?" *Animal Emotions Blog,* May 30, 2016, accessed August 6, 2016, https://www.psychologytoday.com/blog/animal-emotions/201605/why-was-the-gorilla-harambe-killed-the-cincinnati-zoo.

65. Kerry Howley, "What We Mourned when We Mourned Cecil," *The New Yorker*, September 29, 2015, accessed August 6, 2016, http://www.newyorker.com/books/page-turner/what-we-mourned-when-we-mourned-cecil.

66. Rebekah Brammer, "Activism and Antagonism: The *Blackfish* Effect," *Screen Education* 76 (2014): 72–79; Michael Cieply, "SeaWorld Questions Ethics of 'Blackfish' Investigator," *The New York Times,* February 27, 2014, accessed August 6, 2016, http://www.nytimes.com/2014/02/28/business/media/seaworld-questions-ethics-of-blackfish-investigator.html.

67. Moira Gatens, "Feminism as 'Password': Re-thinking the 'Possible' with Spinoza and Deleuze," *Hypatia* 15 (2000): 63.

BIBLIOGRAPHY

Abram, David. *The Spell of the Sensuous: Perception and Language in a More-Than-Human World.* New York: Vintage Books, 1996.

Aristotle. *On Rhetoric: A Theory of Civic Discourse*, 2nd edition. Translated by George A. Kennedy. Oxford: Oxford University Press, 2007.

Ballantyne, Roy, Jan Packer, Karen Hughes, and Lynn D. Dierking. "Conservation Learning in Wildlife Tourism Settings: Lessons from Research in Zoos and Aquariums." *Environmental Education Research* 13 (2007): 367–383.

Bekoff, Marc. "Description and Explanation: A Plea for Plurality." *Behavioral and Brain Sciences* 15 (1992): 269–270.

———. "The Evolution of Animal Play, Emotions, and Social Morality: On Science, Theology, Spirituality, Personhood, and Love." *Zygon* 36 (2001): 615–655.

———. "Are You Feeling What I'm Feeling?" *New Scientist* 194 (2007): 42–47.

———. "The Emotional and Moral Lives of Animals: What Darwin Would Have Said." *Clio Medica: Perspectives in Medical Humanities* 93 (2013): 305–332.

———. "Why Was the Gorilla Harambe Killed at the Cincinnati Zoo?" *Animal Emotions Blog* (May 30, 2016). Accessed August 6, 2016. https://www.psychologytoday.com/blog/animal-emotions/201605/why-was-the-gorilla-harambe-killed-the-cincinnati-zoo.

Biesecker, Barbara A. "Rethinking the Rhetorical Situation through *Différance*." *Philosophy and Rhetoric* 22 (1989): 110–130.

Brammer, Rebekah. "Activism and Antagonism: The *Blackfish* Effect." *Screen Education* 76 (2014): 72–79.

Calarco, Matthew. *Zoographies: The Question of the Animal from Heidegger to Derrida.* New York: Columbia University Press, 2008.

Chaput, Catherine. "Rhetorical Circulation in Late Capitalism: Neoliberalism and the Overdetermination of Affective Energy." *Philosophy and Rhetoric* 43 (2010): 1–25.

Cieply, Michael. “SeaWorld Questions Ethics of ‘Blackfish’ Investigator.” *The New York Times* (February 27, 2014). Accessed June 8, 2016. http://www.nytimes.com/2014/02/28/business/media/seaworld-questions-ethics-of-blackfish-investigator.html.

Conley, Donovan. “Grid and Swerve.” *Critical Studies in Media Communication* 27 (2010): 24–38.

Deleuze, Gilles. *Difference and Repetition.* Translated by Paul Patton. New York: Columbia University Press, 1994.

———. *Spinoza: Practical Philosophy.* Translated by Robert Hurley. San Francisco: City Lights Books, 1988.

Deleuze, Gilles and Felix Guattari. *A Thousand Plateaus: Capitalism and Schizophrenia Volume 2.* Trans. Brian Massumi. Minneapolis: University of Minnesota Press, 1987.

Derrida, Jacques. *The Animal That Therefore I Am.* Trans. David Willis. New York: Fordham University Press, 2008.

Edbauer, Jenny. “Unframing Models of Public Distribution: From Rhetorical Situation to Rhetorical Ecologies.” *Rhetoric Society Quarterly* 35 (2005): 5–24.

Garrett, Mary. “Short Reviews.” Review of George Kennedy, *Comparative Rhetoric: An Historical and Crosscultural Introduction.* New York: Oxford University Press, 1998. *Rhetorica: A Journal of the History of Rhetoric* 16 (1998): 431–433.

Gatens, Moira. “Feminism as ‘Password’: Re-thinking the ‘Possible’ with Spinoza and Deleuze.” *Hypatia* 15 (2000): 59–75.

Greenwalt, Dustin A. “User Generated Videos of Urban Exploration and the Production of Affective Space.” *Explorations in Media Ecology* 14 (2015): 125–139.

Haraway, Donna. *The Companion Species Manifesto.* Chicago: Prickly Paradigm Press, 2003.

Hawhee, Debra. “Language as Sensuous Action: Sir Richard Paget, Kenneth Burke, and Gesture-Speech Theory.” *Quarterly Journal of Speech* 92 (2006): 331–354.

———. “Toward a Bestial Rhetoric.” *Philosophy and Rhetoric* 44 (2011): 81–87.

Howley, Kerry. “What We Mourned when We Mourned Cecil.” *The New Yorker* (September 29, 2015). Accessed June 8, 2016. http://www.newyorker.com/books/page-turner/what-we-mourned-when-we-mourned-cecil.

Kennedy, George A. “A Hoot in the Dark: The Evolution of General Rhetoric.” *Philosophy and Rhetoric* 25 (1992): 1–21.

Lehner, Phillip N. *Handbook of Ethological Methods.* Cambridge: Cambridge University Press, 1998.

Lippit, Akira. *Electric Animal: Toward a Rhetoric of Wildlife.* Minneapolis: University of Minnesota Press, 2000.

Liska, Jo. “The Role of Rhetoric in Semiogenesis: A Response to Professor Kennedy.” *Philosophy & Rhetoric* 26 (1993): 31–38.

Massumi, Brian. *Parables for the Virtual: Movement, Affect, Sensation.* Durham: Duke University Press, 2002.

Nicastro, Nicholas. “Perceptual and Acoustic Evidence for Species-Level Differences in Meow Vocalizations by Domestic Cats (*Felis catus*) and African Wild Cats (*Felis silvestris lybica*).” *Journal of Comparative Psychology* 118 (2004): 287–296.

Parrish, Alex C. *Adaptive Rhetoric: Evolution, Culture, and the Art of Persuasion.* New York: Routledge, 2014.

von Uexküll, Jakob. *A Foray into the Worlds of Animals and Humans with A Theory of Meaning*. Trans. Joseph D. O'Neill. Minneapolis: University of Minnesota Press, 2010.

de Waal, Frans. *The Ape and the Sushi Master: Cultural Reflections by a Primatologist.* New York: Basic Books, 2001.

Wolfe, Cary. "Human, All Too Human: 'Animal Studies' and the Humanities." *PMLA* 124 (2009): 564–575.

Chapter 7

Beyond a Patriarchal Rhetorical Economy

Nonhuman Animals as Agents in Turkic Legends and Political Culture

Iklim Goksel

Historically, women and nonhuman animals have been denied agency and voice in the Western rhetorical tradition. The concept of *alogos* or irrationality played a crucial role in establishing mind, speech, and thought as nonhuman animal's lack and in securing the divide between nonhuman and human animal. Aristotle's assertion that men as rational thinkers are superior to women and nonhuman animals was followed by the philosophy of the Cartesian tradition as "they cannot think, therefore they are not." To point to this philosophy that identified animals as lacking thought, John Cottingham quotes Descartes's letter addressed to philosopher Henry More. In his letter, Descartes explains to More that "the difference between man and animals is that they are incapable of thought: Animals do not penser or cogitare; they are not endowed with a mind (mens, esprit); they lack reason (raison); they do not have a rational soul (âme raisonn- able)."[1]

When René Descartes proclaimed "I think therefore I am," he did not only herald the rise of modern science but also established the parameters within which it would be practiced. As his philosophical statement focused on the human mind, an era that justified ruthless animal experimentations started. As Vera Baumans points out,

> With the Cartesian philosophy in the 17th century, experiments on animals could be performed without great moral problems. The French philosopher René Descartes (1596–1650) stated that living systems could be understood on pure mechanical principles. The difference between man and animals is that man has a mind, which is a prerequisite for awareness and consequently for the capability of feeling pain. Animals cannot think and are more like machines.[2]

Descartes theory of "animal as machine" furthered speculations and doubts about whether animals felt pain or not, and a powerful Cartesian rhetoric that echoed Aristotle's theory of animals as live objects reified the animal-versus-human binary. Building upon the Aristotelian idea that animals have less reasoning ability and exist for the sake of those with more reasoning ability, Descartes described animals as lacking thought, emotion, and consciousness and as merely machines: "And here I specially stayed to show that, were there such machines exactly resembling organs and outward form an ape or any other irrational animal, we could have no means of knowing that they were in any respect of a different nature from these animals" and continues as follows:

> For it is highly deserving of remark, that there are no men so dull and stupid, not even idiots, as to be incapable of joining together different words, and thereby constructing a declaration by which to make their thoughts understood; and that on the other hand, there is no other animal, however perfect or happily circumstanced, which can do the like. Nor does this inability arise from want of organs: for we observe that magpies and parrots can utter words like ourselves, and are yet unable to speak as we do, that is, so as to show that they understand what they say; in place of which men born deaf and dumb, and thus not less, but rather more than the brutes, destitute of the organs which others use in speaking, are in the habit of spontaneously inventing certain signs by which they discover their thoughts to those who, being usually in their company, have leisure to learn their language. And this proves not only that the brutes have less reason than man, but that they have none at all: for we see that very little is required to enable a person to speak; and since a certain inequality of capacity is observable among animals of the same species, as well as among men, and since some are more capable of being instructed than others, it is incredible that the most perfect ape or parrot of its species, should not in this be equal to the most stupid infant of its kind or at least to one that was crack-brained, unless the soul of brutes were of a nature wholly different from ours.[3]

According to Bernard A. Foëx, the question that lies at the heart of the debate about the ethics of animal experimentation is the moral relationship between humans and nonhumans. Foëx points out that

> Western philosophers over the centuries have regarded humans in a different light to the rest of the animal kingdom. For example, Aristotle believed that there was a hierarchy of animals, with humans at the top, as humans could reason and had "rational souls." Even within humans there was a hierarchy: men were more rational than women, and Hellenes were more rational than other races. This made it perfectly acceptable to enslave "barbarians." Descartes considered that non humans were insentient "machines." As such, they could feel no pain, and so could be exploited ruthlessly. Kant, on the other hand, accepted

that animals could suffer, but, by lacking moral autonomy, they also lacked moral status.[4]

Foëx's discussion neatly maps the philosophical foundations of the attitudes toward the status of the nonhuman animal and also addresses the frameworks within which women's status have been relegated to mean "less rational" by patriarchal configurations. These historical constructions of women as substandard, emotional, and irrational have been one of the primary concerns of feminist scholarship that look into the ways in which patriarchal binary systems operate in society. For example, a discussion of Cartesian rationality is found in Susan Bordo's work on "Cartesian Masculinization," a term she uses to closely look at ideals of detachment and lack of empathy within Cartesian logic from a feminist perspective. Genevieve Lloyd's *The Man of Reason: "Male" and "Female" in Western Philosophy* is another classic study that looks at the ways in which reason and rational thought have been traditionally gendered male in Western philosophy.

Descartes's theories not only advanced animal experimentation but also initiated a dualistic thought system, the onset of human versus animal dichotomy, which laid the foundation for the Western patriarchal thought system. In her discussion of vivisection practices among scientists such as William Harvey and Robert Hooke in seventeenth-century England, Anita Guerrini states that "the Cartesian version of the mechanical philosophy, with which these men were familiar, both provided a conceptual framework for their work and also seemed, at least from our perspective, to justify vivisectional practice."[5] Similarly, when Aristotle in *Politics* said that "nature made all animals for the sake of man,"[6] he established human animals as rational beings with the highest form of existence and nonhuman animals as lower beings lacking reason. In Aristotle's male-biased equation, the lower beings were to serve the higher beings. Women, too, were included in this hierarchy as those with fewer reasoning abilities. Womanly and feminine qualities were associated with lack of reason and emotion. For centuries, while reason and rationality have been used to identify male virtues, women and nonhuman animals have been differentiated as lower beings who needed to be silenced.

Hence, male power structures have relied on a binary thought pattern that is based on difference and paved the way for a male-oriented rhetorical tradition that denied voice and language to those who were deemed to lack *logos*. This patriarchal legacy proliferated as male philosophers like Immanuel Kant in the eighteenth century continued to author statements such as "animals are not self-conscious and are there merely as a means to an end. That end is man,"[7] which discursively appropriated the binary constructions at work in patriarchal discourse. Kant's assertion that nonhuman animals are a "means" confers qualities such as rational and masculine in a binary system in stark opposition to femininity and irrationality that includes nonhuman animals as well.

If traditional histories of rhetoric have given rhetorical power to certain groups while denying it to others, we can seek the voice of the Other in alternative intellectual traditions. Given the long history of silenced voices in Western intellectual life, inquiry into less studied contexts can provide us with assumptions and stylistic features of alternative rhetorical traditions where silence is suspended among traditionally silenced groups. This alternative reading is made possible by reading Turkic legends and myths, which depict nonhuman animals as having language and speech. In these tales, birds and wolves speak with their bodies and interact with their voices. They talk and think like human animals. In times of need, they guide human animals to safety or give their khans or rulers advice. With mythical and nationalist overtones, the poetic images of these nonhuman animals are a key rhetorical component in twentieth-century Turkish political culture and continue to captivate the national imaginary. They shape public discourse and materialize as "non-automata" beings in the rhetorical space in Turkish life.

The voice and agency granted to nonhuman animals in Turkic legends and myths is a new model to broaden the definition of rhetoric and to consider critically what a new rhetoric beyond a patriarchal rhetorical tradition might look like. Conceptualizing the nonhuman animal with *logos* is an opportunity to understand rhetoric through nonmasculine forms of communication that traditionally have silenced nonhuman animals and women. In this vein, nonhuman animal agency is also an opportunity to address feminist goals and to conceptualize feminist theory because both women and nonhuman animals have historically been identified as lacking the capacity for reason.

NONHUMAN ANIMAL WITH *LOGOS*

It is in the Turkic Bozkurt legend that rhetorical power moves beyond men (and human animals) to include nonhuman animals, primarily the gray wolf. According to this legend, Nuri Osman Ekiz and Müslim Ergül point out that the first Turks were a Hunnic tribe named Aşine who lived on the western coast of the Caspian Sea. As a result of enemy raids and invasions, their entire population was on the verge of destruction. In the aftermath of ruthless killings, enemy forces discover a little boy who is still alive. The boy is the last Turk on the face of the earth, and the enemy forces decide to spare his life in a gruesome fashion. They cut off his arms and legs and abandon him in a forest wetland. However, upon hearing that the little boy's life is spared, the enemy commander is furious and orders his armies to find him and to make sure that the last remaining Turk on earth is killed. As the enemy forces begin their search, a female gray wolf finds the little boy nestled in agony among tall wetland vegetation. Immediately, she takes him to a cave in the Altai Mountains in central Asia, nurtures him to health with her milk, and marries him when the boy reaches adulthood. Over the years, as they consummate the marriage they have ten children together, all male. When the children reach adulthood, they get married and each of their families extend into becoming a Turkic tribe. One of these tribes is named Aşine, and its leader Asençe becomes the *hakan* (ruler) of all Turkic tribes. In order to remind the future Turkic generations of their ancestry, Asençe plants a flag that has the image of a wolf in front of his yurt (tent-like dwelling). This account of the Turkic origin myth from the Bozkurt legend is the source of the construction of the wolf as a symbol of guidance and life in the face of extinction and the foundation in endowing the nonhuman animal with human qualities such as speech and reason.

In the Ergenekon legend that follows the Bozkurt legend, a discursively powerful image of the wolf (*bozkurt*) emerges. From a position of establishing accountability and trustworthiness and building ethos, the wolf in Ergenekon plays a crucial role in the readjustment of the political make-up of a people and maintains an ethos that legitimately secures a strategic position in Turkic and Turkish intellectual life. The wolf not only gains the trust of the human animal but also gains recognition in faculties, such as speech, traditionally granted to human animals. Hence, it is the wolf's reception in the Ergenekon that publicly and politically recognizes its rhetorical powers. It is this recognition of the wolf with persuasive verbal techniques that positions it as a nonhuman animal with rhetorical voice and presence in the Turkic imaginary.

In the legend, Necati Gültepe explains that bozkurt (the wolf) emerges at a time when Turks find themselves trapped in a mountain range after being defeated by the enemy in a fierce battle. With no alternative escape route, the Turks are trapped within the confines of the mountain range they escape to. In the newly found land, which they called Ergenekon, food is abundant and Turks offer their thanks to Tengri (supreme sky god) and live there for 400 years until they decide to search for an exit way and leave. The Turks in the Ergenekon were surrounded by high mountains that had iron mines underneath them. With the aid of their blacksmiths, the Turks manage to melt the iron and mountains to break free from their confinement. The moment of salvation arrives when two bozkurts (wolves), Börteçine, a male wolf, and Asena, a female wolf, appear from behind the mountain range and tell the Turks that they will lead them to safety. Persuaded by the wolves' leadership abilities and powerful rhetoric, the Turks agree to follow them because they promise leadership, safety, and freedom.

The rhetorical moves of the wolves are impressive in terms of persuading the Turks to listen to them and to trust them. Here, the legend captures the effectiveness of both the voice and the gestures of the wolf and speaks to the rhetorical properties of its abilities. Outside of the traditional interactional characteristics of rhetoric, this communicative act between nonhuman animal and human animal not only recognizes *logos* in nonhuman animals but also portrays new ways of theorizing delivery and authority in speech and rhetoric. The rhetorical abilities of the wolf in the Ergenekon, then, rely on a set of assumptions that supersede the traditional approach of Western understandings of rhetoric in which speech and *logos* are not granted to nonhuman animals. Hence, notions regarding who may speak, communicate, and engage in rhetorical activities are reformulated and redesigned in the Ergenekon legend. With the recovery of the nonhuman animal as a rhetorical being, new possibilities emerge in rhetorical studies. Bozkurt (wolf) as a rhetor subverts the masculine field of the rhetorical tradition and takes into consideration alternative features of rhetorical activity, which is possible only by taking a closer look at the rhetoric of nonhuman animals. What is also at stake is the recognition of the need to go beyond universalized notions of rhetoric to promote efforts in inclusivity and diversity. This inclusion involves issues regarding race, gender, and culture but also broader themes that we yet have to discover. Allowing the nonhuman animal's intervention in the theory and practice of rhetoric is also an opportunity to reflect upon how we may overcome bias about unconventional ways of knowing and meaning making.

Hence, the reversal of qualities such as leadership and rationality from human animal to nonhuman animal makes the move away from dominant forms of knowledge claims a possibility. Because these qualities have also been associated with masculinity, the wolf prompts the human animal to heed a nonhuman animal's reasoning abilities and leadership skills. Recognizing a connectedness to the nonhuman animal's capabilities in leadership and thought also alters the masculine quality of leadership. Finally, by means of the wolf's rhetorical abilities, an emotional connectedness between the nonhuman animal and the animal human is made possible. Recognizing the central role of the nonhuman animal in the legends and human imaginary is not only a means of integrating it to social and political life but also offers new rhetorical formulations. Premised on the notion that human animal is not superior to nonhuman animal, the wolf as a rhetorical animal is a promising new perspective in the study of feminist rhetoric as well.

Implications of Nonhuman Animals with *Logos* for Feminist Theory

The rhetorical underpinnings of nonhuman animal with *logos* are remarkable for feminist theory given the nonhuman animal's absence in rhetorical history. The political and cultural circumstances that have traditionally denied women voice and reason in the discursive field are central to the worldview that have denied the nonhuman animal *logos*. It is this historical narrative that Turkic legends reverse to restore marginalized voices. Central to this worldview is the status of the animal human in harmony with the nonhuman animal and nature. It is characterized by the inclusion of the nonhuman animal in the rhetorical space with intellectual skills and decision-making abilities. A connectedness with nonhuman animals and nature not only erases misogynistic fears and assumptions but also evokes a sympathetic response toward ideals associated with the protection, survival, and success of a community. Conceived in terms of collaboration as opposed to patriarchal competition, rivalry, and the conquest of nature, the inclusion of the nonhuman animal in the discursive field in Turkic legends subverts a patriarchal ideology that seeks to establish dominion over nature. Because it is in itself in collaboration with nature and the nonhuman animal, the Turkic worldview is entwined with the willingness to engage in dialogue with the voice that traditionally has been silenced by Western philosophy to create a sense of community for the well-being of its peoples. Hence, against the ideals that Westernized modern science has historically dictated, there are no limitations and control over listening to the words of the nonhuman animal. The intelligibility of the nonhuman is accepted, and the nonhuman animal is valued for its ability to reason and lead the human animal out of its demise.

The rhetorical influence of the nonhuman animal in Turkic legends is especially remarkable given the fact that Asena emerges as a female wolf in the stories. By placing the female nonhuman animal's voice in a leadership role, Turkic legends reverse the "silent female" patriarchal image and assign speech and agency to the female nonhuman animal who can shape the thoughts and behavior of the human animal. Situating the female voice within the discursive field that traditionally has been reserved for the male voice and listening to it for instructions directly grants it intellectual potency. It is the Turkic cultural climate, then, that extends both the female voice and the nonhuman animal voice a speaking role, agency, and a podium. Having the abilities of achieving public influence and having effects on human thought, behavior, and judgment, the female voice and nonhuman animal voice remain highly visible in the Turkic rhetorical space.

Conceiving the female wolf as a source of help, nourishment, love, respect, and strength and accepting its mental capabilities as sources of awe and admiration serve as the groundwork for identifying the female wolf's vital role engraved in Turkic daily life and function as a governing principle in Turkic legends. The inclusion of the female and the nonhuman in the Turkic rhetorical space brings into sharper focus the ways in which nonpatriarchal forms of speech, language, and oratory can be undermined to have a less privileged status over Othered voices.

Because of the readiness in aligning itself with the female and nonhuman animal voice in legends, Turkic discursive field embodies language and persuasion as both a human animal and nonhuman animal pursuit with a broad public audience. This attitude not only uncovers thoughts associated with language in the Turkic psyche but also signals the many nonhuman animal attributes that are given to language. Representing qualities such as listening and collaboration as governing principles in communicative acts, Turkic legends uncover language as a communal act promoting civic harmony where the nonhuman animal and human animal come together. This subverts the traditionally prominent domineering role of the male human animal in speech and empowers alternative voices and ways of speaking.

THE NONHUMAN ANIMAL'S IMPLICATIONS FOR TURKIC RHETORIC

As Ramilya Yarullina Yıldırım notes, the deep Turkic cultural attachment to the wolf lies at the heart of the *hayvan-ata* (animal male ancestor and guide) and *hayvan-ana* (the female counterpart of hayvan-ata) themes that draw the image of the nonhuman animal as helpers and fellow comrades (*silah arkadaşı*).[8] Both the hayvan-ata and hayvan-ana are metaphors in the legends for the role the nonhuman animal plays in creating and maintaining a medium through which shared communal values are centered in Turkic daily life. Hence, evoking values that are identified as Turkic, both hayvan-ata and hayvan-ana underscore the idealized imaginary of the wolf as a moral and truthful speaker, a comforting listener, and an inspiring helper. This attitude toward language presents legitimate considerations regarding Turkic rhetoric and points to contextual and historical considerations that function as a prelude to theorizing non-Western rhetorical practices. In this vein, the wolf in Turkic legends sketches notions related with language and mirrors Turkic attitudes toward inclusion in the political and social space of all voices as active participants of discourse.

Of course, it is not only the wolf in Turkic legends that plays a crucial role in reworking rhetorical principles in language. Additional nonhuman animals present Turkic ethical principles and thought processes regarding language use. For example, the eagle is endowed with *logos* as well and is viewed as a messenger (*tanrıya gönderilen elçi*) sent to the supreme god Tengri to bring back messages. Very much like the wolf, the eagle plays a leading role in times when armies in warfare need guidance and advice.[9] In the Edigey legend, the elderly wolf and the otter are again interconnected to faculties of wisdom and experience[10] and Toktamış Han, the khan of the Turks in the Edigey legend, expresses his pride in his armies by comparing them to the experience and wisdom of the wolf. In the Oğuz Kağan legend, a gray wolf with sky-blue hair and mane (*gök tuylu ve gök yeleli*) is defined as a savior and a guide (*yol gösterici*) for Oğuz Kağan, the khan of Turks. As noted by Özkartal, because nonhuman animals are considered to be whole in mind and body, legendary heroes, khans, and soldiers are perceived as having the same intellectual qualities as the wolf.[11] In the Orkhon-Yenisey inscriptions,[12] the phrase (*Kağanımın ordusu kurt gibidir/ Kanim Kagan susi bori tek*—The Khan's army resembles the wolf) points to this idea. In other tales and legends, the nonhuman animal manifests human animal intellectual capabilities and aligns empathetically with them.

In the epic of Manas, which is an epic poem composed and sung in oral form, the dog and bird of Manas, a national hero, engage in conversation with human animals as they mourn Manas's passing at his graveside.[13] In the epic of Köroğlu, Kir At, the horse, laments Köroğlu's death and chooses to fast for forty days and forty nights. In another Turkic legend, the Türeyiş, the wolf displays emphatic language and agrees to guide and protect the Turk,[14] and in the Oğuz legend, the wolf is a guide for Oğuz khan to the battlefield.[15] The Orkhon inscriptions emphasize the wolf's abilities in undermining hierarchies of oppression by pointing to its abilities in reproducing relationships that are transformative. In these inscriptions, the wolf is a leader of social transformation that brings about transformation and harmony to a people who are in desperate need of a yurt (nation).[16]

In the book of Dedekorkut, the epic story of the Oğuz Turks, Salur Kazan khan engages in dialogue with the nonhuman animals that he encounters on his way to the battlefield. According to the story, Salur Kazan unexpectedly comes across a wolf and interprets this encounter as an opportunity to engage in a rewarding dialogue that he will learn from.[17] Kazan's conviction that the knowledge possessed by the horse will be beneficial to him is a testament that the nonhuman animal is endowed with the role of a guide, teacher, and wise rational being. In these stories, the horse is also inextricably tied to speech and reason and is a source of wisdom in times of trouble.[18] Hence, the human animal is urged to be receptive to the speech of the horse and to listen to its advice. In the book of Dedekorkut, the khan addresses his horse as *kardaş* (brother/comrade), which, in the Turkic languages, implies bonding and a friendship that is based on trust and exchange of ideas. Similarly, birds are also granted a reputable status because of their abilities to mentor khans. For example, in the book of Dedekorkut, Anka and Huma birds consistently engage in dialogue and are the mentors of the khans due to their ability to reason.[19] It was also the reliance on the reasoning abilities of the Huma bird that had a direct impact on who would become the next khan of the Turks. The Huma bird would literally be in charge of passing the throne to whomever it wished. As Jean-Paul Roux notes, among some Turkic tribes, khans would ascend to the throne if only a Huma bird perched on their heads. To the present day, the Huma bird continues to impact and manage daily affairs in the Turkish national space. Often referred to as the *talih kuşu* (bird of luck) or *devlet kuşu* (bird of the state), a bird poop that lands on the head calls for a trip to a lottery ticket counter.

THE NONHUMAN ANIMAL IN THE TURKISH SOCIAL AND POLITICAL SPACE

Embedded in the twentieth-century sociocultural and political field, the nonhuman animal expands to the everyday and finds expression in grand narratives that remain committed to an understanding of the nonhuman animal with cognition and awareness. This commitment to elevating the status of the nonhuman to human is not only grounded in the assumptions and beliefs of the Turkic psyche but also in the pursuit of elucidating its status in the everyday. Hence, we can begin to consider the nonhuman animal's affiliation with daily life, particularly of the wolf's and the philosophical heritage that endows it with self-consciousness. One of the most tangible gestures toward the wolf's subject positioning is the depiction of Mustafa Kemal Atatürk, the founder of the Turkish republic, as a spectacle resembling the wolf by Turkish poet Nazım Hikmet Ran in his epic poem of the Turkish War of Independence, *Kuvayi Milliye*.[20] Without question, the depiction of Atatürk as a wolf in the poem unmistakably alludes to the Ergenekon Turkic legend and the pivotal role the gray wolf played in the survival of Turkic peoples. More importantly, it enlarges the conditions in which the wolf is imagined to have human faculties especially as a living memory in the Turkish historical and intellectual milieu. Shuttling between legend and life, the wolf stimulates, moves, and occupies the fabric of social life and is inseparable from the realm of the everyday. More importantly, it brings about new models of rhetorical practice in text production, modes of inquiry, and discursive interaction. As illustrated and discussed in the following paragraphs, these models of rhetorical practice do not manifest themselves by assimilating to Western models of rhetoric. Rather, they reveal rhetorical configurations that include the voice of the nonhuman animal. In the Turkish rhetorical space, then, the presence of the wolf is not merely the manifestation of an imagery. Although the wolf's imagery has a notable presence in Turkish culture, what is more significant is the "inclusion" of its voice in the discursive space. The implications of this inclusion offer possibilities for conceptualizing the nonhuman animal as rational. Secondly, the inclusive status of the wolf points to its rhetorical powers in persuasion. It is because of the persuasive abilities of the wolf that a people are led to safety and start a new Turkic nation. Hence, the imageries of the wolf in social and discursive spaces are a testament to its abilities in appealing to the minds and emotions of a people and in establishing its ethos for hundreds of years. It is because of the speech practices, argumentation, and logic of the wolf that its imagery and rhetorical power continue to occupy discursive and nondiscursive spaces in the Turkish context.

The dialogue between the wolf and the poplar tree in a poem written for the fiftieth anniversary of the foundation of the Turkish republic provides a framework for understanding the extent to which the wolf has been a part of Turkish intellectual tradition and how the wolf's affective powers have been instrumental in shaping feelings and sensations associated with the nation. In the poem, the image of the wolf and Atatürk come together as the moral compass of the nation to set the scene in which the poplar tree (the nation) comes to life.[21] The affinity between Atatürk and the wolf not only erases the animal-versus-human duality, but also makes room for nonhuman animal presence in the civic and national space. It associates reason with the nonhuman animal and undermines the distinction between human and nonhuman animal. Eloquence and rhetorical action that is not bound by gender roles but situated beyond the command of the human animal intellect opens up the rhetorical space to all voices. This makes it feasible to imagine what is possible in discourse and to explore alternative cultural assumptions regarding rhetorical behavior and performance. Embracing the nonhuman animal with *logos* and speech is not merely understanding rhetoric as including nonhuman animals. Rather, it is a window through which nonhuman aspects of language and communication can be explored so that the binary of human animal versus nonhuman animal can be approached with suspicion.

Considering the range of attitudes toward the wolf, it is the *hayvan ata* and *hayvan ana* (animal ancestor/elder) symbolism, a key topic in Turkic legends, which is engagingly linked to the Turkish political-cultural space. The theme of the wolf in the Bozkurt anthem dedicated to the republic and the wolf image in the old Turkish Lira notes, in postal stamps, and in the logo of Petrol Ofisi, the Turkish national fuel products distribution company, further point to this symbolism. Furthermore, the following wolf images that occupy the Turkish national space point to the broader framework in which the wolf's importance is understood: the emblem and name of the Turkish national cigarette brand as Bozkurt (the wolf); granting the surname of Bozkurt to Turkish justice minister Mahmut Esat following the surname law of 1934, which required all Turkish citizens to adopt a permanent surname; a painting depicting the exodus from Ergenekon at the entrance of the Ministry of National Education building in the capital city of Ankara; the official name of the cub scouts of the Scouting and Guiding Federation of Turkey as *yavru kurt* (infant wolf); the official emblem of Türkiyat Enstitüsü (Turkish Research Institute) as a bozkurt; and the bozkurt as the official logo of Türk

Ocakları (Turkish Hearth Clubs), founded to promote educational, social, and economic standards in Turkey. In the poetry of Behçet Kemal Çağlar, the wolf motif in "Atatürk'ü Dinlerken"[22] alludes to the "nation-wolf-Atatürk" trilogy, and in Nurettin Özyürek's poem *Mustafa Kemal Paşa*,[23] Atatürk's ascent to the Kocatepe battlefield—also known as the Great Offensive, which was the largest and final military operation during the Turkish War of Independence—is compared with the wolf's guidance of the Turks toward survival from Ergenekon. Vasfi Mahir Kocatürk, in the legend of the Turkish War of Independence,[24] depicts Atatürk as a godly wolf (*tanrısal bozkurt*) that brings hope to the people. In his poem *Ergenekon*, Atatürk resembles the wolf (*mavi kurt*) who is on a mission to save the country from Western colonial ambitions. Oğuz Akkan in his work, also titled *Ergenekon*, draws parallels between the Turkish War of Independence and the exodus from Ergenekon, which are led by Atatürk and the wolf. This echoes the scene in the Oğuz Khan legend where a wolf with a sky-blue coat and exuberant mane tells Oğuz Khan, the khan of the Turks, "I will lead you, follow me." H. C. Armstrong's 1932 biography of Mustafa Kemal Atatürk is titled *Grey Wolf: The Life of Kemal Atatürk.*

Finally, the design of the Turkiyat Institute logo is recounted in many sources in a notable dialogue between Mustafa Kemal Atatürk and Turkologist Fuat Köprülü. Atatürk requests the logo of the Turkiyat Institute to depict a wolf as follows: "I would like to see the wolf (bozkurt) holding a torch by the snow-capped Tian Shan Mountains. Let the torch that the wolf holds signify reason and science in the new Turkish republic. And, let the wolf (bozkurt) who guided us out of Ergenekon be the expression of the founding of the Turkish republic."[25] Additionally, the Turkish national anthem alludes to the wolf's pivotal role in saving a people in the Ergenekon with the lines that translate as "I will tear down the mountains and my accomplishments will be boundless!"[26]

CONCLUSION

Turkic legends offer an alternative way of conceptualizing rhetoric that includes the nonhuman animal. They offer an alternative way of theorizing rhetoric that does not rely on Western patriarchal notions of *logos*, which allows the inclusion of unconventional voices in discourse and communication. They pave the way for formulating a new language that can be included in new rhetorical spaces, new sites for communicating and speaking without excluding women and nonhuman animals. Erasing the silence of both women and nonhuman animals not only revises traditionally accepted understandings of *logos* but also allows the participation of all voices in the rhetorical space.

Promoting communicative acts that are based on notions such as seeking help, negotiating, and listening subvert masculine rhetorical practices that rely on reason, assertiveness, and competition. The animal turn in rhetorical studies, then, is opening up the conversation outside gendered norms, recognizing that the inclusion of the nonhuman in theorizing alternative discursive practices has a broader appeal in considering rhetorical theory. This new turn will bring about new possibilities for ideas related to subjectivity and ethos in language, new ways of seeing who may speak and be a participant in rhetorical practices, and new ways of creating the circumstances under which rhetoric can be practiced. It will also blur patriarchal binaries and gendered biases to expand on ideas of how we may engage in conversations about rhetorical practice, and how traditional canonized concepts such as invention, style, delivery, and audience may be contested. Hence, the inclusion of nonhuman animals in rhetorical performance will generate a host of new questions that will help us formulate new rhetorical categories.

The presence of the nonhuman animal with *logos* in Turkic legends is an unmarked territory that has not been colored by patriarchal ways of knowing and thinking. It is an alternative space for speaking, writing, and listening that is not male-identified. In some ways, giving any nonmale speaker *logos* can be considered a way of co-opting and/or masculinizing rather than subverting the traditional binary. However, the emphatic response exerted by the wolf is geared toward communal ideals such as the protection, survival, and success of a community in need. In contrast to masculine ideals of competition and dominion, the wolf fosters collaboration in the rhetorical space. Hence, the inclusion of the nonhuman animal in the discursive field in Turkic legends works against a masculine binary system and instead subverts patriarchal ideology. Without question, this is appealing for feminist concerns and reinforces alternative rhetorical spaces outside of the commonplace patriarchal narratives. Granting discursive agency to nonhuman animals is a subversive tool against conventional patriarchal configurations that have traditionally associated men with reason and lack of *logos* with women and nonhuman animals. The voice of the nonhuman animal in Turkic legends is a remarkably inclusive social and political economy that is not based on gender binaries. Because "irrational" in the Western tradition has been considered to be womanly within patriarchal power relations, nonhuman animal speech in Turkic legends epitomize nonpatriarchal notions of rhetoric. Hence, the departure from patriarchal speech patterns brings notions such as communal harmony, mutual respect, and collaboration into the rhetorical space. This attitude cuts across rhetorical barriers to articulate alternative forms of communication between speakers and listeners and new patterns in becoming audiences. In the Turkic past, the inclusion of the nonhuman animal in speech is a precondition of rhetoric and it is perceived as an ethical virtue.

Where the nonhuman animal with *logos* is concerned, the twentieth-century Turkish psyche is engagingly preoccupied with it and embraces it in a series of social and political situations. Its impact on Turkish national consciousness operates on a very broad territory and sets the scene in social life and political discourse.

It is in Turkic legends that the animal turn in rhetoric comes to fruition, extending the scope of rhetorical studies. Not bound by traditional notions of persuasion in language, the nonhuman animal's rhetorical agency in Turkic legends appeals to innovative ways of theorizing rhetoric in order to enlarge our understanding of communicative practices.

NOTES

1. John Cottingham, "A Brute to the Brutes? Descartes' Treatment of Animals," *Philosophy* 53 (1978): 554.
2. Vera Baumans, "Use of Animals in Experimental Research," *Gene Therapy* 11 (2004): 64.
3. René Descartes, "Discourse on the Method of Rightly Conducting the Reason, and Seeking the Truth in the Sciences," Part 5.
4. Bernard A. Foëx, "The Ethics of Animal Experimentation," *Emergency Medicine Journal* 24 (2007): 750.
5. Anita Guerrini, "The Ethics of Animal Experimentation in Seventeeth-Century England," *Journal of the History of Ideas* 50 (1989): 391.
6. P. L. P Simpson, *The Politics of Aristotle* (Chapel Hill, NC: University of North Carolina Press, 1997), Book I, Part 8.
7. Immanuel Kant, *Lectures on Ethics*. Translated by Louis Infield (New York: Harper and Row, 1963): 239–40.
8. Ramilya Yarullina Yıldırım, "Edigey Destanında Hayvanların Fonksiyonları/ Functions of Animals in Edigey Epic." *Avrasya Uluslararası Araştırma Dergisi* 1 (2012): 68.
9. Ibid., 67.
10. Ibid., 71.
11. Mehmet Özkartal, "Türk Destanlarında Hayvan Sembolizmine Genel Bir Bakış/ Animal Symbolism in Turkic Epics," *Milli Folklor* 24 (2012): 60.
12. Turkic inscriptions from the eighth century in the Orkhon Valley in Mongolia, erected in honor of two Turkic khans, Bilge Kağan and Kül Tigin.
13. Mehmet Özkartal, "Türk Destanlarında Hayvan Sembolizmine Genel Bir Bakış/ Animal Symbolism in Turkic Epics," *Milli Folklor* 24 (2012): 68.
14. Ibid., 64.
15. Ibid., 65.
16. Ibid.
17. Ibid. The original text in Turkish is as follows: "Bu kez önüne bir kurt çıktı. Kurt yüzü mubarektir, kurtla bir haberleşeyim" dedi.

18. Ramilya Yarullina Yıldırım, "Edigey Destanında Hayvanların Fonksiyonları/ Functions of Animals in Edigey Epic," *Avrasya Uluslararası Araştırma Dergisi* 1 (2012): 68.

19. Abdüsselam Arvas, "Türk Mitolojisinde Hayvan Sembolizmi." In *Türk Mitolojisine Giriş*, edited by Fatma Ahsen Turan and Meral Ozan (Ankara: Gazi Kitabevi, 2015), 617.

20. The original text in Turkish is as follows: Dağlarda tek tek ateşler yanıyordu. ve yıldızlar öyle ışıltılı, öyle ferahtılar ki şayak kalpaklı adam, nasıl ve ne zaman geleceğini bilmeden güzel, rahat günlere inanıyordu ve gülen bıyıklarıyla duruyordu ki mavzerinin yanında, birdenbire beş adım sağında onu gördü. Paşalar onun arkasındaydılar. O, saatı sordu. Paşalar: "üç" dediler. Sarışın bir kurda benziyordu. ve mavi gözleri çakmak çakmaktı. Yürüdü uçurumun başına kadar, eğildi, durdu. Bıraksalar ince, uzun bacakları üstünde yaylanarak ve karanlıkta akan bir yıldız gibi kayarak kocatepe'den Afyon ovası'na atlıyacaktı.

21. Original text in Turkish: Bozkurtla Kavak—ben yabanın kurduyum nazlı kavakgelirim her yılın en karlı günü dumanlı dağlardan Ankara'ya arar Mustafa Kemal'ini susan doruk, bozundan bozundan bir susar kiinilder havalar acısnda. Hadi sen yücesin akıllısın bilirsin bilgeliğini sessizliğinde bana bir yaprak. dinle kardeş kurdum, yellerle gecelerle yaşamakla dolu olduğun kadar düşünve duy:işte özgürlükler, sonsuzluk,işte ölümsüz toprağın gizisen onu aramasanve o sevmese seni ben nasıl yeşil olurdum?

22. The original text in Turkish is as follows:yay yine gerilmede, fırlayacak yine Ok, yine vatanımızın yeryüzünde eşi yok; Bozkurt, Ergenekon'u yeni delmiş gibidir; herkes bir ihtirası seyre gelmiş gibidir.

23. The original text in Turkish is as follows:"ve bir gümüş ay yükseldi yangınlıklar üstüne kuru dallar çiçek verdi çıkıp kocatepe'ye bozkurt misâli Kemal Paşa yol gösterdi."

24. The original text in Turkish is as follows: Istiklâl Savaşı Destanı—"güneşin çizdiği izden canlı bir umut geliyor; ufku dumanlı denizden tanrısal bozkurt geliyor"

25. İbrahim Candan. *Seni Anlasaydık Bu Hale Gelmezdik*. (Ankara: Akasya Kitap, 2006). The original text in Turkish is as follows: "Karlı Tanrı Dağları'nın önünde meş'ale tutan bir bozkurt olsun, bu meş'ale genç Türkiye Cumhuriyeti'nin ilminin ifadesi olsun. Ergenekon'dan çıkmamızın klavuzu olan bozkurt, Türklüğün Anadolu topraklarındaki yeni devletinin kurtuluşunu ifade etsin."

26. Ekiz, Nuri Osman and Müslim Ergül, *İslamiyet Öncesi Türk Edebiyatı* (İstanbul: Toker, 1986). The lines of the Turkish national anthem, *İstiklal Marşı*, composed by Mehmet Akif Ersoy, are as follows: "Yırtarım dağları, enginlere taşarım!"

BIBLIOGRAPHY

Akkan, Oğuz. *Ergenekon*. Bursa: Uygun Basımevi, 1946.

Armstrong, H. C. *Grey Wolf—Mustafa Kemal*. London: Methuen, 1932.

Arvas, Abdüsselam. "Türk Mitolojisinde Hayvan Sembolizmi." In *Türk Mitolojisine Giriş*, edited by Fatma Ahsen Turan and Meral Ozan, 183–196. Ankara: Gazi Kitabevi, 2015.

Baumans, Vera. "Use of Animals in Experimental Research." *Gene Therapy* 11 (2004): 64–66.

Bordo, Susan. "The Cartesian Masculinization of Thought." *Signs: Journal of Women in Culture and Society* 11 (Spring 1986): 439–456.

Candan, İbrahim. *Seni Anlasaydık Bu Hale Gelmezdik*. Ankara: Akasya Kitap, 2006.

Cottingham, John. "A Brute to the Brutes? Descartes' Treatment of Animals." *Philosophy* 53 (1978): 551–559.

Cottingham, John, Robert Stoothoff, and Dugald Murdoch, trans., *The Philosophical Writings of Descartes*. Cambridge: Cambridge University Press, 1985.

Çağlar, Behçet Kemal. "Atatürk'ü Dinlerken." In *Çağdaş Türk Şiiri Antolojisi*, edited by Mehmet Fuat. İstanbul: İkinci Adam, 2013.

Descartes, René. "Discourse on the Method of Rightly Conducting the Reason, and Seeking the Truth in the Sciences." Last modified December 17, 2014. https://ebooks.adelaide.edu.au/d/descartes/rene/d44dm/index.html. Accessed January 27, 2017.

Ekiz, Nuri Osman and Müslim Ergül. *İslamiyet Öncesi Türk Edebiyatı*. İstanbul: Toker, 1986.

Foëx, Bernard A. "The Ethics of Animal Experimentation." *Emergency Medicine Journal* 24 (2007): 750–751.

Guerrini, Anita. "The Ethics of Animal Experimentation in Seventeenth-Century England." *Journal of the History of Ideas* 50 (1989): 391–407.

Gültepe, Necati. *Türk Mitolojisi*. İstanbul: Resse, 2015.

Kant, Immanuel. *Lectures on Ethics*. Translated by Louis Infield. New York: Harper and Row, 1963.

Kocatürk,Vasfi Mahir. "Atatürk'ü Dinlerken." In *Çağdaş Türk Şiiri Antolojisi*, edited by Mehmet Fuat. İstanbul: İkinci Adam, 2013.

Lloyd, Genevieve. *The Man of Reason: "Male" and "Female" in Western Philosophy*. Minneapolis: University of Minnesota Press, 1984.

Özyürek, Nurettin. "Mustafa Kemal Paşa." "Atatürk'ü Dinlerken." In *Çağdaş Türk Şiiri Antolojisi*, edited by Mehmet Fuat. İstanbul: İkinci Adam, 2013.

Özkartal, Mehmet. "Türk Destanlarında Hayvan Sembolizmine Genel Bir Bakış/ Animal Symbolism in Turkic Epics." *Milli Folklor* 24 (2012): 58–71.

Ran, Nazım Hikmet. "Kuvayi Milliye." In *Çağdaş Türk Şiiri Antolojisi*, edited by Mehmet Fuat. İstanbul: İkinci Adam, 2013.

Roux, Jean-Paul. *Eski Türk Mitolojisi*. Translated by Musa Yaşar. Ankara: Bilgesu, 2015.

Simpson, P. L. P., trans. *The Politics of Aristotle*. Chapel Hill, NC: University of North Carolina Press, 1997.

Yıldırım, Ramilya Yarullina. "Edigey Destanında Hayvanların Fonksiyonları/ Functions of Animals in Edigey Epic." *Avrasya Uluslararası Araştırma Dergisi* 1 (2012): 67–77.

Chapter 8

Humans, Dolphins, and Other People

Alex C. Parrish

ON NAMING

Between the years of 1847 and 1857, an African American slave named Dred Scott attempted a feat that would be unthinkable in the modern day: he attempted to turn a piece of property into a living, breathing human being. By what means could such a magical transformation occur? Polymorphism? Reincarnation? Would he buy this piece of property from an ancient Chinese sage and proceed to get it wet and feed it after midnight?

It was none of these things. In fact, the conversion from thing to person that Scott attempted relied on a much more powerful magic—an ancient magic that permeates almost all aspects of life. Scott channeled the power of naming. Much like Isis, who was able to divine the true name of Ra and thus hold complete power over him, Dred Scott hoped to hold power over his own future, and that of his wife and daughter, by changing their status from some*thing* to some*one*. That is, he wanted to give them a true name, which is something most people do not grant to mere objects.

His justification for this alteration of name or status? He went somewhere that did not allow biological humans to be turned into property. This sounds simplistic (and I mean it to), perhaps even glib (and I do not), but it demonstrates aptly the fantastical nature of antebellum American thought, as well as the awesome power behind name and status: crossing the border from one state to another turned an object into a person or back again. Scott's case depended heavily on his four-year residence in the free state of Illinois and the territory of Wisconsin (in the area around Fort Snelling, which in 1858 became part of the free state of Minnesota), and it bears repeating that the same living, breathing creature could be property in one part of our country and a person with rights in another.

It is important for nations to formally recognize personhood, as Daniel Hoffman argues, and to do so inclusively. This will help ensure that incidents like the Dred Scott case are not repeated; a person should not turn into an object merely because he crosses an arbitrary boundary between states.[1] And yet, thinking, feeling individuals in the present day are held captive across the world because they happen to have been captured somewhere where they are not people, transported to enclosures where they have no rights. If they had remained in other countries, they would be free. Worse yet, immoral humans, who are protected by the states in which they operate, use these legal inconsistencies to traffic living, breathing "goods" across international boundaries, even in some cases through nations that do not allow the capture or display of certain species.

Some people might quail at the comparison between race-based slavery in the United States and the enslavement of nonhuman animals, but this depends entirely on a similarly uncaring human exceptionalism that allows feelings, thoughts, and desires to human animals but no other. It is akin to quibbling with the word "genocide" when used to describe the murder of millions of unwanted cats and dogs every year—not because they are unsuitable for home placement, but because there are too many and they have become inconvenient (despite a large percentage of them originating from for-profit puppy and kitten mills). We forget the power behind such words, power that could be used for good if only people would allow it to happen. That is why I choose to open this discussion of personhood with an example of someone who was denied it, but almost everyone today would feel he truly deserved that status. Just as the unenlightened in nineteenth-century America might have denied that some human animals were people, many in the United States of the twenty-first century still feel that even the most intelligent nonhuman animals should not qualify, either.

This is a dangerous viewpoint to anyone concerned about the state of humanity. We are mere decades from the last major international event that featured an attempt to dehumanize one or more ethnic groups en masse by *animaling* them—that is, by treating them as if they were some perceived lesser beast, which in some people's minds would allow them to be murdered without consequence or moral reservation. While the idea of "animaling" might seem, *prima facie*, contradictory to my argument, it is in fact the other side of the same hierarchical coin: we use the bestial to oppress both human and nonhuman animals, so that we can separate ourselves from those beings of any species that we want to harm. In order to understand the precariousness of denying a being the right to personhood, we need only consider the many other individuals we would harm by withdrawing the rights we only assume they hold. That is why the topic of this chapter, while being *about* one genus of nonhuman animals, has far-reaching philosophical consequences beyond the well-being of any one species. Human animals have power over the lives of every other species on our planet, whether we deserve our place of supremacy or not. So the question that any group of people with power must ask is, "How can I create benefit for my own people without harming those who are weaker than I am?"

While it is true that some people would not care, those sorts of people do not read essays such as this. They have already self-selected out of humanistic, let alone humane, considerations. For those who do care what happens to the weaker, to the other, or to the underdog, I hope to guide you toward a new way of thinking about the other intelligent species on our planet, using one genus of cognitively gifted creatures as a case study in support of nonhuman personhood. And, despite my firm belief that this is a noble goal, it will not be easy. The idea of personhood is fragmented at times and spongy at others; personhood implies rights and responsibilities in some cases, abilities in others, and even nationality or residency in cases like the scenario above, in which one man's family was ultimately denied personhood by the Supreme Court of the United States of America (and not by a small margin, either—the vote was 7–2 against Dred Scott). Thus, it will be important to examine some of the ways personhood has been defined in the West, and to orient ourselves in a more inclusive direction before we can move on to a discussion of dolphins' personhood through their sophisticated methods of communication, their incontrovertible individuality, and even those abilities that make them superhuman in some regards, for it has long been the case that humans have assumed that all of their cognitive and communicative abilities are better than those of all other species, when our brains have in fact evolved many deficiencies and blind spots. All of this should indicate that humans need to adopt a more inclusive worldview that allows for difference without hierarchy—something our species has a long history of being unable, or at least unwilling, to do.

ESTABLISHING PERSONHOOD

In 1983, when discussions of personhood were continuing to gain political importance for several groups of individuals, Ruth Macklin was already complaining that the literature on personhood in the young field of bioethics was quite vast, yet there was little evidence that researchers were approaching consensus.[2] Thirty-four years later, the current research dwarfs that of 1983, and we have taken only very small steps toward defining what a person might be. Personhood studies are often performed under three branches of consideration: whether it is a status question, a metaphysical question, or a moral one.

As a status concern, there is often a differentiation between a biological person and a legal person; being one of these things does not, in the minds of many scholars, entail the other. Plato was perhaps the first to attempt a biological definition of personhood, when he suggested that people are featherless bipeds. Fellow philosopher Diogenes of Sinope did not regard this definition as highly as the rest of Athens seemed to at the time, so in order to point out one of the flaws of this definition, he plucked a chicken, released it during one of Plato's lectures at the Academy, and shouted, "Behold: a man!"[3] Philosophers would struggle with this issue throughout history, and many fine attempts have been made to nail down what makes a biological person. John Locke's famous definition of a person, for example, brought us no closer to satisfying all parties and, some would argue, did not even satisfy his own internal logic—according to his theory, a person may be a being or a state of mind, or something in between.[4]

A more modern biological definition might sound something like the following: *humans are all people and all people are humans*. This is at once too restrictive and too inclusive, however, and lends itself to several debates. I will address the idea of extrahuman personhood in several places below, but there are many ways to question whether this idea is too inclusive. Is a baby a person? What about someone who has suffered a horrible accident and is only being kept alive by machines that make that body breathe, despite its "brain dead" status? Is a dead human a person, or was it? What about a human who is missing the parts most other humans have—possibly there are gradations where, if I am only missing my tonsils, I am still a person, but I really have to hope I don't get appendicitis or I'd be a borderline case. Cyborg humanism, among other materialist research programs, discusses human-synthetic hybridization when we add on to our bodies with devices like pacemakers and artificial limbs, so I will not delve too deeply into that here; it is important, however, to remember that these debates about biological humans and personhood do exist. A person cannot merely be the sum of its parts.

If the idea of biological personhood seems somewhat arbitrary and easily muddled, the concept of legal personhood will be completely unhelpful in clearing up the controversy. One problem with legal theories of personhood are that they often include completely nonbiological agents, such as financial corporations. There comes a time in the life of an evolving language that a metaphor becomes so common that people begin to take it literally. This seems to be the case with the concept of incorporation—several financial interests coming together to form a single "body" with which they will carry out their business. Corporate personhood is especially insidious in modern capitalist systems because it allows for the rights of personhood without most of the responsibilities. For instance, while a corporation can own property, enter into contracts, and in at least one case be married to a human being, corporations cannot be sent to jail or drafted into military service.

The 2009 Supreme Court decision *Citizens United v. The Federal Election Commission* expanded some constitutional protections given to corporations, which made them legal persons.[5] In fact, corporations in the United States now have legal rights to "speech" that transcend that of a biological person. In response to this result, one of the dissenting judges, Justice Sonia Sotomayor, quipped that the court just "gave birth" to corporate babies.[6] It is difficult to describe the frustration this development causes for people who have been fighting for decades to liberate sensate beings from the shackles of unpersonhood—shackles that allow animals to be caged, tortured, and killed at the whims of humans and their corporations. Groups like the International Forum on Globalization have launched a campaign to re-dehumanize what many call the "corporate monsters" that conservative politicians have created, but the battle for the next four to eight years will be an uphill struggle.

Other troublesome debates that legal personhood definitions inspire often have to do with ability and disability. This is an area of concern that several rhetoricians have covered more completely. Jenell Johnson, as just one example, following discussions by Cynthia Lewiecki-Wilson, argues convincingly against a person's rights being granted based solely on the fact that they have certain abilities in common with normally developing humans.[7] The problem with this idea should be fairly obvious—not all humans are able to do the same things, either physically or mentally. If this were the case, the Olympics would be a ho-hum affair, and everyone could be poet laureate (which is not to say all people must excel at all human activities, but that we would be hard-pressed to justify defining the boundary of personhood based on pole-vaulting ability). Steven Wise reminds us that the examples people use when asserting human ability creates personhood are always extraordinary—Albert Einstein, Ella Fitzgerald, William Shakespeare. If our "fundamental rights turned on the ability to write prose like Shakespeare," he continues, "few humans would have them."[8] Even when we attempt to avoid drawing distinct ability-based lines between human and nonhuman people, most efforts at employing a gradient are lacking in justification or consistency. Arbitrariness becomes the rule, as any "human trait" can be difficult for certain individuals.

For the rhetorician working in a field that has in the last century largely attempted to rally against the defining away of rights based on shoddy reasoning or a lack of identification with others, it should be obvious that when we deny the personhood of another, it is our own personhood that suffers. The troubling trend, however, is that there seems to be more and more forces working against an inclusive humanitarian view. The Romans would ask *cui bono*, or who stands to benefit, when a troublesome legal question would arise; it might be wise to keep this question in mind as we consider why a human, a politician, or a corporation[9] would want to challenge the personhood of another. One constitutional law expert tells us that

> Arguing for constitutional rights on behalf of non-human beings, which some people viscerally oppose, shouldn't be confused with giving certain non-human interests absolute priority over conflicting human claims. Recognizing rights is fully consistent with acknowledging circumstances in which such rights might be overwritten, just as human rights themselves sometimes come into conflict.[10]

Legal personhood has increasingly had to deal with challenges in the form of moral rights that certain groups assert for themselves or on behalf of others who are unable. Moral rights (such as the anti-abortionist claims that fetuses are people with the right to life)[11] often require special-case thinking to expand right-holder definitions for religious or political reasons. Rather than require an individual to have certain abilities, moral cases often stand on a foundation of potential. A fetus, to an anti-abortionist, is nonetheless a partial person, who is awarded some rights, which by necessity abridge the rights of other, fully formed people. The definition of human, person, and child in these debates is so slippery and so legally and emotionally fraught as to make this author want to avoid the discussion altogether, but it is important to at least remind the reader of the vitriol politicians can generate by asking a seemingly simple medical question like "when does life begin?" Whether it is at birth, when the heart begins beating, or as early as conception, the child will not have most rights and entitlements of adult humans until it reaches eighteen years of age (with some things, like drinking alcohol or renting an automobile remaining difficult or illegal until human people are in their twenties).

The legal concept of moral rights at first blush sounds similar to the separate, and broader, question of moral personhood, but this is a category almost wholly unrelated to questions of legal personhood and should be treated separately. Moral personhood is a concept often restricted to philosophical discussion. One of the most famous definitions of a moral person is Daniel Dennett's. He states that a person is "an agent who is accountable," one who has rights and responsibilities.[12] Working from a discourse community separate from the one most rhetorical theorists occupy, Dennett is likely not as liberal in his definition of agency as rhetoricians are, with our love of Latour's talking train and the interesting scholarship such arguments have inspired about everyday objects.[13] Nonetheless, even with expansive definitions, the idea of moral personhood does rely on taking an intentional stance, as Dennett would frame it, toward a body. That is, morality does not spring forth from an unreasoning agent, and because humans have in the past proven loathe to allow other entities thoughts or feelings, the definition cannot be complete or even representative.

A major problem arises when people of various species fall through the cracks of moral personhood's definition. Many who are accountable have responsibilities but no (or few) rights. Some who have full or limited rights—like children—have no or few moral responsibilities compared to an adult. The disabled, children, animals, and even the intelligent space aliens of Dennett's *gedanken* experiment that informs his definition of moral personhood, could easily be defined out of moral personhood, whether or not that is a moral argument.

Another common way the question of personhood is approached is through the concept of metaphysical personhood, definitions of which can be limited or fairly inclusive, depending on one's metaphysical beliefs (or lack thereof). One commonality that cuts across most definitions is that a metaphysical person can be seen as a conscious, intelligent, feeling agent. There are, however, many additional conditions that can inform this definition. Michael Goodman lists several such conditions of metaphysical personhood, which enhance the older idea of biological personhood: consciousness, rationality, the ability to take or reciprocate an attitude toward another being, complex communication, self-consciousness, and self-motivation.[14] With the influence of organized religion, other groups have insisted on the person in question having a soul. Many of these qualities are impossible to prove, and most require subscription to a certain religious or idealist philosophy.

Discussions of metaphysical personhood, like those of biological personhood, are assuming "normal" development for humans and other animals to qualify for personhood, which is a flaw of most modern definitions. Even though one can easily imagine humans and other intelligent creatures that do not fulfill all of the terms of metaphysical definitions, it is the most basic and inclusive among those of philosophers who currently debate the matter because most people are willing to concede that animals at least have feelings, and thus deserve some level of compassion, if not the rights of a moral person, which is a status almost always restricted to human adults of able mind and body.

To anyone concerned with rights and responsibilities beyond the able-bodied adult human (and the corporations who, oddly, hold many of the same rights), these definitions largely do not satisfy. To move beyond these types of definition that serve mainly to enable asymmetries of power or even oppression, it is important to explore a concept of rhetorical personhood. What I argue relies heavily on Eugene Schlossberger and others like him, who conceive of personhood as a moral question, not a factual one. There is no empirical evidence for personhood that would satisfy a scientific method of verification. Personhood is a status we extend to others because we ought to care about their well-being for unselfish reasons (and it is normally for purely selfish reasons that anyone would attempt to deny another feeling creature personhood).

Schlossberger argues that "what one says about personhood must be compatible with the ways in which we treat (or ought to treat) persons; an explanation of what it is to be a person must take account of the role that the concept of a person plays."[15] This may seem open-ended, and it is meant to be. Personhood should be assumed until proven otherwise, in order to reduce the harm assuming the opposite can and does cause. Allowing a creature personhood if it is not a person does relatively little harm, by comparison.

Richard Rorty makes a similar argument in his *Philosophy and the Mirror of Nature*, relating that personhood is "a matter of decision rather than knowledge, an acceptance of another being into fellowship rather than a recognition of a common essence."[16] While I use the term rhetorical personhood to suggest that we should be searching for "fellowship" rather than seeking reasons to exclude, there is indeed a decision to be made. Must there be a solid boundary between the human and nonhuman? The human and the cyborg? Even the human and the differently-abled human? (Or the different-looking human!) The core of my argument for rhetorical personhood is that we should subvert the hierarchy-in-progress of personhood defining entirely, in order that we can escape harmful separations based on difference. How much better it would be to embrace similarities for a while. To that end, I do not suggest dolphin intelligence as a sole criterion for personhood nor their individuality; instead, I use the discussions of these traits merely to point out that their similarities are impressive, and that their differences do not make them inferior, which is the running assumption in many fields. To define my conception of rhetorical personhood more explicitly, it is an invitation to discuss what makes others special or interesting, rather than falling into the same old trap of identifying difference for the purpose of placing them lower in an arbitrarily defined hierarchy. Rhetorical personhood is not a right or an ability but a celebration. Sadly, it is difficult for the human mind to accept such ideas.

A mostly benign example of the fuzzy thinking it takes to systematically exclude in order to create a hierarchy of persons comes from Goodman's discussion, mentioned previously. In it, he uses (what I assume are domestic) cats as an example of conscious animals that have neither rights nor responsibilities. There are several flaws with this theory, as rights are a purely human concept that likely contain no meaning in any form of cat thought or communication. Not even all human cultures have concepts of legal rights—at least not in the strictly Western sense Goodman invokes. As to the idea that cats do not have rights and responsibilities, or restrictions on either, ask any pet guardian about this and they will not believe a thinking, feeling being made such a naïve claim.

Cat rights may seem pretty mundane, such as where they can climb, or defecate, or be fed (or at what time, or in what order, ad nauseam), but these are what we might call the rights of shared living that all gregarious, cohabiting creatures negotiate on a daily basis. More formally, we might speak of the legal rights of pets, which are fewer and less rigidly enforced than human rights, but several times per year there are incidents where one neighbor will walk into another neighbor's yard in order to murder a pet that has in some way bothered them. Not only are the offenders subject to animal cruelty laws, but also to trespassing, brandishing a weapon, and so on. These are insufficient protections for what many people feel are family members, but pets do seem to at least have the right not to be treated cruelly when they are owned by a legal person, especially when on that legal person's property. These laws do seem primarily concerned with the destruction of belongings (pets) on private land, however; if pets were not legal property, they would likely have fewer rights.

Cats have few rights, if any, by law, but in many households, they have secured certain freedoms. That cats, or other animals for that matter, have no responsibilities is also often untrue. Whether we mean moral responsibilities, like caring for their young, or responsibilities to the humans who feed them (consider the plight of the horse, who has had to work for his food for at least four millennia under human control), and especially the growing numbers of service animals who live out their lives for the benefit of their humans, it is not difficult to imagine animals having responsibilities to our species, let alone their own. These responsibilities should be obvious to anyone who cares to acknowledge them, but it is likely that Goodman and I are not in agreement about what constitutes a responsibility. In order to exclude cat responsibilities, one has to have already defined cats out of personhood, and so there is no sense in dealing with contrary examples. They are thought mere inconvenient anecdotes when brought to bear on definitions meant to exclude.

This sort of wishful thinking pervades personhood discussions among scientists and humanists alike. A further example from Goodman is his suggestion that we'll know who is and who is not a person because "a being who is considered to be a person by another person is a person."[17] Once one's mind is done being boggled, statements like this lead to the very obvious question, "Well, who considered the first person a person, then?" Clearly it was a person.

Inconsistencies like all of the above examples are why we must pass the test of personhood through a more inclusive medium—we can choose to invite comparison in order to celebrate shared traits and unique abilities, rather than draw lines in the sand for exclusion. It is far too easy to allow fuzzy thinking and hierarchical self-aggrandizement to dehumanize biological humans, let alone to exclude the thoughts and feelings of nonhuman people. Moreover, it is not difficult to see how other animals react to pain, and very good arguments have been put forward that we include emotional suffering in this category.[18] That there are other thinking, feeling creatures on this planet is not a matter for debate any longer. What is a matter for debate is why we should allow others to treat them without respect, to harm them, to keep them captive, and to strip them of their dignity, all because of a trick of language that makes some humans feel special.

In the rest of this chapter, I want to explore the idea of nonhuman personhood in one specific case. I have intentionally chosen to discuss a nonprimate group of animals, whose common ancestor with humans is very distant, but nonetheless demonstrates a staggering number of impressive physical, social, communicative, and cognitive traits—some of which put humans to shame. Understanding the abilities of dolphins is important for rhetoricians concerned with identity, not least because, as Baruch Brody argues, personhood is not something to be defined so much as something to be discovered.[19] That personhood needs to be invented (in the Latin sense of *invenire*, i.e., "to find"), much like any new species must be argued to belong in a certain taxonomic category, suggests it should be a matter of interest to rhetoricians. It is a discussion, not a hierarchy or legal code. Consensus and a reduction of harm should always be the goal, not the maximization of profit or the punishment of Others. Thus is it important to begin a conversation about nonhuman animals that demonstrates to the reader why they should be considered persons, too. My case for rhetorical personhood (that is, my invitation for you to share your personhood with dolphins) stands on four legs: that dolphins are sophisticated communicators; that they are individuals with their own identities; that, though they do not have all of the abilities of humans, they have some unique skills of their own that outstrip us in many ways; and therefore, that difference is not a case for inferiority.[20]

DOLPHIN COMMUNICATION

Dolphin communication is rich and varied. Much like humans, they have strong vocal skills and a broad range of meaningful noises they can produce, while also relying on myriad forms of nonverbal communication. When dolphins and humans interact, both in captivity and the wild, it is across several media. Dolphins are sophisticated listeners, who can be easily persuaded to complete tasks whether a food reward is offered or if the dolphins are simply in a playful mood.

Humans are unable to hear some dolphin vocalizations, which makes us comparatively poor listeners, even if we have the will to try. Of those we can hear, whistles are the most frequent. This goes especially for a specific type, called signature whistles (which will be discussed below) that make up more than half of the instances of dolphin signals that we can hear.[21] Whistling can serve several social or informational functions for dolphins, and young dolphins seem to learn to make their individual sounds in similar ways to human children. Juvenile dolphins participate in an epideictic mimicry and elaboration process much like human toddlers who point, name objects, and seek confirmation from a trusted adult.[22] While pointing with fingers is a very human activity, dolphins do show several types of gestures that can supplement speech, although we do not know if these aid the learning process. Vocal learning has been demonstrated in both wild and captive populations, and can be confirmed through cross-species communication experiments.

In addition to having listening skills that outstrip human abilities at present, at least in terms of number of other-species words and phrases either species can comprehend, dolphins are able to understand gestures and symbolic communication when interacting with humans. This includes the use of objects to stand for other objects; words, gestures, or images that act like linguistic verbs; and novel translation. An instance of the latter is found in the easy use of metaphorical body language dolphins can perform and understand when, for instance, a trainer waves one hand or foot and the dolphin is able to mimic this through the use of the corresponding flipper or one portion of a fluke beyond the median notch (as in "left foot" or "right foot").

Dolphins understand and attend to both pointing and gaze direction as signals of intent.[23] They are both quick and accurate in their formation of mental representations that allow them to react to brief pointing or gazing at objects a trainer wishes to indicate. In the case of one dolphin experiment by Pack and Herman, for instance, a captive dolphin was shown in sequence a surf board, a hoop, and then the "fetch" signal, which requires the dolphin to store in mind images about each object and then construct the relationship between them that the action signal demands.[24] While this is impressive, considering that this animal has to use a second language negotiated with a member of another species in order to form these linguistic relationships and the corresponding images in their minds, Pack and Herman's experiments also point out one area where human senses far outstrip those of the dolphin. That area is far-sightedness above water. In tests of attentiveness to gestures to objects with possibilities both near and far, the dolphin tested only responded to gaze by interacting with the near objects, whereas pointing, which is a far more obvious bodily movement, elicited the correct response far more frequently.[25] The failures of dolphins in long-range gaze recognition portions of this test are quite similar to the performances of human infants, possibly again because of visual limitations and the more dramatic lateral displacement of a pointing arm as opposed to a head.

Among conspecifics—and the vast majority of dolphin studies have come from only one species, *Tursiops truncatus*, or the common bottlenose dolphin—these animals elicit a wide variety of vocalizations that are meaningful to their own kind. These include several types of narrowband communication, like whistles, which can carry identity markers or provide contextual information, such as "there's food over here," or "maintain relative position."[26] Other narrowband vocalizations include what scientists have dubbed: squawks, blats, mewings, and brays. We are only beginning to understand the meanings of these sounds, but it is likely that brays, at least, serve a social function, like we know whistles do. It is difficult to know exactly what we are missing vocally, contextually, and socially, since even as recently as the 1990s, most studies were being performed above water.[27] Listening well often requires listening three-dimensionally so that researchers can isolate which individual in a school is speaking. However, water noise and triangulation are traditionally complicating factors when attempting underwater recordings.[28]

In addition to narrowband communication—much of which humans can hear—dolphins are able to employ broadband emissions. Echolocation clicks are one type, and as the reader can likely guess, these vocalizations are used primarily in exploring the underwater environment. Burst pulses are another broadband strategy, but these are more likely to be explicit communication efforts, again, possibly, for social reasons.

We cannot yet fathom the full communicative abilities of dolphins and so must reserve judgment of their overall prowess. We know that in many ways, their communication outstrips our own—especially in underwater environments. Their senses are very different from humans', which fundamentally alters the way they interact with the world. For instance, while their eyesight out of water might have a shorter range than ours, they are much cleverer in sensing and using magnetic fields. In Hui's study of short-beaked common dolphins (*Delphinus delphis*), he found that these animals could actively use magnetic fields for guidance when their other senses failed them in navigating and they became stranded.[29] Free-ranging dolphins did not resort to this tactic, however, which may signal that reading bottom topography is the more useful navigational tool when it is possible to use it. While this is an interesting addition to our knowledge of dolphin abilities, it does not approach completing our understanding of how their senses align with their communication and behavioral strategies. What we do know so far about dolphin communication is that they are individuals with high levels of cognition. These two claims (that they are individuals, and that they are intelligent) show through especially in the development and use of signature whistles, which are an important consideration when discussing the potential for dolphin personhood.

DOLPHIN INDIVIDUALITY

Many readers likely grew up watching the television show *Flipper*, or reruns of the show on Nickelodeon and other cable networks. The program is about an intelligent dolphin who assists a park ranger and his two sons in protecting the natural environment in a fictional marine preserve in southern Florida. As the audience for *Flipper* grew over the years, the idea of dolphins as intelligent assistants and friends to humankind planted itself in the collective subconscious of several generations of Western viewers. It should be no surprise, then, that scientific experiments continue to show that there is much exciting information we can learn about the cognitive abilities of these creatures, who perform very well on tests normally geared toward evaluating human intelligence (which therefore contain elements of human exceptionalist bias).

In addition to tests measuring vocal learning, mental representation, metacognition or theory of mind, behavioral innovation, and tool use, several experimenters have found that dolphins are very comfortable with individuality and self-recognition tasks.[30] One potent example is the Gallup Test. The procedure for this test is to leave a mark on an animal's face or head that would be easily noticeable if the individual was presented with a mirror but would not be felt on the skin or hair of the test subject. While this is a very human-centered task, as humans are often visually able, and those who are able are preoccupied with visual stimuli, dolphins readily pass this test, turning their heads to view the reflections of their new mark from different angles or attempting to rub it off on other objects or individuals. Many animals are either unable to pass this test, or the test is irrelevant to them despite human researchers' attempts to make them respond.[31] Human children can take several years to develop this sort of mirror self-recognition, although more than half will achieve this understanding by the time they are two years old.

Cultural transmission plays a large part in both human and dolphin learning. It also helps shape many of the interactions between the species.[32] Without cultural learning, dolphins likely would not be able to find advantages in the many ways they work with human cultures, such as cooperative fishing—an activity which is not as rare as we once thought, being active in human-dolphin shared communities in Australia, India, Burma, Mauritania, and Brazil and in the Mediterranean.[33] Further evidence for social learning or culture in dolphins can be found in studies of vocal learning and whistle convergence in wild animals, inter- and intrapopulation behavioral variation without genetic or ecological explanations, and correlation between degrees of social association and similarities in behavioral traits.[34]

Because the development of culture demands the recognition, if not the celebration, of individuality, the tests listed above and the cultural evidence researchers have observed should be more than enough to establish that dolphins have a firm sense of individuality. Nonetheless, there is one communicative behavior that should be of special interest to rhetoricians and communication scholars who are concerned with the dolphin's sense of self and how they relate vocally to others. I refer to dolphins' signature whistles. An important way that signature whistles signal or send information is through their frequency contour.[35] Isolated from other acoustic qualities, the pattern of frequency over time is unique for each dolphin, so it acts like a human signature or fingerprint. Like a signature, it can be mimicked with practice, but it is ultimately a means of identifying an individual with a measurable output of sound over time. While humans often use pitch or voice cues to identify who is speaking, a dolphin will broadcast their identity by means of these idiosyncratic whistle contours.

Adding to the frequency contour information, signature whistles may be altered to also convey status information.[36] In human terms, this could be useful if I wanted to know more than the simple statement "meet immediately." If I knew, rather, that "Bill is saying we should meet immediately," I could potentially know a whole lot more through that contextual clue. Perhaps Bill likes to borrow money, or maybe a group of people had all agreed to meet at a later time but Bill is impatient. It may even signify that Bill is the type to cry wolf, so I am probably safe to ignore Bill.

The signature portions of what signature whistles convey—that is, the identifying information—are traits dolphins develop as calves. Until around their first birthday, calves produce a variety of whistles. Over time, and in conversation with others in their community, the calves settle on a predominant whistle contour.[37] I speak of "others" generally, rather than the more specific "other dolphins," because captive dolphins seem to be influenced by human communication as well. Captive dolphin calves develop flatter whistle contours because of the influence of the flat tones of the whistles their human trainers make. That dolphins require vocal learning in order to communicate makes them rare in the animal kingdom, but what is even more interesting is that the most influential individuals in the development of young dolphin whistles are often dolphins of the same sex who associate only very rarely with the juvenile dolphin's usual groups.[38] Dolphins, like the young of at least one primate species, seem to want to identify with the mysterious celebrities of other pods, or at least they want to avoid sounding too much like their own parents and relatives. Nonetheless, each signature whistle is unique, so models will influence their eventual sound over time pattern but are not copied outright.

Signature whistling is common generally, but its use is ramped up in at least two particular situations. When dolphins are isolated or captured, they will call out their "names" more frequently, likely in order to locate other members of their group. This practice of reaching out to others when alone is a common behavior among many species of mammal. Another situation where signature whistling increases markedly is when groups of dolphins are joining together. This suggests several potential epideictic functions of signature whistles, including group cohesion, alliance formation, and individuality assertion within the context of a group.[39]

Signature whistles, once selected by the individual, are generally static. One thing that can cause these whistles to change, however, is the lasting social relationships of male dolphins. Female dolphins have highly variable associations that change based on kinship and by reproductive state. Because reproductive states influence relationships between male and female dolphins so heavily, females will often congregate based on where they stand in relation to reproduction. For instance, mothers with newborn calves will often congregate for mutual defense, leaving females in other reproductive states to form groups of their own. Dolphin societies are fission fusion, so groups may form and split based on other factors, like preferred foraging methods or time of day. However, the female groups are a driver of dolphin cultural learning, which, as a result, is highly matrilineal.[40]

Despite much learning being passed from mother to calf, dolphin males are not immune to the urge to appropriate from other cultures. Sometimes this even happens across species lines, like in the case of Lusseau's observation of a dolphin population in New Zealand that has adopted a unique agonistic display[41]—unique to dolphins, not to the many sheep that populate the New Zealand countryside. Having apparently observed the dramatic running head butting that several ungulate species perform, these dolphins have begun the practice of swimming quickly toward one another, leaping out of the water, and ramming their heads together violently. One can only assume that animals as adept at social learning as dolphins are must have been influenced by witnessing two rams fighting near shore at some point, and as a consequence decided that this looked like an interesting new pastime.

A more mundane example of male dolphin culture is in the way that they do not settle on their final signature whistle form until they have paired off into the lasting social relationships they form with other male dolphins. Most often, we see males form dyadic relationships that remain stable, as opposed to the changefulness of female relationships. While juveniles of either sex will mix frequently, male dolphins tend to pick a partner who will stick with them for long periods and who will join other groups of dolphins in order to maximize breeding chances.[42] Some male alliances have been maintained for up to 18 years, as observed in Shark's Bay in Australia.[43] These bonds being so strong, they even seem to alter male dolphin identity markers. When males form a dyad, their two signature whistles converge, which promotes social cohesion and helps to form a group identity.[44] They will also, when separated, call out the signature whistle of their partner in addition to their own.[45]

DOLPHIN SUPERHUMANS

While the discussion of dolphin abilities that surpass those of humans has taken part piecemeal above, it is worth taking a moment to consider one additional way that dolphins are superhuman. Their neurosensory equipment is far different from that of a human, as they have at separate times evolved as both land and sea mammals, and in some ways that equipment is far more sophisticated. For instance, not only can dolphins perform fine-tuned sensory abilities, such as echolocation, researchers have found that dolphins also hear the clicks of that system through an array of two separate receivers.[46] This allows them to coordinate what in humans we might consider two hearing senses in order to better triangulate information from environmental sources. However, since hearing is the primary sense for dolphins, as opposed to vision for humans, dolphins could be thought to "see with their ears."

What is more astounding about this ability is that, much like a doctor performing an ultrasound to see the developing fetus inside a pregnant human woman, dolphins can use their ultrasonic clicks to see inside one another, or to locate fish below the sandy surface of the ocean. Fat, muscle, and other soft tissue shows up as mere outlines in ultrasound, with teeth, bones, and full gas bladders showing more solid images. But here is where dolphin senses meet dolphin communication to create something truly impressive: dolphins need not describe echolocation images to one another vocally. Instead, they are able to communicate precise acoustic images to one another, much like if humans were able to place picture prints in the brains of other humans, rather than trying to describe the visual attributes of a moment frozen in time.[47] In science-fiction parlance, this is like something out of an H. G. Wells novel, in that we have intelligent alien species swimming around in our oceans who can see inside our bodies and communicate the results telepathically. If dolphins were not so friendly to humans, we might consider an H. P. Lovecraft metaphor more appropriate, as this would be a frightening power in the hands of a species that counts humans as prey. (It would be a pretty frightening power for humans themselves to have, too, frankly).

The ability to see inside one another, for humans at least, would present dramatic opportunities for both comfort and mischief. Implications for the latter should be obvious, but the opportunities for the former are far more exciting. For instance, being able to casually detect the initial stages of tumors forming in other bodies, long before any symptoms appear, would drastically cut the mortality rates of most forms of cancer. Finding out about a new pregnancy much sooner after conception could be beneficial to new mothers (although, once again, the opportunity for mischief arises when information about the reproductive states of others is automatically always public knowledge).

While doctors have limited forms of ultrasound equipment at present, the potential applications of this technology are nearly endless. This is a further reason why we need to continue studying natural echolocation systems in a variety of animals, like dolphins and bats, and it points to the importance of studying communication under the lens of evolved neurosensory apparatus. Every species is subject to the conditions of its evolved mind and senses, and how we express our needs and desires is constrained by both. That dolphins can perform feats of communication that far outstrip the sophistication of anything humans can do should not be used as the sole criterion for considering them people (and this should remind us of the dangers of abilities-based definitions of personhood discussed in the first section of this chapter), but it is suggestive that we need to look beyond our own biases about human superiority in order to judge other animals more fairly (or, better yet, to stop judging them in hierarchical terms).

DIFFERENT, NOT BETTER

In May of 2010, scientists and conservationists gathered in Helsinki, Finland, to discuss the future of cetaceans in an Anthropocene largely characterized by extinction and destruction of habitat by humans. The conference concluded with a declaration of rights for cetaceans, which insists upon the following points:

1. Every individual cetacean has the right to life.
2. No cetacean should be held in captivity or servitude; be subject to cruel treatment; or be removed from their natural environment.
3. All cetaceans have the right to freedom of movement and residence within their natural environment.
4. No cetacean is the property of any State, corporation, human group or individual.
5. Cetaceans have the right to the protection of their natural environment.
6. Cetaceans have the right not to be subject to the disruption of their cultures.
7. The rights, freedoms and norms set forth in this Declaration should be protected under international and domestic law.
8. Cetaceans are entitled to an international order in which these rights, freedoms and norms can be fully realized.
9. No State, corporation, human group or individual should engage in any activity that undermines these rights, freedoms and norms.
10. Nothing in this Declaration shall prevent a State from enacting stricter provisions for the protection of cetacean rights.[48]

The articles of this declaration take as a starting point much of the modern scientific literature on dolphin and whale communication and cognition. Thomas White, a noted animal ethicist, claims to set the bar higher than most when considering whether an animal is a person. In his book *In Defense of Dolphins* (2007), he claims that, to be convinced himself, a genus or family of animals would have to demonstrate the following traits: being alive and aware; having the ability to experience positive and negative sensations (pleasure and pain); having demonstrable emotions; being self-conscious and possessing a personality; being capable of self-controlled behavior; recognizing and treating other persons appropriately; and being capable of a series of higher-order intellectual abilities such as learning and communication.[49] Cetaceans, he says, fit all of these criteria.

Evidence for dolphin personhood is mounting in all corners of animal studies, in both the sciences and humanities. Yet, if this is so, how can we continue to argue that dolphins are not people of one sort or other? To Daniel Hoffman, it seems that we often accept our inability to deal with a different creature as proof of its lack of personhood.[50] But often all this signals is our own failing to understand and empathize. Humans are not yet smart enough to understand animal cognition and communication. In fact, several species of animals understand us far better than we understand them.

Dog guardians: think for a moment about how many words your dog understands compared to how many dog vocalizations you understand. Cat guardians: don't even try if you don't want to feel silly—while dogs only make about ten vocalizations, cats make around 100 that we've identified so far. And while composer David Teie has recently kickstarted a CD of "Music for Cats" based on what scientists think might be cat-specific sounds of comfort from shortly after their birth (suckling, purring, birds chirping in the distance), our ability to "speak their language" is yet in question. Humans and animals clearly negotiate the functioning of their shared living spaces, whether these conversations are explicit or if they are simply habits that are "fallen into" over time. However most pet guardians feel about their own nonhuman animals, though, these important processes of cocreation are rarely acknowledged in conversations of rights or responsibilities. In legal situations, humans are almost always treated as more important than other animals, even in the most egregious cases containing asymmetries of harm.

Worse yet, there is something in the psyche of many humans that causes them to adopt the most cowardly positions at the mere suggestion that humans are not special and better than other animals in every way. Frans de Waal relates in his recent book on animal intelligence the fact that linguists have on several occasions attempted to ban the study of language origins and have even wanted to stop researchers from teaching nonhuman animals human languages. "Such measures reflect intellectual fear rather than curiosity. What are linguists afraid of? They had better pull their heads out of the sand, because no trait, not even our beloved linguistic ability, ever comes about de novo."[51] To paraphrase Lucretius, no ability, no matter how special, comes from nothing—de Waal's point is not only one about studying animals as origins, but of shared ontogeny or even convergence (for, despite the common error of thought, humans do not "come from" monkeys; rather we share an ancestor with monkeys and apes).

Even among members of our own species, we in the West have tended to ignore the ways others communicate with and about animal persons. If one were a Buddhist, the standard view in the United States would seem parochial, bordering on barbaric, since many Eastern philosophies and religions believe that some animals could once have been human and may be able to become human again someday. But it does not require metaphysical belief to recognize that animals feel pain. And it does not take religion to understand that we shouldn't cause pain when we don't have to, or that if there are ways to reduce pain, a moral person would do so.

This is why we need to begin a conversation about rhetorical personhood. Even if one does not agree with all of the elements of the above argument, even if one still thinks humans are special, or that their god has given humans all the animals to watch over, then one has to ask oneself: "How good of a job are we doing?" I know of no god or philosophy that would approve of the way we treat animals in the modern West and in many other places around the world. Some ersatz holy men and women may twist their beliefs to justify oppression—we have seen this over and over in human history, including in the case with which I opened this chapter—but these are not actions an empathetic person should support.

The activist scholar has to overcome both hate and ignorance, which recent political trends have taught us are still powerful forces in the world. Moreover, we need to find a way to raise awareness among the vast majority who are kind, caring, empathetic humans and people. In the case of dolphin personhood, there is much work to be done, but the tide is turning. In Hungary, Costa Rica, and Chile, dolphins have officially been recognized as nonhuman persons who are afforded the legal rights to self-determine. India, among other countries, has ordered the closure of all dolphin shows and suggested that cetaceans should be thought of as nonhuman persons (even if there are no legal protections to back this up, as yet). The idea of personhood is expanding in other nations, and it is time for the West to catch up.

That personhood could indeed be a rhetorical negotiation—something that must be named and defined before what was once property can now be transformed into people—means this should be a debate of especial interest to rhetoricians and communications scholars. In order to expand our definition of what makes people people, we need to understand what it is we are arguing for. This will allow us to look inside ourselves and figure out why it is that we might want to prevent other thinking, feeling creatures from having the protections and rights that people of any species are due. Because if other animals are intelligent, feeling, unique individuals, it becomes extremely difficult to continue finding new ways to define them out of personhood, and history will not look kindly on those who continue to try.

NOTES

1. Daniel N. Hoffman, "Personhood and Rights," *Polity* (1986): 82.
2. Ruth Macklin, "Personhood in the Bioethics Literature," *The Milbank Memorial Fund Quarterly. Health and Society* (1983): 35.
3. Diogenes Laertius and Robert Drew Hicks, *Lives of Eminent Philosophers*, 2 vols. (London, New York: W. Heinemann; G. P. Putnam's sons, 1925). VI.40.
4. Jenny Teichman, "The Definition of Person," *Philosophy* 60, no. 232 (1985): 175.
5. Joshua Barkan, *Corporate Sovereignty: Law and Government under Capitalism* (University of Minnesota Press, 2013). 65.
6. Ibid., 66.
7. Jenell Johnson, "Disability, Animals, and the Rhetorical Boundaries of Personhood," *JAC: A Journal of Rhetoric, Culture & Politics* 32, no. 1/2 (2012): 372–73.
8. S. Wise and J. Goodall, *Rattling The Cage: Toward Legal Rights For Animals* (Da Capo Press, 2014). 179.
9. I see these as distinct categories, and I will leave it to the reader to determine if any of these groups should be strictly considered people.
10. Laurence H. Tribe, "Ten Lessons Our Constitutional Experience Can Teach Us About the Puzzle of Animal Rights: The Work of Steven M. Wise," *Animal L.* 7 (2001): 2.
11. Eugene Schlossberger, *Moral Responsibility and Persons* (Temple University Press, 2010). 22.
12. Daniel Clement Dennett, *Brainstorms: Philosophical Essays on Mind and Psychology*, 1st MIT Press ed. (Cambridge, MA: MIT Press, 1981). 176.
13. For Latour's chatty train, see Bruno Latour, *Aramis, or, The Love of Technology* (Cambridge, MA: Harvard University Press, 1996). For one good example of everyday objects as agents in new materialist accounts, see Scot Barnett and Casey Boyle, eds., *Rhetoric, Through Everyday Things* (2016).
14. Michael F. Goodman, "A Sufficient Condition for Personhood" (paper presented at the The Personalist Forum, 1992), 75.
15. Schlossberger, *Moral Responsibility and Persons*: 22.
16. Richard Rorty, *Philosophy and the Mirror of Nature* (Princeton University Press, 2009). 38.
17. Goodman, "A Sufficient Condition for Personhood," 77.
18. Barbara J. King, *How Animals Grieve* (University of Chicago Press, 2013).

19. Baruch Brody, "On the Humanity of the Fetus," in *What Is a Person?* ed. Michael F. Goodman (Totowa, NJ: Humana Press, 1988), 229.

20. Several readers of this chapter have asked what *would* make a case for inferiority, which only serves to strengthen my argument that humans are obsessed with ranking and ordering others. While I would rather not answer the question, as it seems to me that little good can come of it (as searching for inferiority will either enable further harm or further ignorance about the suffering of animals), I will instead make an honest attempt to suggest some ways we can abandon the idea of animal inferiority and make our interactions with animals more humane. If an animal can feel pain, we should not harm it. If we must harm it, we should try to do so quickly and with as little suffering as possible. We should prioritize the care and treatment of animals, especially those of us who eat them. If we eat them, we should prefer animals that have lived full lives. Of animals that have had full lives, we should prefer those who lived in the most natural environments possible. (Local wild game should be eaten before meat sourced from factory farms that must be shipped to supermarkets). Eat less of it. Unless we are sorely confused, vegetables do not seem to feel pain. Therefore, if pressed, I would argue that broccoli is an inferior animal.

21. Arik Kershenbaum, Laela S. Sayigh, and Vincent M. Janik, "The Encoding of Individual Identity in Dolphin Signature Whistles: How Much Information is Needed?" *PloS One* 8, no. 10 (2013): 6.

22. Heidi E. Harley, "Whistle Discrimination and Categorization by the Atlantic Bottlenose Dolphin (Tursiops truncatus): A Review of the Signature Whistle Framework and a Perceptual Test," *Behavioural Processes* 77, no. 2 (2008): 246.

23. Adam A. Pack and Louis M. Herman, "The Dolphin's (Tursiops truncatus) Understanding of Human Gazing and Pointing: Knowing What and Where," *Journal of Comparative Psychology* 121, no. 1 (2007): 35.

24. Ibid., 35–6.

25. Ibid., 41.

26. Vincent M. Janik, Laela S. Sayigh, and R. S. Wells, "Signature Whistle Shape Conveys Identity Information to Bottlenose Dolphins," *Proceedings of the National Academy of Sciences* 103, no. 21 (2006): 8293.

27. Kenneth S. Norris and Randall S. Wells, "Observing Dolphins Underwater," in *The Hawaiian Spinner Dolphin* (University of California Press, 1994), 54.

28. Ibid., 57.

29. Clifford A Hui, "Lack of Association between Magnetic Patterns and the Distribution of Free-Ranging Dolphins," *Journal of Mammalogy* 75, no. 2 (1994): 399.

30. Brooke L. Sargeant and Janet Mann, "From Social Learning to Culture: Intrapopulation Variation in Bottlenose Dolphins," in *The Question of Animal Culture* (2009): 152.

31. For a discussion of using human-centered tests inappropriately on animals that the tests may not apply to, see Frans de Waal, *Are We Smart Enough to Know how Smart Animals Are?* (WW Norton & Company, 2016).

32. Hal Whitehead and Luke Rendell, *The Cultural Lives of Whales and Dolphins* (Chicago: University of Chicago Press, 2015), 99.

33. Ibid., 100–10.

34. Sargeant and Mann, "From Social Learning to Culture: Intrapopulation Variation in Bottlenose Dolphins," 161.

35. Harley, "Whistle Discrimination and Categorization by the Atlantic Bottlenose Dolphin (Tursiops truncatus): A Review of the Signature Whistle Framework and a Perceptual Test," 243.

36. Ibid., 245.

37. Ibid., 245–6.

38. Deborah Fripp et al., "Bottlenose Dolphin (Tursiops truncatus) Calves Appear to Model their Signature Whistles on the Signature Whistles of Community Members," *Animal cognition* 8, no. 1 (2005): 17.

39. Kershenbaum, Sayigh, and Janik, "The Encoding of Individual Identity in Dolphin Signature Whistles: How Much Information is Needed?" 1–5.

40. Sargeant and Mann, "From Social Learning to Culture: Intrapopulation Variation in Bottlenose Dolphins," 153.

41. Related in ibid., 167.

42. S. Gero et al., "Behaviourally Specific Preferred Associations in Bottlenose Dolphins, Tursiops spp," *Canadian Journal of Zoology* 83, no. 12 (2005): 1566.

43. Ibid., 1571.

44. Sargeant and Mann, "From Social Learning to Culture: Intrapopulation Variation in Bottlenose Dolphins," 161.

45. Harley, "Whistle Discrimination and Categorization by the Atlantic Bottlenose Dolphin (Tursiops truncatus): A Review of the Signature Whistle Framework and a Perceptual Test," 247.

46. David E. Bain, Birgit Kriete, and Marilyn E. Dahlheim, "Hearing Abilities of Killer Whales (Orcinus orca)," *The Journal of the Acoustical Society of America* 94, no. 3 (1993): 1829.

47. Ralph Strauch, "Do Dolphins Think without Language?" *Sea Frontiers* 30, no. 2 (1984).

48. The Helsinki Group, "Declaration of Rights of Cetaceans: Whales and Dolphins," WDC, Whale and Dolphin Conservation, http://www.cetaceanrights.org.

49. T. I. White, *In Defense of Dolphins: The New Moral Frontier* (Wiley, 2008). 175.

50. Hoffman, "Personhood and Rights," 86.

51. De Waal, *Are We Smart Enough to Know How Smart Animals Are?*, 109.

BIBLIOGRAPHY

Bain, David E., Birgit Kriete, and Marilyn E. Dahlheim. "Hearing Abilities of Killer Whales (Orcinus Orca)." *The Journal of the Acoustical Society of America* 94, no. 3 (1993): 1829.

Barkan, Joshua. *Corporate Sovereignty: Law and Government under Capitalism*. University of Minnesota Press, 2013.

Barnett, Scot, and Casey Boyle, eds. *Rhetoric, through Everyday Things*. Tuscaloosa, AL: University of Alabama Press 2016.

Brody, Baruch. "On the Humanity of the Fetus." In *What Is a Person?* edited by Michael F. Goodman. 229–50. Totowa, NJ: Humana Press, 1988.

de Waal, Frans. *Are We Smart Enough to Know How Smart Animals Are?* New York: WW Norton & Company, 2016.

Dennett, Daniel Clement. *Brainstorms : Philosophical Essays on Mind and Psychology*. 1st MIT Press ed. Cambridge, MA: MIT Press, 1981.

Fripp, Deborah, Caryn Owen, Ester Quintana-Rizzo, Ari Shapiro, Kara Buckstaff, Kristine Jankowski, Randall Wells, and Peter Tyack. "Bottlenose Dolphin (Tursiops Truncatus) Calves Appear to Model Their Signature Whistles on the Signature Whistles of Community Members." *Animal cognition* 8, no. 1 (2005): 17–26.

Gero, S., L. Bejder, Hal Whitehead, J. Mann, and R. C. Connor. "Behaviourally Specific Preferred Associations in Bottlenose Dolphins, Tursiops Spp." *Canadian Journal of Zoology* 83, no. 12 (2005): 1566–73.

Goodman, Michael F. "A Sufficient Condition for Personhood." Paper presented at The Personalist Forum, 1992.

Group, The Helsinki. "Declaration of Rights of Cetaceans: Whales and Dolphins." WDC, Whale and Dolphin Conservation, http://www.cetaceanrights.org/.

Harley, Heidi E. "Whistle Discrimination and Categorization by the Atlantic Bottlenose Dolphin (Tursiops Truncatus): A Review of the Signature Whistle Framework and a Perceptual Test." *Behavioural Processes* 77, no. 2 (2008): 243–68.

Hoffman, Daniel N. "Personhood and Rights." *Polity* (1986): 74–96.

Hui, Clifford A. "Lack of Association between Magnetic Patterns and the Distribution of Free-Ranging Dolphins." *Journal of Mammalogy* 75, no. 2 (1994): 399–405.

Janik, Vincent M., Laela S. Sayigh, and R. S. Wells. "Signature Whistle Shape Conveys Identity Information to Bottlenose Dolphins." *Proceedings of the National Academy of Sciences* 103, no. 21 (2006): 8293–97.

Johnson, Jenell. "Disability, Animals, and the Rhetorical Boundaries of Personhood." *JAC: A Journal of Rhetoric, Culture & Politics* 32, no. 1/2 (2012): 372–82.

Kershenbaum, Arik, Laela S. Sayigh, and Vincent M. Janik. "The Encoding of Individual Identity in Dolphin Signature Whistles: How Much Information Is Needed?" *PloS One* 8, no. 10 (2013): e77671.

King, Barbara J. *How Animals Grieve*. University of Chicago Press, 2013.

Laertius, Diogenes, and Robert Drew Hicks. *Lives of Eminent Philosophers*. 2 vols London, New York: W. Heinemann; G. P. Putnam's sons, 1925.

Latour, Bruno. *Aramis, or, the Love of Technology* [in English]. Cambridge, MA: Harvard University Press, 1996.

Macklin, Ruth. "Personhood in the Bioethics Literature." *The Milbank Memorial Fund quarterly. Health and Society* (1983): 35–57.

Norris, Kenneth S., and Randall S. Wells. "Observing Dolphins Underwater." In *The Hawaiian Spinner Dolphin*. 54–65. University of California Press, 1994.

Pack, Adam A., and Louis M. Herman. "The Dolphin's (Tursiops Truncatus) Understanding of Human Gazing and Pointing: Knowing What and Where." *Journal of Comparative Psychology* 121, no. 1 (2007): 34–45.

Rorty, Richard. *Philosophy and the Mirror of Nature*. Princeton: Princeton University Press, 2009.

Sargeant, Brooke L., and Janet Mann. "From Social Learning to Culture: Intrapopulation Variation in Bottlenose Dolphins." In *The Question of Animal Culture* (2009), 152–73. Cambridge, MA: Harvard University Press.

Schlossberger, Eugene. *Moral Responsibility and Persons*. Philadelphia: Temple University Press, 2010.

Strauch, Ralph. "Do Dolphins Think without Language?" *Sea Frontiers* 30, no. 2 (1984): 110–4.

Teichman, Jenny. "The Definition of Person." *Philosophy* 60, no. 232 (1985): 175–85.

Tribe, Laurence H. "Ten Lessons Our Constitutional Experience Can Teach Us About the Puzzle of Animal Rights: The Work of Steven M. Wise." *Animal L.* 7 (2001): 1.

White, T. I. *In Defense of Dolphins: The New Moral Frontier*. New York: Wiley, 2008.

Whitehead, Hal, and Luke Rendell. *The Cultural Lives of Whales and Dolphins*. Chicago: University of Chicago Press, 2015.

Wise, S., and J. Goodall. *Rattling the Cage: Toward Legal Rights for Animals*. Boston: Da Capo Press, 2014.

Part III

FURTHER EXPANSION: CROSS-SPECIES AND ACROSS CULTURES

Chapter 9

Learning to Howl

An Exercise in Internatural Abduction

Emily Plec and Susan Hafen

Almost two decades ago as doctoral students, we professed our desire to study human-animal communication to two different professors—one supportive, one not so much.[1] Here, briefly, are the stories of those germinal instructional moments.

> Susan's story: "The professor shook his head, patted my hand, and explained that our discipline defined communication as occurring between humans—*intra*species, not *inter*species. Years later, tenured, I presented several conference papers on the subject, to the chagrin of panel participants whose presentations on unrelated topics were ignored in the barrage of questions (many hostile) directed at my paper."
>
> Emily's story: "I wrote the paper for my dialogue class on an interaction I had with our dog, Lila. After evaluating my paper, my professor suggested I revise and submit it to an academic conference. I followed his advice and the paper received a 'Top Student Paper' award from the division to which it had been submitted. Almost all of the audience members responded positively and several approached me after the presentation to show their interest and support and share their own stories of meaningful interspecies communication."

These experiences took place during roughly the same time period, the mid-1990s, when most scholars in the discipline of communication still held staunchly to the idea that humans were unique and special animals because of our capacity to engage in complex communication behaviors. Yet many of these same people had directly experienced the very phenomenon they denied in their scholarship. That perhaps explains why Susan's experience was negative; her paper made an argument for animal communication as a kind of semiotic language. It is no accident that the latter community of support consisted largely of feminist communication scholars who saw complex relationships between human women and other animals within patriarchal systems, a critical landscape already well-mapped by that time by ecofeminists. Today, the tide of supportive, interdisciplinary opinion may have turned enough to convince the arbiters of appropriate communication studies that human-animal communication is not only relevant to our discipline but also spans interpersonal, family, health, organizational, intercultural, instructional, legal, mass mediated, spiritual, and—as this book demonstrates—rhetorical communication.

This chapter challenges the communication discipline's adherence to a humans-only paradigm of communication by briefly tracing developments in anthrozoology and zoosemiotics, and arguing for an abductive orientation to interspecies encounters. Abduction, as C. S. Peirce first introduced the term in his 1903 Harvard lectures, was a kind of "logical goodness" that brought together both instinct and inference to account for discovery.[2] Abduction begins with guessing, according to Peirce, and then "abductive inference shades into perceptual judgment without any sharp line of demarcation between them."[3] After discussing Peircean semiotics, which argues that abductive instincts should not be denied to nonhuman animals, we describe such an orientation by examining working relationships between humans and canines (on sled teams, in particular) and conclude by illustrating some of the possibilities provoked by an expanded view of what we call "internatural communication." Internatural communication is a subfield of communication that, like intercultural communication, is concerned with relationships and interactions across lines of difference; it has been used to explore human-animal interactions in a variety of contexts.[4]

THE TRANSHUMAN TURN IN COMMUNICATION STUDIES

Writing in 1992, rhetorical scholar George Kennedy argued that "rhetorical energy ... is present in physical actions, facial expressions, gestures, and signs generally."[5] He also asserted, in a chapter of *Comparative Rhetoric* entitled "Rhetoric Among Social Animals," that "The rhetorical study of animal communication primarily seeks to identify basic principles and formal aspects of communication that are fundamental to all rhetorical structures and used by both human and nonhuman animals."[6] Kennedy's 1992 essay (and public presentations) prompted a trickle of responses, including a 2011 forum published in *Philosophy and Rhetoric* featuring landmark essays by Diane Davis, Debra Hawhee, and John Muckelbauer. Described as an effort to "locate and unsettle the rhetorical-philosophical terrain of animal studies,"[7] the forum probes the possibilities of conceptions of rhetoric that take animality into account. Likewise seeking to place rhetoric and biology into closer conversation, Alex Parrish's *Adaptive Rhetoric* advances a biocultural communication paradigm that positions persuasion as an evolutionary capacity across species lines.[8]

As mentioned previously, "internatural communication" is defined as the exploration of "interaction among and between natural communities and social groups that include participants from what we might initially describe as different classifications of nature.... [It] includes the exchange of intentional energy between humans and other animals as well as communication among animals and other forms of life."[9] The need for a more-than-human theory of communication is evidenced in the work of environmental, rhetorical, and cultural studies scholars. Communication theorists are increasingly moving beyond the boundaries of the human communication paradigm when conceptualizing interaction. John Durham Peters offers an expansive idea of communication that occasionally takes an internatural perspective for analogical and metaphorical purposes, illustrating the potential of internatural communication as a thought experiment to illuminate dimensions of experience. In *Speaking Into the Air*, he recounts Emerson's view of the universe as "a constant transmission to those who have ears to hear,"[10] a view that resonates with the transhuman turn. In this context, the transhuman turn refers to intellectual efforts to transcend the limitations imposed by our very definition of what it means to be human. Though these efforts have yielded some utopian and dystopian technological visions of a posthuman condition, transhumanism also refers to a decentering of what we know as "the human" and the permeation of the boundaries that once circumscribed it.

Beginning in the late 1990s, a number of communication scholars began to challenge the boundaries of the human communication paradigm by introducing interspecies comparisons, animal studies, and transhuman theories of communication. Illustrating the comparative approach, Nieva and Hickson take a communibiological look at how language enhances humans' ability to deceive other humans beyond the attempts of nonhuman animals to deceive other nonhumans.[11] In an article in the *Journal of Business Communication*, Barker describes the potential organizational benefits of companion animals in the workplace, providing opportunities for interdisciplinary research among management, medical, psychosocial, and communication realms.[12] In that same issue, Dieterich responds with the benefits of bringing his golden retriever to his business writing class: Students are more relaxed, stimulated, and interactive with the instructor and each other.[13] In her examination of the rhetoric of organizations and facilities dedicated to orangutans, Stacey Sowards uses the Burkean ideas of perspective by incongruity and consubstantiality to demonstrate how nature/culture dualisms that have "persisted for centuries" can be deconstructed through identification, fostering an "animalcentric anthropomorphism."[14] Clearly, compared to only a decade or two ago, the landscape of communication research has changed considerably.

Furthering the internatural agenda, Donal Carbaugh examines the relationships between humans and nature by focusing on listening, place, and dwelling; Tema Milstein writes about human-animal interactions in the contexts of the rhetoric of whale-watching and zoo tours; Natasha Seegert offers critical readings of disruptive animals as media spectacles; and Richard Rogers investigates the rhetoric of petroglyphs and other forms of rock art from a dialogic perspective.[15] In fact, it was Rogers' essay calling for a transhuman, materialist theory of communication that offered one of the most significant early contributions to the disciplinary literature, sparking a fire that continues to grow with the present book.[16] But it was not always so.

In the very first volume of *Communication Yearbook*, Klaus differentiates three types of communication based on reflexivity: human, animal, and mass.[17] His categorization appears to follow the lead of George Herbert Mead, the oft-cited social psychologist who contrasted humans' use of symbols with the "conversation of gestures" by animals, theorizing that humans are intentional and animals merely instinctual, lacking any consciousness and thereby incapable of communication even within species.[18] Klaus takes a moderate approach, according animals the nonverbal communication and reflexivity needed for play and attentiveness. Charles Darwin was more generous, perhaps due to his extensive observations of animal behavior, proposing in *The Expression of the Emotions in Man and Animals* that humans and animals experience emotions and cognition on a continuum of degree, not in categorical exclusions.[19]

Today, the dualisms of mind versus body, emotion versus cognition, and humans versus animals are increasingly challenged, with many leading scientists now arguing that animals do indeed communicate, manipulating symbols and at times demonstrating a sense of self and other. These scientists include ethologists such as Miklosi and Griffin; zoologists like Dawkins; veterinarians such as Dodman, McMillan, Schoen, and Timmel; primatologists such as Fouts and Goodall; psychiatrists and psychologists like Masson, Horowitz, Wells, and Burghardt; sociologists like Alger and Alger or Franklin; geographers such as Philo and Wilbert; biologists like Bekoff and Pierce, Markowitz, and Timmel; ecologists such as Abram and Shepard; animal scientists such as Grandin and Johnson; biochemists such as Sheldrake; and experts in a relatively new field, anthrozoologists, represented by the likes of Serpell and Herzog.[20]

ANTHROZOOLOGY, ZOOSEMIOTICS, AND PEIRCEAN ABDUCTION

Anthrozoology, explains Hal Herzog, clearly cuts across many academic boundaries and "includes the study of nearly all aspects of our interactions with other species."[21] He decries that, until recently, psychologists concentrated on behavioral processes such as motivation, perception, and memory, neglecting "facets of daily life such as food, religion, and how we spend our leisure time" and other things that psychologists ignore but everyday people care about, such as our relationships with pets.[22] He names reasons why psychologists should care about attitudes and behaviors toward other species, including the fact that two-thirds of Americans live with pets, and many Americans describe deep, communicative relationships with their companion animals. These relationships affect family dynamics and their health (both positively and negatively), and debates are deepening not only about the ethical treatment of animals but about their environments as cultures. More than 150 colleges and universities have courses in human-animal interaction, and in 2011, Carroll College in Montana offered the first state-approved, undergraduate anthrozoology major in the United States.

The term zoosemiotics was first coined in 1963 by Thomas Sebeok to designate the study of animal communication as semiosis within and across species.[23] The influential French semiologist Umberto Eco excluded animals from language and thus from semiotic communication in a manner echoing rhetorical theorist Kenneth Burke, stating that "languages are what constitute human beings as such, that is as signifying animals."[24] American semiotician, Charles Peirce, on the other hand, "socialized with his horse and his canary" and described his interactions with them as communication.[25] Using abduction, he affirmed semiosis between humans and other animals: "You would never persuade me that my horse and I do not sympathize, or that the canary bird that takes such delight in joking with me does not feel with me and I with him; and this instinctive confidence of mine that it is so, is to my mind evidence that it really is so."[26]

The science of signification, as set forth by Peirce, operates within a field of meanings in order to study the life of signs within cultures. If this sounds like General Systems Theory, there is historical coincidence. Peirce was just twenty years old when Darwin's *Origin of Species* was published; and, toward the end of Peirce's life, Albert Einstein presented his theory of general relativity.[27] Peirce's fascination with logic, and how animals (including humans) regard inferences in relation to social factors, points to the influence of the ideas of his time:

> The catholic consent which constitutes the truth is by no means to be limited to men in this earthly life or to the human race, but extends to the whole communion of minds to which we belong, including some probably whose senses are very different from ours, so that in that consent no predication of sensible quality can enter, except as an admission that so certain sorts of senses are affected.[28]

For Peirce, the capacity to understand particulars from a more expansive view (a "universal community of observers") was "not about being as such, but about the conditions in the act of seeing imbued with reflexivity."[29] Peircean semiotics refers to the real world, which then "opens up an undiscovered country, populated with diverse indigenous signs, which it is the task of communication theory to explore."[30] Indeed, Peirce's pragmatist semiotics provides a mode of inquiry and investigation that is particularly well suited to our project of expanding communication studies: the theoretical inference of abduction. Abduction is perhaps an even more appropriate form for inquiry into human-animal communication due to its inherent disruption of hierarchical dualisms, despite Thora Margareta Bertilsson's suggestion otherwise. She describes abduction as "at the same time more primitive and more advanced than induction, as it helps in giving meaning to the perception of a special case."[31] The concept also helps to explain why, as she points out, "We are often utterly sure of ourselves when we 'see' things, especially so when we are part of a culture of seeing, and when there is no one around to challenge our percepts and beliefs."[32] Her subtly speciesist commitment to human exceptionalism aside, Bertilsson's point about Peirce's notion of abduction is provocative:

> The only mode of inference that can truly give rise to surprises is that of abduction. Deduction does not really "see" anything at all as it serves a control office. Induction does not see anything new either for its main purpose is that of helping us to form habits of expectations. Both these operations are absolutely essential for human living. We probably share these inferential modes with other mammals, although they are refined and explicated as logical thought operations in the case of humans. However, the case of abduction may be different. The act of seeing something anew, to connect a case with an unexpected rule, demands mental faculties of some considerable degree. It might be that a blind evolutionary process (of natural selection) provides other mammals with similar abilities. New circumstances demand that any living creature can change and accommodate their behavioural patterns accordingly.[33]

In fact, Peirce regarded science as kind of open system, revisable and "radically empirical" in that it must return ultimately to experience for validation.[34] At the heart of Peirce's semiotics is his threefold categorizing of conceptions. "Firstness" (or *icon*) refers to the arousal produced by sensations, or feelings that are an immediate, prereflexive consciousness, often experienced vaguely. "Secondness" (or *index*) is the observation of an other, related to "brute facts."[35] In secondness, a firmness is achieved and "[t]he horizon of contingencies is no longer indefinitely open; a process of selection has set in and closed the possibilities contained in firstness; this is 'it,' and it can be no other. Indices stand for brute connections: seeing something 'as'; experiencing something 'as.'"[36] "Thirdness" (or symbol) is the necessary interpretive process between experience (firstness) and observation (secondness) "where signs are set in motion."[37] It goes beyond the individual and particular to become a category of law, habit, structure, relatedness. Our ability to understand things in common is a function of the capacity of signs to extend meaning and interpretation to others in time and space. In this way, "thirdness" aligns with deduction as a mode of reasoning.

Peirce's account of semiotics can be related to internatural communication through his firstness, secondness, and thirdness categories. Anderson summarizes:

> Now, if we return to Peirce's description of perception and feeling, we can see that for him there is no barrier in principle or in experience to communication or semiosis between humans and other animals. Moreover, although some other animals may engage us with fully articulate signs, such as facial expressions and whimpers, even such forms of communication are made possible by the shared feeling of differential perceivers .... That is, we humans can sympathize or feel with other animals.[38]

By establishing a sensate "firstness" (the sensory experience of an interspecies encounter) and linking this to the observable world in "secondness" (the human's perception that the other is, for instance, a dog, or vice versa), Peirce's "thirdness" does not foreclose the possibility of *shared* signification between human and nonhuman animals (the shared pleasure of a human and canine engaged in play). Thirdness here is interpreted within the social structures signifying play for both dog and human—including vocalizations, open mouths in "grins," wiggling bodies in approach postures, and the solicitation of tactile behaviors.

Exploring the implications of human-animal communication with its "diverse indigenous signs" requires critical engagement with the languages or sign systems of other communicators, and we might as well begin at the level of abduction, with those iconic inferences that tell us why things are the way they are. They are, after all, both insightful and extremely fallible. As Bertilsson points out, "An abductive inference gives us a clue to reality, but it always remains on the level of a *maybe*.... Abduction, Peirce says, is closest to reality for we cannot form any judgment at all if it were not for the power of abduction; we now see what we did not see before."[39] Abduction is also essential to any phenomenological study, and Merleau-Ponty's environmental philosophy was grounded in relationality and "interanimality."[40] Thus, to better understand animals and their communication, particularly as it pertains to our relationships with them, we need to begin with close observers who experience those relationships in ways that encourage such reflection and judgment.

Humans with close working relationships with animals, especially those with relations of mutual dependency, may be better able to identify and understand modes of communication that work outside the (current) bounds of the human communication paradigm yet fruitfully expand and enrich it. To that end, we now turn to the case of canine-human communication in the context of the work, love, and play required to be a successful sled team. We choose to look at canine communication for many reasons, including their adaptive relationship with humans throughout history and across cultures and a symbiotic relationship between the two species that relies on intentional and reciprocal communication. We also are lucky to have lived with canine companions and are motivated by a desire to listen and communicate more effectively with them. Our desire to better understand the rhetorical lives of canines is also a function of our conviction that they have much to teach us about assistance, awareness, affection, teamwork, and a host of other relational interactions.

WORKING WITH DOGS

Humans' reliance on canines precedes the animals' domestication. In early hunter-gatherer groups, people found the presence of dogs useful in two primary ways: they signaled the approach of other animals or humans and cleaned up the scraps and waste left behind by early human societies. While dogs long served the purposes of cleanup and providing alarm, as they became increasingly domesticated, dogs also were seen as useful providers of heat and, of course, as excellent hunters and herders. Depending upon location and purpose, dogs have been variously trained as shepherds, retrievers, and setters, and a few dogs have even specialized in capturing live tigers, though more recently they have been used to track tigers and other endangered animals for conservation purposes.[41]

During both World Wars, dogs were used to carry messages and accompany troops. Dogs have also served as vital members of tactical units during the US wars in Korea, Vietnam, Afghanistan, and Iraq.[42] As Noyes summarizes, "Dogs and humans, of course, have a long cooperative history. In return for food and shelter, dogs have served as beasts of burden in war and peace, as guards, messengers, hunters, herders, and finders of lost children."[43] Other roles played by dogs in the lives of humans include family members/pets, personal protection, seeing-eye dogs, assistance for people with physical disabilities, seizure alerters for epileptics, cancer-sniffers, search and rescue, bomb and drug detectors, plant identifiers, venomous snake alerters, calf birth indicators (in ranching), insect and rare plant alerters (in forestry), therapy dogs, and sporting dogs (such as racing, fighting, and sledding).

In recent years, tremendous attention has been paid to the powerful role of dogs in therapy and disability contexts. Sakson describes the four types of assistance dogs provided by an organization called Canine Companions for Independence (CCI). "Service dogs" are for adults with physical disabilities. "Hearing dogs" are for those who are deaf or hard of hearing. "Facility dogs" work permanently in a health-care setting such as a physical therapy or retirement center. Finally, "skilled companions" is the term used to designate "dogs going to children or adults with developmental or cognitive disabilities along with physical disabilities. In these cases, the dog, the person with a disability, and the human assistant are trained together as a team."[44]

Like most people with animal companions, the clients of CCI find their human-animal relationship to be an important source of affection, love, self-esteem, and stability. In addition to these valuable benefits, the relationship can also improve human-human communication.

> People with service dogs reported time after time that one of the biggest benefits that came with their dog was the public's willingness to speak to them ... . [S]tudies confirmed that when they have a dog with them, members of the public are ten times more likely to smile and say hello. They don't feel like a person with a disability; they are a person with a dog, just like anybody else.[45]

As for dogs "who don't meet the demanding requirements" of working with CCI clients, they are adopted by other agencies, organizations, or individuals (including the people CCI calls "Puppy Raisers" who have already bonded with and trained the dogs). As Sakson notes, "These dogs continue to serve by using their confidence, ultra-sensitive noses and willingness to please, functioning as bomb searchers, border patrol, customs inspectors, and therapy dogs."[46] Some researchers have become interested in working with dogs because of their remarkable capacity to sniff out cancerous tissue or identify urine samples from patients with cancer. Sometimes referred to as "dognoseis," the story of the dog who smelled cancer has received much popular attention and stimulated new ways of approaching disease.[47]

A final manifestation of human-canine professional cooperation, and one that helped to inspire this chapter, is that of the sled team. The practice of using dog sleds for checking trap lines and fulfilling other work-related functions has given way in recent decades to two overwhelming forces—the mechanization of snow travel and the popularity of dogsled races such as the *Iditarod* and *Yukon Quest*. Today, most sled dogs raised in kennels are trained principally for racing rather than for work on trap lines or for basic transportation. Dogsled teams offer a unique organizational formation for consideration of communication dynamics. All teams have their leaders, but what makes for good leadership is, in part, a function of the attitudes and interactions of the team as a whole—a dimension of adaptive leadership that goes beyond mere pack order. Mushers offer a unique perspective on canine-human communication in work contexts. As Oregon musher Liz Parrish points out, "the communication/trust between a musher and team has lots of nuances" that are "above and beyond" the relationship between a human and a single dog.[48]

At Muktuk Kennels in Whitehorse, Yukon Territory, Frank Turner, along with his partner Ann Taylor, raises, cares for, and races with the huskies for whom the place is named. An Inuktitut word, *muktuk* refers to the outer layer of whale fat that is rendered into a culinary delight for the sled dogs, all of whom have names and houses of their own. Introductions to the dogs include a sustained howling session; hence, the title of this chapter. Howling, as a communication act, varies by context and company: "Wolves howl when separated from the group, and also when setting out with the group for a hunt or in reunion afterward. A howl when alone is a communication seeking company; howling together may be simply a rallying cry or celebration of the group. It has a contagious component … "[49] Howling with the sled dogs at Muktuk carries with it a powerful sense of celebration of community. It is not a mournful experience, as it is often cast in popular media, but rather like a group hug accomplished through vocalization. Learning to howl is just one of the many lessons humans learn from sled dogs.

The safety and success of a sled team is a function of the effectiveness of interspecies communication. In this fashion, communication on sled teams could provide insights relevant to other high-risk professional interspecies teams, such as bomb squads and firefighters. What is perhaps most interesting about the sled context is the mushers' awareness that it is incumbent upon them, the humans, to learn as much as possible about canine communication in order for the team to work together well and adapt to circumstance. The dogs are engaging in communication with each other, no doubt, but it is their efforts to communicate with their human musher that is the focus of this case illustration.

Often, the focus in literature about work dogs focuses upon how to train the dogs to respond effectively to human communication signals, with little consideration given to adapting human communication to match aspects of canine communication. When such consideration is given, it is frequently part of a larger discussion of establishing appropriate master-subject (read: human-dog) dynamics or understanding canine behavior for the purposes of discipline and training. Though these dynamics are certainly present in a musher's relationship with her dogs, what is more striking is the desire and ability to come to know the dogs on their own terms. In Gail Kuhl's study of the stories of mushers, a common and recurring theme had to do with "the musher's responsibility to understand and communicate with dogs … it was the musher who must adapt to the dogs to both understand and read dog 'language' and communicate in a way their dogs would understand."[50] Sled dogs live at the intersections of the categories of canine literature. They are closer to their wild canine ancestors and cousins than some other domesticated breeds. They are also often handled and treated with the affection, attention, and familial interest bestowed upon pets and therapy animals, as Liz Parrish's household approach to raising puppies demonstrates.[51]

Sled dogs do not function as official assistance animals, but they are employed. They are runners. How does a human come to understand, embrace, and communicate with them in terms of their complex, multiplicitous identities and relationships? In contrast with other contexts in which dogs and humans work together, the sled team is an environment in which the human trainer may need to be significantly more knowledgeable about canine communication—and capable of adapting to it—than the other way around. Humans may teach their domesticated dogs how to "heel," but sled dogs teach us to howl.

CANINE COMMUNICATION

Numerous studies have examined the behavior and interaction of canines such as wolves and domesticated dogs.[52] Some studies assert that wolves are smarter than domesticated dogs because of their capacity to problem solve in efficient and independent ways. Vilmos Csányi pinpoints the cause of the misconception: "The wolf is independent, while the dog prefers to work under human direction, which is the source of the differential assessment of the intelligence of the two."[53] After observing two groups of dogs—those identified as outdoor dogs or guard dogs and those identified as indoor dogs or family pets—Csányi and his colleagues found that the dogs' behavior was much more strongly influenced by the presence or absence of a meaningful communication relationship than it was a marker of their intelligence or capacity for problem solving.

In the study, dogs and their owners were placed in a setting where the dog had to solve a problem involving the presence of a fence in order to get to a morsel of food. The outdoor dogs solved the problem relatively quickly with little regard for the presence of the human owner. The indoor dogs, on the other hand, took longer to solve the problem and often looked to their human companion for permission or assistance. After careful observation, the researchers found that the family dogs were not less capable of obtaining food but that their relationships with their owners had a stronger influence on how and to what extent the dogs obtained food. In short, they were acknowledging secondness (the presence of humans with whom they had a relationship) and thirdness (through attempted symbolic interaction with human interlocutors).

The outdoor dogs' efforts were significantly less mediated by the presence of their owners, whereas indoor dogs used eye gaze, vocalizations, and gestures to gain attention, permission, or assistance from the humans with whom they have relationships. Similarly, in Kuhl's study, the mushers "often emphasized the importance of reading dogs' body language," with one musher saying, "you can see if someone has any issues just by the way they're holding their ears or the way they're holding their head or the way, you know, their gait."[54]

Continuing their interest in the variations in rules and signaling behaviors in bonded pairs, researchers videotaped seeing-eye dogs with their human companions and analyzed behavior, particularly in the moment when "a significant change takes place or in which a major decision is made ... . It seemed useful to examine who initiated the decision: the master or the dog?"[55] They found that the dogs made more than half of the decisions, with the balance by the blind person, and that there was significant variation among pairs as well as within certain contexts. In other words, some pairs indicated the dog making the decision 80 percent of the time and others indicated that the dog decided only 20 percent of the time, though the latter group included mostly people who had some partial vision and may have therefore been more directive with their dogs. More importantly, the decision or significant change in course frequently appeared to be a negotiated reality. For example, in some cases the dog would yield to the human's decision, even if it initially wanted to turn a corner, move in a different direction, or stop; but in other instances, the dog was more persistent or less yielding, suggesting the blind person should heed the dog's cues.[56] Kent Allen, a musher from Colorado, describes his dogs' decision-making ability as a strength of the team: "I always let the dogs determine the pace going out, I only ask them to run my speed when I call them up to go home."[57]

Of course, dogs are relational beings with each other as well as with humans. Much canine research focuses on pack behavior and the establishment of rank order. Bonding is seen as the basis for cooperation among dogs and, while "the alpha individual does its very best to induce the pack to follow it … it will not go off on its own" if unable to establish its will as dominant in a particular instance.[58] Pack behavior has been extensively studied among both domesticated dogs and wolves. Canine researchers note that understanding pack behavior, rank order, and aggression "is greatly assisted by the wolves' rich body language, which always needs to be interpreted within the particular context."[59]

Research on pack interactions (and canine communication generally) has the potential to call attention to alternate, nonverbal means of negotiating power, influence, and identity. As Csányi later notes, "The 'top-down' approach, which compares animal accomplishments to those of humans, is heavily burdened with ideology ... . Of course, there are other avenues as well. We could examine, for example, how animals, or even humans, *understand* how one should behave in a small community ... . A true evolutionary characterization would adopt such an approach."[60] What communicative possibilities inhere in the enforcement and transgression of pack relations? How does a single dog, not to mention one lighter in weight and lesser in strength than its teammates, nonetheless succeed in holding an entire team in check in the presence of compelling distractions? Paying attention to the ethological literature, abduction leads to multiple possible explanations, including trust and confidence—expressed through scent, eye contact/glance, and immobility, among others. One young musher explains that "the most important accomplishment you can get from training" is a team "that will work together and respect each other."[61]

Peirce, describing his dog's senses, noted canines' remarkable communicative capacities and compulsions in relation to information conveyed by scent as being akin to his human dependence upon sight:

> But when I reflect to how small a degree he thinks of visual images, and of how much *smells* play a part in his thoughts and imaginations analogous to the part played by *sights* in mine, I cease to be surprised that the perfume of roses or orange flowers does not attract his attention at all and that effluvia that interest him so much, when at all perceptible to me, are simply unpleasant. He does not think of smells as sources of pleasure and disgust but as sources of information.[62]

Obviously, scent is one of the most powerful ways dogs receive information, but what role does it play in their self-awareness and the sending of messages to others? Self-awareness has for many years been assessed in infants and a few animals such as elephants, chimpanzees, dolphins, and magpies using the mirror recognition test (MSR). Canines, however, fail this test. Marc Bekoff hypothesized that the cause was that dogs, unlike the other animals mentioned, use their noses, not their eyes, to navigate their world and recognize each other. Bekoff used his own dog as a research subject, tracking Jethro's recognition of his own urine versus the urine of other known dogs in the snow. Bekoff concludes that Jethro's self-awareness is better "proven" by this method than by the MSR, designed by humans based on our own tendency to privilege sight over scent.[63]

In another personal example, Elizabeth Marshall Thomas recounts the story of the death of one of her dogs. Upon returning home from the veterinarian's office, she showed the dog's collar to his pack mates and allowed them to sniff her extensively.

> Then, as we stood there together in the cold, bright room, just gazing at one another, they suddenly began to emit an odor. It was the odor of dog, of wet dog, musky and penetrating, and it was rising like a cloud of cold steam from one or both of the two dingo sisters, seeping through their skins, growing stronger and stronger until the room was filled with it. … As voices travel where their makers can't go, so odors cling where their makers can't stay. The odor was a dog thing. I couldn't pretend to understand it. But interestingly enough, my body responded to it: the hair rose on my skin.[64]

As Thomas' experience (exemplary of Peirce's notion of firstness) suggests, and studies of canine physiology show, scent is a powerful communication medium for dogs, and they do not appear to use it exclusively with other canines. It is likely that, due to their relatively underdeveloped olfactory receptors, most humans are ill prepared to receive and process the many messages dogs send via scent. Whereas humans have approximately six million olfactory receptors, dogs have between two and three hundred million.[65] Dogs have a remarkable ability to smell and differentiate among odors, and scent is just one of their many modes of engaging in motivated communication. Despite limited experience by many of us, humans *are* capable of overcoming our "trained incapacities," to borrow a Burkean phrase borrowed from Veblen.[66] For instance, the practice of *honi*, a greeting that entails the inhalation of breath, enabled early Hawaiians to distinguish between native Hawaiians and foreigners, who were called *haole*, either because of their different scent or their failure to participate in the traditional practice at all. As Kennedy suggests, scent is a kind of rhetoric.[67]

Eye contact is yet another major form of canine communication. Dogs use eye contact to "gain information about others' perceptions and intentions, and they appear to use eye contact to initiate and emphasize communication."[68] Indeed, many canines, like humans (and unlike monkeys) have a sclera (white) in their eyes, which makes it easier to give eye signals. According to Csányi, "Among dogs, the genetic changes accompanying domestication also started to develop the signaling role of the color pattern of the eyes. We have good reason to suppose that dogs use this in their collaboration and information exchange with humans."[69] Having eyes that show their whites allows others to read their "whale eye," when the head and eyes aren't pointing in the same direction, so you see the whites of the eyes on one side or the other. According to zoologist and animal behaviorist Patricia McConnell, this is a nervous expression seen in dogs that want to look away, but can't take their eyes off the anxiety-producing object or person. Humans can be seen doing this at a horror movie, when they turn their heads away, but still look up at the screen.[70]

As with humans, the same nonverbal signal can mean different things for different dogs. In one of the photographs in McConnell's book, her beloved

Luke has his head turned while looking at the camera, appearing to be amused, as if sharing a joke. Our own dogs, in our respective households, seem to display such a sense of humor by "playing jokes"—dropping a ball, then grabbing it away at the last minute and showing this "whale eye." Other authors have also commented on the canine sense of humor. Vicki Hearne, horse and dog trainer, describes her pointer Salty first bringing her feed dish to Vicki, and then, when Vicki absentmindedly put it on the desk, "bringing a wastepaper basket, wriggling so hard with gleeful appreciation of her own wit that she drops it on her way."[71] Mushers also note their dogs' senses of humor, whether that is manifested in tricks or in play, and emphasize the importance of maintaining a positive attitude and social environment. Yes, the dogs on the team are working, but they should also be having fun. Musher Ann Mariah Cook describes the interconnectedness of work and play for the canine members of the team: "As we pulled up to the kennel yard, I felt the sense of mystery that I always felt when I saw sled dogs run. It amazed me that they worked this happily and this hard for food, shelter, and human affection. Each day was an adventure to them, each practice a new challenge. What secret, held deep inside them, made them *want* to do this work?"[72]

Other prominent canine communication behaviors studied by ethologists are also meaningful to members of a sled team. For instance, tail wagging can show that the wagger is paying attention to you, and nudging with the nose shows that the nudger is seeking your attention. Wagging tails are not always a signal of friendliness or question asking, unless it is a "loose relaxed tail that wags its hindquarters with it" or what McConnell calls, "wagging from the shoulders back."[73] A tail wagging only from its midsection tells you this dog is not necessarily feeling friendly or asking a friendly question. On this latter point, Csányi elaborates on the observation that dogs ask questions and accept answers: "The most exciting thing is when dogs ask questions. It is strange that it has not been reported in the scientific literature that this type of behavior exists at all. … One must pay very close attention to the requests of and questions posed by dogs, because if one does not react appropriately, they soon stop asking."[74]

Musher Liz Parrish confirms this observation in a memorial to former team leader Rope, who she describes as "challenging." In one race, Parrish wished to react differently to Rope's choice to take the team down off the trail "toward the shelter of a small group of trees" but chose instead to make an immediate change to the team structure for practical purposes. She later worked with Rope during a break so that his question did not remain unanswered, and thus his doubts confirmed:

> I knew I'd have to deal with Rope later, to work him fully through his next "I can't" episode which was sure to happen since I'd set him up so nicely with this

> one. That happened overnight a few days later, on the long run between Cripple and Ruby, and we did the training we needed to do. Didn't matter it was in the middle of the race … my commitment was to Rope, and to showing him what he could do, not leaving him with a failure. And, of course, he proved to himself that he could do it.[75]

Similarly, musher Shelley Ramsay describes the benefit she and her husband find in working with the "trouble dogs," saying they have "been our best teachers when we take the time to figure them out. There is also a sense of relief we get from the dog when we finally address him to his proper place and solve his puzzles."[76]

Another canine nonverbal mode of communication is bodily movement, particularly what McConnell calls being "frozen like a popsicle." She warns dog owners to note that "a stiff body and a closed mouth are signs of a dog on high alert, whose next move might be to lunge forward," either to snap at the dog groomer or a dog approaching. Dogs will "micro-freeze" with each other, even while playing, as a quick warning not to mount them or to back off now. They will also micro-freeze with humans, turning their heads in a slight "turn to face," with a closed jaw, to warn us that whatever we are doing should not be repeated.[77]

Finally, the impression that dogs "smile," is not mere anthropomorphic wishing. When humans smile, our lower jaws relax and our teeth do not touch; the same is true for dogs. McConnell writes extensively about the emotional expressions dogs use to communicate with us and each other. The "open-mouth play face" shows relaxation and pleasure (unless, of course, the open mouth is panting in heat or exhaustion). McConnell instructs her clients to watch the open mouth/closed mouth of their dogs, which communicate their emotional states: "Dogs will close their mouths when they go on alert, and watching a mouth go from open to closed is a good way to know your dog has begun to concentrate on a change in the environment. Dogs will also close their mouths when they're serious about something."[78]

Whether smiling, standing still, asking questions, wagging their tails, or smelling the air, a wide variety of canines are engaging in what humans call communication. They express themselves in ways that require us to abduce meaning from likely contextual clues. Abduction provides a platform for the proposition of new hypotheses about animal communication, ones that have the capacity to capture some likely explanations for utterances previously confined to behaviorism. Canines are also able to abduce clues from humans, as indicated by research that shows dogs are superior to chimpanzees in their ability to "infer" from humans' pointing gestures and retrieve the correct objects. The researchers concluded that "dogs' skills with human communication are an adaptation to life with humans."[79]

The mounting evidence that many animals have intelligence, emotions, consciousness, and intentionality should alter the field of communication in at least three important ways. First, as this book demonstrates, we need to abandon the mythology foreclosing the possibilities of human-animal communication. Instead, we should extend language to animals acting as subjects with a semiotics based not on referentiality to other signs but to the natural world and the systems of meaning they engage. Second, we need to respect and study the lived experience ("local knowledge," both phenomenological and ethnographic) of humans who interact with animals to share understanding and exchange meaning. In this way, our initial foray into the communication of mushers and their teams of sled dogs charts a prospective course. Finally, as an outcome of these two endeavors, we need to use the insights from studies of internatural communication to expand our understanding of what it means to be, and communicate as, humans.

Our inconsistent and contradictory attitudes toward animals reflect what James Serpell calls "a fundamental conflict of interests":

> If we regard animals as unfeeling and uncaring objects [to be used and eaten], then their apparent interest in us—their affection and companionship—must just be an illusion. ... Conversely, if we regard animals as equivalent to friends and family, we cannot expect to exploit them harmfully without experiencing moral anxiety in the process.[80]

Hal Herzog is not bothered by these contradictions, by the "flagrant moral incoherence" he encountered when he began studying human-animal interactions."[81] He no longer feels like a hypocrite when he eats beef but not veal, or because he has a cat but believes pet ownership is unethical. Instead, he embraces these paradoxes as what makes us human, because neither logic nor intuition can be relied upon when "resolving the moral conundrums in our dealings with other species."[82] Thus, something that resides in between the two—namely, abduction—may be how we go about resolving such conundrums, by accepting the reasonableness of the conclusion that we live within a wide, wide world of rhetorical animals.

NOTES

1. Portions of this paper were previously presented by the first author at the Conference on Communication and the Environment in 2009 and 2011 and by the second author at the 2012 International Communication Association Conference.

2. Chihab El Khachab, "The Logical Goodness of Abduction in C. S. Pierce's Thought," *A Quarterly Journal in American Philosophy* 49, no. 2 (2013): 157.

3. Peirce quoted in Mark Tschaepe, "Guessing and Abduction," *A Quarterly Journal in American Philosophy* 50, no. 1 (2014): 117.

4. Emily Plec, Ed. *Perspectives on Human–Animal Communication: Internatural Communication* (New York: Routledge, 2013).

5. George Kennedy, "A Hoot in the Dark: The Evolution of General Rhetoric," *Philosophy & Rhetoric* 25 (1992): 3–4.

6. George Kennedy, *Comparative Rhetoric: An Historical and Cross-cultural Introduction* (New York and Oxford: Oxford University Press, 1992), 12.

7. Erik Doxtader, "Addressing Animals," *Philosophy & Rhetoric* 44 (2011): 79.

8. Alex C. Parrish, *Adaptive Rhetoric: Evolution, Culture, and the Art of Persuasion* (New York: Routledge, 2014).

9. Plec, *Perspectives on Human–Animal Communication,* 5–6.

10. John Durham Peters, *Speaking Into the Air: A History of the Idea of Communication* (Chicago: University of Chicago Press, 1999), 155.

11. Eduardo Nieva and Mark Hickson, "Deception and Honesty in Animal and Human Communication: A New Look at Communicative Interaction," *Journal of Intercultural Communication* 32 (2003), 23–35.

12. Randolph T. Barker, "On the Edge or Not?" *Journal of Business Communication* 42 (2005): 299–315.

13. Dan Dieterich, "A Response to Randy Barker's Call for Research," *Journal of Business Communication* 42 (2005): 316–317.

14. Stacey Sowards, "Identification through Orangutans: Destabilizing the Nature/Culture Dualism." *Ethics & The Environment* 11 (2006): 46.

15. See Donal Carbaugh, "Just Listen: 'Listening' and Landscape Among the Blackfeet," *Western Journal of Communication* 63 (1999): 250–270; Donal Carbaugh, "Naturalizing Communication and Culture," in *The Symbolic Earth: Discourse and our Creation of the Environment,* ed. Christine L. Oravec and James G. Cantrill (Lexington, KY: University Press of Kentucky, 1996), 38–57; Donal Carbaugh and Tovar Cerulli, "Cultural Discourses of Dwelling: Investigating Environmental Communication as a Place-based Practice," *Environmental Communication: A Journal of Nature and Culture* 7 (2013): 4–23; Tema Milstein, "When Whales 'Speak for Themselves': Communication as a Mediating Force in Wildlife Tourism," *Environmental Communication* 2 (2008): 173–192; Tema Milstein, "Banging on the Divide: Cultural Reflection and Refraction at the Zoo," in *Perspectives on Human–Animal Communication: Internatural Communication*, ed. Emily Plec (New York: Routledge, 2013); Natasha Seegert, "Queer Beasts: Ursine Punctures in Domesticity," *Environmental Communication: A Journal of Nature and Culture* 8 (2014a): 75–91; Natasha Seegert, "Play of Sniffication: Coyotes Sing in the Margins," *Philosophy & Rhetoric,* 47 (2014b): 158–178; Richard A. Rogers, "Overcoming the Objectification of Nature in Constitutive Theories: Toward a Transhuman, Materialist Theory of Communication," *Western Journal of Communication*, 62 (1998): 244–272; Richard Rogers, "Overcoming the Preservation Paradigm: Toward a Dialogic Approach to Rock Art and Culture," in *American Indian Rock Art* 33 (2007): 53–56; Richard Rogers, "'Your Guess is as Good as Any': Indeterminacy, Dialogue, and Dissemination in Interpretations of Native American Rock Art." *Journal of International and Intercultural Communication* 2 (2009): 44–65.

16. Rogers, "Overcoming the Objectification of Nature," 268.

17. Merten Klaus, "Reflexivity in Human Communication," *Communication Yearbook* 1 (1978): 121–131.

18. George Herbert Mead, *Mind, Self, and Society* (Chicago: University of Chicago, 1934), 43.

19. Charles Darwin, *The Expression of Emotion in Man and Animals* (Chicago: University of Chicago Press, 1872/1965).

20. See bibliography for representative writings by the authors listed above.

21. Hal Herzog, *Some We Love, Some We Hate, Some We Eat* (New York: Harper Perennial, 2010), 16.

22. Ibid., 16.

23. John N. Deely, *Introducing Semiotic: Its History and Doctrine* (Bloomington, IN: Indiana University Press, 1982), 107.

24. Umberto Eco, "How Culture Conditions the Colours We See," in *The Communication Theory Reader*, ed. Paul Cobley (New York: Routledge, 1968/1996), 212.

25. Rulon S. Wells, "Criteria for Semiosis," in *A Perfusion of Signs*, ed. Thomas A. Sebeok (Bloomington, IN: Indiana University Press, 1977), 6.

26. Peirce quoted in Douglas R. Anderson, "Peirce's Horse: A Sympathetic and Semiotic Bond," in *Animal Pragmatism: Rethinking Human–Nonhuman Relationships*, ed. Erin McKenna and Andrew Light (Bloomington, IN: Indiana University Press, 2004), 88.

27. Ibid.

28. Peirce quoted in E. Valentine Daniel, *Fluid Signs: Being a Person the Tamil Way* (Berkeley: University of California Press, 1984), 17.

29. Thora Margareta Bertilsson, "The Elementary Forms of Pragmatism: On Different Types of Abduction," *European Journal of Social Theory* 7, no. 3 (2004): 373.

30. Paul Cobley, "Introduction," in *The Communication Theory Reader*, ed. Paul Cobley (New York: Routledge, 1996), 30.

31. Bertilsson, "The Elementary Forms of Pragmatism," 372.

32. Ibid., 373.

33. Ibid., 377.

34. Jay Zeman, "Peirce's Theory of Signs," *A Perfusion of Signs*, ed. Thomas A. Sebeok (Bloomington, IN: Indiana University Press, 1977), 22.

35. Bertilsson, "The Elementary Forms of Pragmatism," 373.

36. Ibid., 374.

37. Ibid.

38. Douglas R. Anderson, "Peirce's Horse: A Sympathetic and Semiotic Bond," in *Animal Pragmatism: Rethinking Human–Nonhuman Relationships*, ed. Erin McKenna and Andrew Light (Bloomington, IN: Indiana University Press, 2004), 87–88.

39. Bertilsson, "The Elementary Forms of Pragmatism," 377.

40. Susan Hafen, "Listening with the Third Eye: A Phenomenological Ethnography of Animal Communicators" in *Perspectives on Human–Animal Communication: Internatural Communication*, ed. Emily Plec (New York: Routledge, 2013), 196.

41. Deborah Noyes, *One Kingdom: Our Lives with Animals: The Human–Animal Bond in Myth, History, Science, and Story* (Boston: Houghton Mifflin, 2006); Vilmos

Csányi, *If Dogs Could Talk: Exploring the Canine Mind*, trans. Richard E. Quandt (New York: North Point Press, 2000/2005); Center for Conservation Biology.

42. Sharon Sakson, *Paws & Effect: The Healing Power of Dogs* (New York: Alyson Books, 2007).

43. Noyes, *One Kingdom*, 95–96.

44. Sakson, *Paws & Effect*, 105–106.

45. Ibid., 105.

46. Ibid., 111.

47. Noyes, *One Kingdom*, 96.

48. Liz Parrish, in discussion with the first author, September 2009.

49. Alexandra Horowitz, *Inside of a Dog: What Dogs See, Smell, and Know* (New York: Scribner, 2009), 103.

50. Gail Kuhl, "Human–Sled Dog Relations: What Can We Learn from the Stories and Experience of Mushers?" *Society & Animals* 19 (2011): 29.

51. Liz Parrish and Jan Kelley, *Crimp! On-By! The True Story of a Most Unlikely Iditarod Lead Dog* (Mattoon, IL: United Graphics, 2009b), 9.

52. See David Mech, *The Wolf: The Ecology and Behavior of an Endangered Species* (New York: Natural History Press, 1970); John Paul Scott and John L. Fuller, *Genetics and the Social Behavior of the Dog* (Chicago: University of Chicago Press, 1965).

53. Vilmos Csányi, *If Dogs Could Talk: Exploring The Canine Mind*, trans. Richard E. Quandt (New York: North Point Press, 2000/2005), 104.

54. Gail Kuhl, "Human–Sled Dog Relations: What Can We Learn from the Stories and Experience of Mushers?" *Society & Animals* 19 (2011): 29.

55. Csányi, *If Dogs Could Talk,* 114.

56. Ibid., 116.

57. Sled Dog Central, "Interviews," accessed December 12, 2015, http://www.sleddogcentral.com/meet.htm.

58. Csányi, *If Dogs Could Talk*, 14.

59. Ibid., 21.

60. Ibid., 167.

61. Sled Dog Central.

62. Peirce quoted in Anderson, "Peirce's Horse," 87.

63. Marc Bekoff (2001). "Observations of Scent-marking and Discriminating Self from Others by a Domestic Dog (Canis familiaris): Tales of Displaced Yellow Snow," *Behavioural Processes* 55, no. 2 (2001): 75–79.

64. Elizabeth Marshall Thomas, *The Hidden Life of Dogs* (New York: Houghton Mifflin, 1993), 136–137.

65. Alexandra Horowitz, *Inside of a Dog: What Dogs See, Smell, and Know* (New York: Scribner, 2009), 71.

66. Kenneth Burke, *Permanence and Change: An Anatomy of Purpose* (Berkeley CA: University of California Press, 1984), 7–9.

67. Kennedy, *Comparative Rhetoric.*

68. Dale Peterson, *The Moral Lives of Animals* (New York: Bloomsbury Press, 2011), 181.

69. Vilmos Csányi, *If Dogs Could Talk: Exploring The Canine Mind*, trans. Richard E. Quandt (New York: North Point Press, 2000/2005), 119.

70. Patricia McConnell, *For the Love of a Dog: Understanding Emotion in You and Your Best Friend* (New York: Ballantine Books, 2005/2007), 60.

71. Vicki Hearne, *Adam's Task: Calling Animals by Name* (New York: Skyhorse Publishing, 2007), 75.

72. Ann Mariah Cook, *Running North: A Yukon Adventure* (Chapel Hill, NC: Algonquin Books of Chapel Hill, 1998), 129.

73. McConnell, *For the Love of a Dog,* 49.

74. Csányi, *If Dogs Could Talk,* 140.

75. Liz Parrish, "Challenges are the Good Stuff," May 29, 2015, *Be the Lead Dog,* http://www.betheleaddog.com/, accessed January 27, 2016.

76. Sled Dog Central.

77. McConnell, *For the Love of a Dog*, 46.

78. Ibid., 44.

79. Katharina C. Kirchhofer, Felizitas Zimmermann, Juliane Kaminski, Michael Tomasello, "Dogs (Canis familiaris), But Not Chimpanzees (Pan troglodytes), Understand Imperative Pointing," *PLoS ONE* 7, no.2 (2012): n.p., accessed June 20, 2016.

80. James Serpell, "Having Our Dogs and Eating Them Too: Why Animals are a Social Issue," *Journal of Social Issues* 65 (2009): 642.

81. Hal Herzog, *Some We Love,* 279.

82. Ibid., 259.

BIBLIOGRAPHY

Abram, David. *The Spell of the Sensuous: Perception and Language in a More-Than-Human World.* New York: Pantheon Books, 1996.

Alger, Janet M. and Steven F. Alger. "Beyond Mead: Symbolic Interaction between Humans and Felines," *Society and Animals* 5 (1997): 65–82.

Anderson, Douglas R. "Peirce's Horse: A Sympathetic and Semiotic Bond," in *Animal Pragmatism: Rethinking Human–Nonhuman Relationships*, edited by Erin McKenna and Andrew Light, 86–94. Bloomington: Indiana University Press, 2004.

Barker, Randolph T. "On the Edge or Not?" *Journal of Business Communication* 42 (2005): 299–315.

Bekoff, Marc. "Social Communication in Canids: Evidence for the Evolution of Stereotyped Mammalian Display," *Science* 197 (1997): 1097–1099.

———. "Observations of Scent-marking and Discriminating Self from Others by a Domestic Dog (Canis familiaris): Tales of Displaced Yellow Snow," *Behavioural Processes* 55, no. 2 (2001): 75–79.

———. *The Emotional Lives of Animals.* Novato, CA: New World Library, 2007.

Bekoff, Marc and Jessica Pierce. *Wild Justice: The Moral Lives of Animals.* Chicago: University of Chicago Press, 2009.

Bertilsson, Thora Margareta. "The Elementary Forms of Pragmatism: On Different Types of Abduction," *European Journal of Social Theory* 7, no. 3 (2004): 371–389.

Burghardt, Gordon M. "Ethics and Animal Consciousness: How Rubber the Ethical Ruler?" *Journal of Social Issues* 65 (2009): 499–521.
Burke, Kenneth. *Language as Symbolic Action.* Berkeley: University of California Press, 1966.
———. *Permanence and Change: An Anatomy of Purpose.* Berkeley: University of California Press, 1984.
Carbaugh, Donal. "Naturalizing Communication and Culture," in *The Symbolic Earth: Discourse and Our Creation of the Environment,* edited by Christine L. Oravec and James G. Cantrill, 38–57. Lexington: University Press of Kentucky, 1996.
———. "Just Listen: 'Listening' and Landscape Among the Blackfeet," *Western Journal of Communication* 63 (1999): 250–270.
Carbaugh, Donal and Tovar Cerulli, "Cultural Discourses of Dwelling: Investigating Environmental Communication as a Place-based Practice," *Environmental Communication: A Journal of Nature and Culture* 7 (2013): 4–23.
Cobley, Paul. "Introduction," in *The Communication Theory Reader*, edited by Paul Cobley, 1–36. New York: Routledge, 1996.
Cook, Ann Mariah. *Running North: A Yukon Adventure.* Chapel Hill: Algonquin Books of Chapel Hill, 1998.
Csányi, Vilmos. *If Dogs Could Talk: Exploring The Canine Mind*, translated by Richard E. Quandt. New York: North Point Press, 2000/2005.
Daniel, E. Valentine. *Fluid Signs: Being a Person the Tamil Way.* Berkeley: University of California Press, 1984.
Davis, Diane. "Creaturely Rhetorics," *Philosophy & Rhetoric* 44 (2011): 88–94.
Darwin, Charles. *The Expression of Emotion in Man and Animals.* Chicago: University of Chicago Press, 1872/1965.
Dawkins, Marian Stamp. "The Science of Suffering," in *Mental Health and Well-Being in Animals*, edited by Franklin D. McMillan, 47–56. Ames: Blackwell Publishing, 2005.
Deely, John N. *Introducing Semiotic: Its History and Doctrine.* Bloomington: Indiana University Press, 1982.
Dieterich, Dan. "A Response to Randy Barker's Call for Research," *Journal of Business Communication* 42 (2005): 316–317.
Dodman, Nicholas H. *Dogs Who Love too Much: Tales, Treatments, and the Psychology of Dogs.* New York: Bantam Books, 1996.
Doxtader, Eric. "Addressing Animals," *Philosophy & Rhetoric* 44 (2011): 79–80.
Eco, Umberto. "How Culture Conditions the Colours We See," in *The Communication Theory Reader*, edited by Paul Cobley, 209–224. New York: Routledge, 1968/1996.
El Khachab, Chihab. "The Logical Goodness of Abduction in C. S. Pierce's Thought," *A Quarterly Journal in American Philosophy* 49, no. 2 (2013): 157–177.
Franklin, Adrian. *Animals and Modern Cultures: A Sociology of Human–Animal Relations in Modernity.* New York: Routledge, 2000.
Fouts, Roger. *Next of Kin: What Chimpanzees Have Taught Me About Who We Are.* New York: William Morrow and Company, 1997.

Goodall, Jane. *Through a Window: My Thirty Years with Chimpanzees of Gombe.* Boston: Houghton Mifflin, 1990.

Grandin, Temple and Catherine Johnson. *Animals in Translation: Using the Mysteries of Autism to Decode Animal Behavior.* Orlando: Harcourt Books, 2005.

———. *Animals Make Us Human: Creating the Best Life for Animals.* Orlando: Houghton Mifflin, 2009.

Griffin, Donald R. *The Question of Animal Awareness: Evolutionary Continuity of Mental Experience.* Los Altos: William Kaufmann, 1981.

———. *Animal Thinking.* Cambridge: Harvard University Press, 1984.

———. *Animal Minds.* Chicago: University of Chicago Press, 1992.

Hafen, Susan. "Listening with the Third Eye: A Phenomenological Ethnography of Animal Communicators," in *Perspectives on Human–Animal Communication: Internatural Communication*, edited by Emily Plec, 185–206. New York: Routledge, 2013.

Hawhee, Debra. "Toward a Bestial Rhetoric," *Philosophy & Rhetoric* 44 (2011): 81–87.

Hearne, Vicki. *Adam's Task: Calling Animals by Name.* New York: Skyhorse Publishing, 2007.

Herzog, Hal. *Some We Love, Some We Hate, Some We Eat.* New York: Harper Perennial, 2010.

Horowitz, Alexandra. *Inside of a Dog: What Dogs See, Smell, and Know.* New York: Scribner, 2009.

Kennedy, George. "A Hoot in the Dark: The Evolution of General Rhetoric," *Philosophy & Rhetoric* 25 (1992): 1–21.

———. *Comparative Rhetoric: An Historical and Cross-cultural Introduction.* Oxford: Oxford University Press, 1992.

Kirchhofer, Katharina C., Felizitas Zimmermann, Juliane Kaminski, and Michael Tomasello. "Dogs (Canis familiaris), But Not Chimpanzees (Pan troglodytes), Understand Imperative Pointing," *PLoS One* 7, no. 2 (2012), accessed June 20, 2016.

Klaus, Merten. "Reflexivity in Human Communication," *Communication Yearbook* 1 (1978): 121–131.

Kuhl, Gail. "Human–Sled Dog Relations: What Can We Learn from the Stories and Experience of Mushers?" *Society & Animals* 19 (2011): 22–37.

Markowitz, Hal and Gregory B. Timmel, "Animal Well-Being and Research Outcomes," in *Mental Health and Well-Being in Animals*, edited by Franklin D. McMillan, 277–284. Ames: Blackwell Publishing, 2005.

Masson, Jeffrey M. *Dogs Never Lie about Love: Reflections on the Emotional World of Dogs.* New York: Crown, 1997.

———. *The Emperor's Embrace: Reflections on Animal Families and Fatherhood.* New York: Pocket Books, 1999.

Masson, Jeffrey M. and Susan McCarthy. *When Elephants Weep: the Emotional Lives of Animals.* New York: Delta, 1995.

McConnell, Patricia. *For the Love of a Dog: Understanding Emotion in You and Your Best Friend.* New York: Ballantine Books, 2005/2007.

McMillan, Franklin D. "Do Animals Experience True Happiness?" in *Mental Health and Well-Being in Animals*, edited by Franklin D. McMillan, 221–234. Ames: Blackwell Publishing, 2005.

———. "Stress, Distress, and Emotion: Distinctions and Implications for Mental Well-Being," in *Mental Health and Well-Being in Animals,* edited by Franklin D. McMillan, 93–112. Ames: Blackwell Publishing, 2005.

Mead, George Herbert. *Mind, Self, and Society*. Chicago: University of Chicago, 1934.

Mech, David. *The Wolf: The Ecology and Behavior of an Endangered Species.* New York: Natural History Press, 1970.

Milstein, Tema. "When Whales "Speak for Themselves": Communication as a Mediating Force in Wildlife Tourism," *Environmental Communication* 2 (2008): 173–192.

———. "Banging on the Divide: Cultural Reflection and Refraction at the Zoo," in *Perspectives on Human–Animal Communication: Internatural Communication,* edited by Emily Plec, 162–181. New York: Routledge, 2013.

Miklósi, Ádam. *Dog Behaviour, Evolution, and Cognition.* Oxford: Oxford University Press, 2007.

Muckelbauer, John, "Domesticating Animal Theory," *Philosophy & Rhetoric* 44 (2011): 95–100.

Nieva, Eduardo and Mark Hickson, "Deception and Honesty in Animal and Human Communication: A New Look at Communicative Interaction," *Journal of Intercultural Communication* 32 (2003): 23–35.

Noyes, Deborah. *One Kingdom: Our Lives with Animals: The Human-Animal Bond in Myth, History, Science, and Story*. Boston: Houghton Mifflin, 2006.

Parrish, Alex C. *Adaptive Rhetoric: Evolution, Culture, and the Art of Persuasion.* New York: Routledge, 2014.

Parrish, Liz (musher/owner of Life Through Dogs LLC), in discussion with Emily Plec, September 2009.

———. "Challenges are the Good Stuff." May 29, 2015, *Be The Lead Dog,* http://www.betheleaddog.com/, accessed January 27, 2016.

Parrish, Liz and Jan Kelley. *Crimp! On-By! The True Story of a Most Unlikely Iditarod Lead Dog*. Mattoon: United Graphics, 2009.

Peirce, Charles Saunders. *Selected Writings (Values in a Universe of Chance),* edited by Philip Wiener. Stanford: Stanford University Press, 1958.

Peters, John Durham. *Speaking into the Air: A History of the Idea of Communication.* Chicago: University of Chicago Press, 1999.

Peterson, Dale. *The Moral Lives of Animals.* New York: Bloomsbury Press, 2011.

Philo, Chris and Chris Wilbert. *Animal Spaces and Beastly Places: New Geographies of Human–Animal Relations.* New York: Routledge, 2000.

Plec, Emily. "Introduction," in *Perspectives on Human–Animal Communication: Internatural Communication,* edited by Emily Plec, 1–13. New York: Routledge, 2013.

Rogers, Richard A. "Overcoming the Objectification of Nature in Constitutive Theories: Toward a Transhuman, Materialist Theory of Communication," *Western Journal of Communication* 62 (1998): 244–72.

———. "Overcoming the Preservation Paradigm: Toward a Dialogic Approach to Rock Art and Culture," *American Indian Rock Art* 33 (2007): 53–56.

———. "'Your Guess is as Good as Any': Indeterminacy, Dialogue, and Dissemination in Interpretations of Native American Rock Art," *Journal of International and Intercultural Communication* 2 (2009): 44–65.

Sakson, Sharon. *Paws & Effect: The Healing Power of Dogs.* New York: Alyson Books, 2007.

Scott, John Paul and John L. Fuller. *Genetics and the Social Behavior of the Dog.* Chicago: University of Chicago Press, 1965.

Schoen, Allen M. *Kindred Spirits: How the Remarkable Bond Between Humans and Animals Can Change the Way We Live.* NewYork: Broadway Books, 2001.

Seegert, Natasha. "Play of Sniffication: Coyotes Sing in the Margins," *Philosophy & Rhetoric* 47 (2014): 158–178.

———. "Queer Beasts: Ursine Punctures in Domesticity," *Environmental Communication: A Journal of Nature and Culture* 8 (2014): 75–91.

Serpell, James. *In the Company of Animals: A Study of Human–Animal Relationships.* New York: Blackwell, 1986.

———. *The Domestic Dog.* Cambridge: Cambridge University Press, 1995.

———. "Having Our Dogs and Eating Them Too: Why Animals are a Social Issue," *Journal of Social Issues* 65 (2009): 633–644.

Sheldrake, Rupert. *Dogs that Know When Their Owners are Coming Home.* New York: Crown, 1999.

Shepard, Paul. *Thinking Animals: Animals and the Development of Human Intelligence.* New York: The Viking Press, 1978.

———. *The Others: How Animals Made Us Human.* Covelo, CA: Island Press, 1996.

Sled Dog Central, "Interviews," http://www.sleddogcentral.com/meet.htm. Accessed December 12, 2015.

Sowards, Stacey. "Identification through Orangutans: Destabilizing the Nature/Culture Dualism," *Ethics & The Environment* 11 (2006): 45–61.

Tschaepe, Mark. "Guessing and Abduction," *A Quarterly Journal in American Philosophy* 50, no. 1 (2014): 115–138.

Thomas, Elizabeth Marshall. *The Hidden Life of Dogs.* New York: Houghton Mifflin, 1993.

Turner, Frank (musher/owner of Muktuk Kennels), in discussion with Emily Plec, November 2010.

Wells, Deborah L. "The Effects of Animals on Human Health and Well Being," *Journal of Social Issues* 65 (2009): 633–644.

Wells, Rulon S. "Criteria for Semiosis," in *A Perfusion of Signs*, edited by Thomas A. Sebeok, 1–21. Bloomington: Indiana University Press, 1977.

Zeman, Jay. "Peirce's Theory of Signs," in *A Perfusion of Signs*, edited by Thomas A. Sebeok, 22–40. Bloomington: Indiana University Press, 1977.

Chapter 10

Touring the Sixth Persona

Dodos and the Rhetorical Effects of Missed Communication

T. Jake Dionne

A bout of foul weather ushered in the end of time for the dodos, a species of flightless birds endemic to the island of Mauritius. There the dodos lived without grave threat until, on August 28, 1598, rough winds and waters caused Dutch sailors to veer off course and drop anchors next to the landmass isolated in the Indian Ocean.[1] This landing meant certain death for the birds, as little to nothing prepared them to protect themselves against foreign invasion. Over the next one hundred years, a fatal concoction of conquest, predation, and habitat loss triggered the extinction of the dodos.[2]

In the wake of the untimely annihilation of the dodos, various communicators in overlapping spheres of influence (comparative anatomists, curators, environmental historians, naturalists, ornithologists, policy makers, etc.) sought to document the tumultuous relationship between humans and the birds,[3] particularly asking how an entire avian collective could find themselves victim of an unplanned genocide.[4] Whereas conquest, predation, and habitat loss made the list of key reasons for the demise of the dodos, these explanations often accompanied a popular yet inaccurate understanding of the birds as "sitting duck[s]"—a "fat and flightless, clueless and clumsy" grouping of creatures far too trusting and stupid for their own survival.[5]

Were the dodos cognizant of the dangers posed to them by foreign invasion? Were the birds too dumb to foresee impending doom? Did the dodos attempt to convince crews to leave them alone? Did the birds even possess the capacity to persuade? Taking concern with an "anthropocentric worldview" that outwardly blames the dodos for their own extinction,[6] this chapter picks up more than 300 years after the eradication of the birds to ask and answer the following question: If rhetoricians presumed rather than denied that the dodos made earnest attempts to convince sailors to leave them alone, then how would scholars conceptualize the human in the study of persuasion? In other words, what would become of the human if rhetoricians started their theorizations and criticisms from the perspective that the birds composed texts to be read by others both within and beyond their species?

In her analysis of how coyotes in Chicago challenged anthropocentric worldviews of persuasion as an art practiced solely by humans, Natasha Seegert divorced herself from a definition of rhetoric as symbolic action in favor of an unfixed, animate approach.[7] "I define rhetoric as the relational force of signals interacting in the world," advanced Seegert, "A focus on signal stresses a dissemination model of communication and reminds humans that rhetoric is a relational construction of how individuals—human and more-than-human—interpret the signals that have been sent."[8] Concerned with anthropocentric beliefs in "human exceptionalism" that flourish in the humanities,[9] Seegert moved to block rhetoricians from relying upon "anthropocentric logics" to uphold dangerous hierarchical notions of nonhumans as inferior to humans due to their so-called incapacity to produce easily recognizable symbolic texts.[10] In turn, Seegert further paved the way for rhetoricians to study the rhetorical effects of communication between different species by asking scholars to move beyond solely questioning *if* nonhumans possessed the capacity to persuade,[11] a path-breaking task previously taken up by several other rhetoricians.[12]

The reverberations of Seegert's unruly howl—a loud call for rhetoricians to consider "the possibility that everything *might* be speaking"[13]—guide this chapter by prompting the following demonstration of how speculation can be a heuristic for rhetoricians studying the rhetorical effects of nonhuman rhetoric, especially in contexts where signals are not readily locatable. As fleshy members of a more-than-human world long denied "an affectab*ility* or persuad*ability*" by rhetoricians faithful to anthropocentric worldviews,[14] the dodos are no longer alive to testify about their massacre. Rather than the birds and their signals, what survives the dodos are piles of bones and descriptive entries in journals—scraps that do little to demonstrate that the birds possessed the capacity to persuade, but nonetheless find themselves on display in museums of natural history across the globe. This chapter suggests that rhetoricians can bypass anthropocentric worldviews that deny the dodos their due rhetoricity by simply rejecting mechanist and correlationist portrayals of nonhumans as popularized by philosophers like René Descartes and Immanuel Kant in favor of outlooks that unapologetically grant that the birds sought to persuade Dutch sailors to leave them alone. That is, rhetoricians can begin their theorizations and criticisms from the point of view that the dodos, like all nonhumans, possessed the capacity to signal and hence produced texts to be read by others; thus, when crews failed to notice those signals, certain rhetorical effects ensued.

Henceforth, this chapter speculates that the dodos did not remain aloof during the onset of catastrophic violence brought on by Dutch sailors, and what likely happened was that crews failed to observe and thus react to the birds' pleas for autonomy and survival. Consequently, this missed communication between the dodos and sailors—a failed alignment in signals disseminated between nonhuman rhetors and human audiences—initiated an ongoing relationship between the birds and humans whereby the self-proclaimed superior species embodied what I call *the sixth persona*: a human audience who failed to recognize and respond to nonhuman rhetoric. As my analysis will reveal, tourists and museums of natural history perform a readily identifiable contemporary strand of the sixth persona, the audience responsible for further perpetuating a series of violent rhetorical effects regarding the dodos.

Already conceptualizing the way rhetoricians understand the human in the study of persuasion is a host of theories—known as *rhetorical personae*—that characterize the relationship(s) between rhetors and audiences. Beginning with Edwin Black's naming of rhetors as *the first persona* and their implied audiences as *the second persona*,[15] discussions about the formation and implication of rhetorical personae have been generative for rhetoricians.[16] Take, for example, Philip Wander's and Dana L. Cloud's respective theorizations of *the third persona* and *the null persona*. Whereas Wander described *the third persona* as the audience negated in a rhetorical situation,[17] Cloud framed *the null persona* as the negation of the rhetor by the rhetor.[18] Together, these theories prompted rhetoricians to (re)consider the ways in which various systems of power determined who talked, who could talk, and about what they talked. Given the institutionalized link between persuasion and power in rhetorical studies, stabilizing the assorted rhetorical personae in various rhetorical situations, even if only briefly, helped rhetoricians to better understand *how* and *to what effect* intersecting identities (race, class, gender, sexuality, etc.) converged at the site of communicative transmissions among humans.

In line with the other rhetorical personae, the existence of the sixth persona posits a return to the most essential feature of any given rhetorical situation: the relationship between rhetors and audiences within the context of internatural communication. Here nonhumans peck, tweet, sniff, and bite at humans, each referential of a process of signaling that denaturalizes "the human as the unchallenged center of rhetoric."[19] Defined against the backdrop of Seegert's definition of rhetoric as "the relational force of signals interacting in the world,"[20] the existence of the sixth persona presumes that nonhumans disseminate signals imbued with meaning, but that humans—whether by choice or sensory in/ability—fail to pick up on those texts. In turn, nonhumans fail to communicate internaturally and thus to persuade humans to behave in ways desired by said nonhumans, an outcome whereby missed communication spins into existence a series of undesirable rhetorical effects.

In the case of the dodos, if rhetoricians speculate that the birds produced signals aimed at curtailing conquest, predation, and habitat loss, then what emerges is a situation in which scholars can observe that Dutch sailors overlooked—*they missed*—texts produced by the dodos that might have prevented their domination and extinction. Through presumption, whether the birds possessed the capacity to persuade becomes an irrelevant inquiry; what matters becomes the rhetorical effects of missed communication. Because those rhetorical effects are plentiful, the forthcoming critical vignettes will focus on how missed communication and its constructed sixth persona transformed the dodos from autonomous beings into zoographic strands of what Barbara Kirshenblatt-Gimblett detailed as "ethnographic artifacts," or artifacts that performatively became artifacts "by virtue of being defined, segmented, detached, and carried away" from their originating location.[21] Forever trafficked, the dodos find themselves far from extinct as they perform as ethnographic artifacts for tourists of museums of natural history.

As such, the objective of this chapter is not to definitively prove that dodos disseminated signals intended to persuade Dutch sailors to leave them alone. Such a task is an unfair request for any rhetorician, as the process of piecing together texts is always incomplete and relies on a fallible critic to build context. Instead, the goal is to signal a high-pitched chirp pleading for rhetoricians to begin their theories and criticisms from the position that nonhumans possess the capacity to persuade, but that humans miss the fine textures of nonhuman rhetoric.

In what follows, I espouse three rhetorical effects of the sixth persona—*of humans failing to recognize and respond to signal-based texts produced by the dodos*—by recreating the scenes in which I toured three museums of natural history that display the remnants of the birds. In the tradition of rhetoricians who have aptly taken up the study of museums,[22] I "exerted physical and temporal effort" to make myself present at the following three museums within a one-year span of time:[23] the Zoological Museum in Copenhagen, Denmark; the Oxford University Museum of Natural History in Oxford, England; and the American Museum of Natural History in Manhattan, New York. There I acted as an "in situ rhetorical critic" by embodying the role of tourist-rhetorician to engage with the dodos on display.[24] Reflecting on my experience as a tourist-rhetorician, I offer thick descriptions of each institution and proceed by oscillating between moments *on tour* and moments of critical reflection *at home* to showcase the rhetorical effects of the sixth persona as they emerge through museums.[25]

TOURING THE SIXTH PERSONA

On Tour: Copenhagen, Denmark

With cellphones in hand, my teenage brother and I locate free Wi-Fi as we approach a freight elevator that doubles as a lift for guests of the

Zoological Museum in Copenhagen. This oversized elevator, a product of human ingenuity, triggers a productive iteration of dehumanization by reminding us of our shared connection with the dodos: *We are flightless.*

Taking advantage of our unfettered access to social media, we burst through the sliding metal doors of the elevator and immediately begin uploading images and videos of the museum's vast collection of artifacts. In our search for the dodo, we come across several nonhumans that differentiate this museum of natural history from others, including a rare spider called "The Lonely Spider," the fossilized remains of a dinosaur named "Misty," and a majestic narwhal with unicorn-like tusk still intact. We unabashedly pause for selfies in front of each artifact.

"It's much less packed in here than other museums," I comment to my brother. As we slink through the institution, we encounter only a handful of other tourists. No more than fifty guests seem to frequent the entire museum.

"Over here," bellows a voice from an undisclosed location. By happenstance, our zealous mother has stumbled upon the dodo. We follow the echoes of her voice around a corner only to find her standing in front of the museum's prized display, "The Dodo Skull."

"The Dodo Skull" showcases a single, off-white-colored cranium separated from tourists by an unbreakable glass barrier (see Figure 10.1). With the descriptive textures of skin long decomposed, an elongated beak announces

Figure 10.1 The Dodo Skull.

that the exhibited artifact is of avian origin as a nearby didactic further specifies that herein lies "*Raphus cucullatus*," the scientific name for the dodos.

"This is the dodo. Obviously, it's just her head," I record for a video that I share with friends witnessing my travels unfold via social media. An electric-like current runs through my body as the experience of touring extinction suddenly becomes all too real.

"Look at this keypad. I don't see other displays with keypads. This is so weird," I remark. "They keep her locked in a high security display—*like a bank vault!*" Unlike other displays, curators rely on advanced technologies to protect the dodo. Whereas my brother and I could touch Misty, we could not rub our fingers against the equally extinct dodo.

To test the thickness of the display, I knock on the dense glass and surrounding metal framing and proclaim with certainty, "No one is breaking into this sucker!"

As I survey the display, I make note of a feature that warrants contemplation. Catching my attention are the bird's eye sockets. These holes do not appear to give way to a forward-leaning line of sight, a fact that corroborates scientific claims that dodos did not prey on other nonhumans.[26]

"This dodo looks just like the dodo on *American Dad*!" squeals my brother as he eyes the hollow skull. In true millennial fashion, he dials up the episode of the popular adult cartoon in which Stan Smith, a case officer at the Central Intelligence Agency, brings home from work a cloned dodo.

"Ah, the dodo. *Raphus cucullatus*," laments Stan to his son. "These magnificent creatures would still be around today if not for the blood thirsty Dutch hunting them to extinction in 1662."[27] As Stan delivers this monologue, a cartoon dodo hops onto the kitchen counter, rolls around in olive oil and flour, and stumbles into a hot frying pan.

Although captivated by the cartoon, we sense an approaching group of tourists and quickly regain a respectful composure as they draw near. We turn our attention toward a lengthy didactic housed in the display. Rather than a panoramic backdrop of Mauritius that would relocate, albeit remotely, the skull of the dodo back to her rightful final resting place, composing the scene of the display is a charcoal-gray panel that reads as follows:

> The skull of the extinct bird, the dodo, is one of the museum's greatest treasures. The only other existing specimen in the world is found in Oxford, England.
>
> The dodo was first described in 1501, when Portuguese mariners landed on the island of Mauritius—the only place the species ever lived. They described it as big and clumsy and noted that the meat was not very tasty. Many years later, in 1843, Johannes Theodor Reinhardt, from the Royal Museum of Natural History, declared that the lumbering creature was some kind of pigeon. Although met with great scepticisim at the time, his view is now acknowledged as being correct. The closet living relative of the dodo is the Nicobar pigeon.

> The story of the dodo skull and its long journey from Mauritius to Copenhagen is shrouded in mystery. It may have been part of the Dutch doctor Bernardus Paludanus' collection of curiosities and specimens from all over the world, many of which ended up at Gottorp Castle in Schleswig-Holstein around 1650. When the palace and all its treasures fell to Frederik VI during the Great Northern War, the dodo skull was among the booty, and became part of the Royal Chamber of Curiosities. When the chamber closed in 1825, the dodo was transferred to the Royal Museum of Natural History.

The word "booty," terminology that denotes stolen goods, catches my eye. Museums of natural history are places of abduction. Not only do they kidnap and kill nonhumans, but also they take possession of other institution's artifacts. The dodo in front of me, here reduced to nothing more than a floating head, experienced an unfortunate iteration of trafficking in both her life and afterlife.

Feeling an odd mixture of awe and sadness, we say goodbye to the dodo. We move toward the elevator and do something that the dodos, dead but not gone, once did before Dutch sailors colonized Mauritius: *We venture outdoors without fear of extinction.*

At Home: Reflections on the Zoological Museum

The Dutch colonization of Mauritius transformed Copenhagen, a city located more than 5000 miles away from the Indian Ocean, into the unlikely final resting place for one of only a few dodos currently possessed by humans. Speaking to the powers of being recognizable or obtaining what Pierre Bourdieu named "symbolic capital,"[28] tourists from around the globe traveled from their homelands to the Zoological Museum to witness firsthand the authentic remains of the birds. In turn, tourists embodied the collective ethos of Dutch sailors, the group that originated the sixth persona. More specifically, when crews failed to recognize and respond to the dodos' pleas for autonomy and survival, this missed communication initiated an ongoing relationship in which humans dominated the birds by way of trafficking them for purposes of human pleasure.

Almost immediately after Vice Admiral Wybrand van Warwijck of the Netherlands ordered his crew to abandon their predetermined course and sail to Mauritius, Dutch sailors began *bothering* the dodos. Consumption, perhaps, was the first type of pain inflicted on the birds for purpose of human pleasure. Longing for some time on land and a meal, the crew sunk their teeth into the plump dodos. "[W]e also found large birds, with wings as large as those of a pigeon, so that they could not fly," described a steersman named Heyndrick Dirckz Jolinck. "These particular birds have a stomach so large that it could provide two men with a tasty meal and was actually the most delicious part of the bird."[29] Fortunately for the birds, such practices of consumption were short-lived as the dodos tasted gamey and sailors, as the rules of colonization would have it, had jobs to perform.

As happens when a nation locates unclaimed territory, Dutch sailors needed to establish a channel between the Netherlands and Mauritius. For the next one hundred years, crews sailed back and forth between the two locales often accompanied by trafficked dodos. Sailors captured the birds, carried their protesting bodies aboard ships, and transported them to territories never meant to be seen by the flightless dodos. They did so in hopes that others—from dignitaries to common folk—would have the pleasure of tangling with the exotic birds. The dodo on display at the Zoological Museum, as the didactic noted, experienced a "long journey from Mauritius to Copenhagen," one that likely included pit stops in the Netherlands and Germany before finally roosting in Denmark. Each of these stops—brief pauses along "acquisition routes" for nonhumans[30]—spoke to the pleasure of capturing and owning the birds, whether kept in cabinets of curiosity, private collections, or museums.

Such trafficking did not end with the extinction of the dodos in the latter half of the seventeenth century. As years passed, the birds gained clout and became celebrities in their own right—an avian collective that indexed one of the first times in history that humans realized they could eliminate an entire species of nonhumans from existence.[31] With that came the curatorial practice of displaying remnants of the birds and the touristy desire to see and witness firsthand the bones of the dead and gone dodos. As my performance as a tourist of the Zoological Museum highlighted, the birds found themselves trafficked in a way inconceivable by those who originally abducted the dodos. Far from their homeland, museums showed remains of the dodos to tourists who gazed upon the likes of skinless craniums before dispersing photographs—*images taken without permission*—to audiences of friends and family around the world. Whereas this practice did not appear harmful, it was behavior that ignored the birds' pleas for autonomy and survival by perpetuating an understanding of the dodos, as the Zoological Museum described, as "booty."

On Tour: Oxford, England

A near two-hour commute separates the British Museum in London from the Oxford University Museum of Natural History in Oxford. Emblematic of English transportation, a double-decker bus arrives just before lunch to transport me between the two cities. I board the vessel, the Oxford Tube, and rest my eyes.

I awake in Oxford, a quaint town home to cobblestone streets and one of the most famous universities in the world. Although endless rows of whimsical edifices surround me, what enchants me most are the thousands of tourists traversing the uneven pavements of this historic town.

As I venture a short distance from the bus station to the museum, it becomes more and more apparent that the dodos are Oxford icons. I pass a myriad of gift shops that sell, in addition to Oxford University shirts, merchandise featuring the image of the birds, including key chains, tote bags, and playing cards. Wanting to get my hands on a souvenir, I meander through several gift shops in search of the perfect trinket, but leave empty handed in favor of purchasing a keepsake from the museum itself.

Although the architectural features of the buildings in Oxford change little as I approach the museum, I know that I have arrived when I see a sign featuring an outline of a dodo. Like the Zoological Museum, the Oxford Museum thinks of their dodo as one of their "greatest treasures."[32] From advertisements to merchandise to brown paper sacks that accompany any purchase in the museum gift shop, the bones of their bird are far more than artifacts on display: They are mascots!

"Where is the dodo?" I ask an employee as I walk through the old wooden doors separating the exterior and interior of the museum. He points toward a glass case conveniently located less than twenty feet from my rear end. Whereas the Zoological Museum hid their booty amidst a plethora of closely positioned displays, this museum clearly makes better use of their so-called "trophy" by displaying her remains near the entrance of the institution.

Contained in a display aptly titled "The Oxford Dodo" are the corporeal fragments of the flightless entities that continue to tangle with humans, albeit in a far less mortal state (see Figure 10.2.). These priceless scraps—much of which the museum derived from a single dodo—include a foot, leg bone, and skin-covered cranium. To that end, a small didactic tells me that elsewhere in safekeeping "is a small amount of skin from the foot."

Cold chills take over my body. I can't believe my eyes. I can see her hair follicles!

This cranium—covered in dark, discolored skin—lays ear-down upon a small crown pillow. Because previous owners of this dodo did not protect her remains from fleas and other scavengers, the now extinct bird appears to tourists as a mistreated, bruised, and tattered victim of domestic violence.

Due to the detailed textures of the cranium, I feel underwhelmed by the nearby foot and leg bone. Maybe it's the fact that I can see the skin of this dodo's eyelid that draws my line of sight to her skull—a reminder that the dodo saw her abductors. Maybe it is the sheer presence of skin, an apparatus of life, that piques my interest. Whatever it may be, I feel less compelled to tend to this dodo's foot and leg, two artifices that communicate that the birds had the capacity to flee, yet failed to effectively do so.

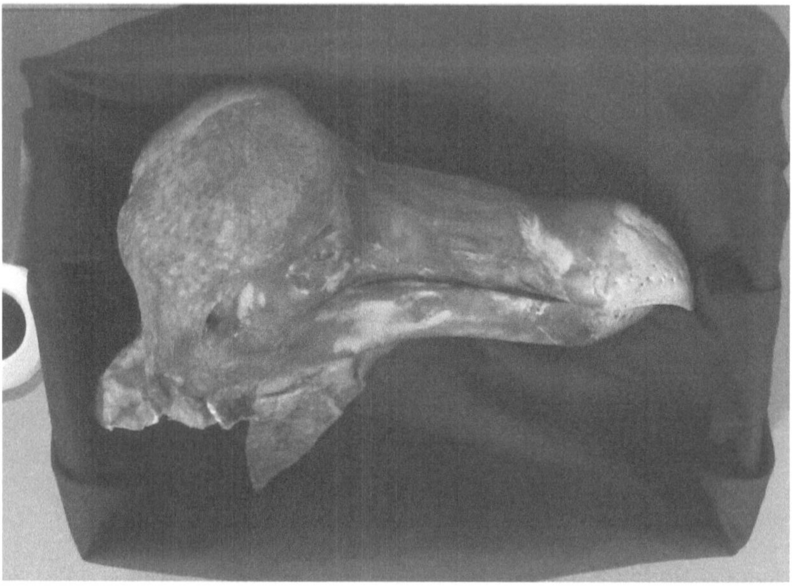

Figure 10.2 The Oxford Dodo.

As the awe subsides, I read a few of the display's didactics.

"This specimen has had its DNA analysed and it was recently CT scanned," reports a didactic. "The DNA analysis has identified that the closet living relative of the Dodo is the Nicobar pigeon (*Caloenas nicobarica*) from Southeast Asia. The CT scan revealed some interesting results, which are currently confidential."

This latter detail causes pause. The dodos are a species representative of the first time that humans realized their potential to eliminate nonhumans from existence.[33] The dodos have already told humans so much. It is inspiring to know that she has more to share in years to come.

Hungry for knowledge, I survey the display for more information about the dodos. Unlike other museums that house parts of the birds, this institution builds identification between tourists and the dodos by way of pop culture. Charles Lutwidge Dodgson, better known as Lewis Carroll, frequented the halls of this museum. The Dodo, a character known to be the self-insertion of Dodgson himself, appeared in *Alice's Adventures in Wonderland.* Situating artwork from this book against the dodo on display, the museum capitalizes on their former patron's claim to fame.

Whereas tourists around me take the bait and *ooh* and *ahh* about Dodgson, I find myself more invested in the display's lack of security than its celebration of the author and his artistically rendered dodo. What shocks me most about this display is not its silence about abduction and murder, two forms of violence that preceded the exhibition of this bird, but rather that the museum does little to protect its finite archive: the evidence that the so-called mythical dodos once roamed the beaches of Mauritius. Unlike the Zoological Museum, it seems that only thin glass and the fear of getting caught prevents me from grabbing ahold of these treasures and running toward the exit.

After browsing the vast collection of species on display, I stop at the museum gift shop and purchase a magnet, two postcards, and a book titled *The Oxford Dodo: The Sad Story of the Ungainly Bird that Became an Oxford Icon*. Due to rainy skies, I huddle under an archway. I take to social media to upload a selfie with the dodo on display at the museum.

"You weren't meant to be forgotten," I caption the photo.

At Home: Reflections on the Oxford Museum

The Dutch sailors who discovered the dodos were no strangers to commerce. When bad weather compromised their safety, crews were accompanying Admiral Jacob Cornelius van Neck on the Second Dutch Expedition to Indonesia. Given that their sole purpose for sailing was the spice trade, it should come as no surprise that when sailors stumbled upon Mauritius, among their goals was the assessment of *who* or *what* would be of financial worth to the Netherlands. In trafficking the birds, crews were not just interested in supplying other humans with pleasure. They were looking for *things* that might be of worth to their government. This practice of understanding the dodos within the context of economics did not go extinct with the birds, as the Oxford Museum demonstrated. When Dutch sailors failed to recognize and respond to the dodos' pleas for autonomy and survival, this missed communication initiated an ongoing relationship in which humans dominated the birds by categorizing them as *things* to be bought and sold.

As evidenced by the Zoological Museum and Oxford Museum, humans have routinely failed to regard the dodos as anything more than things. Whereas Dutch sailors suppressed any secretly harbored cannibalistic desires as they traveled between Indonesia and the Netherlands, they did not mind eating the birds. The incapacity to eat one another while simultaneously being able to eat the dodos pointed toward how the birds occupied a lower status in the hierarchy between humans and nonhumans. Trafficking only added fuel to this fire, as crews demonstrated a lack of consideration for how the birds, for example, might experience a type of diaspora or long to be among the dodos with whom they were familiar. To that end, how cabinets of curiosities, private museums, and, especially in the contemporary moment, museums of natural history came to possess a species of birds that once roamed an island at will indexed how the birds found themselves objectified by humans and labeled as *things* rather than beings worth of dignity.

The Oxford Museum, an internationally renowned tourist destination, merely amplified this process of objectification. Take, for instance, how the institution relied upon the image of the dodos to attract tourists to their facilities. Even before tourists arrived at the museum, the institution enticed them with their mascot via promotional material on their website. "The dodo is the most famous of all creatures to have become extinct in historical times," touted the website for the museum, "The remains of the dodo at Oxford are one of the greatest treasures of the Museum."[34] Upon arrival, the museum greeted tourists with signage and paperwork featuring a black outline of the birds. Of course, in a place like Oxford, tourist destinations must do something to set themselves apart from other sites. Nonetheless, like a sports franchise deploying Native Americans as their mascot, the Oxford Museum kept stable the downgraded status of the dodos as things rather than beings by using their image to attract tourists, a class of individuals best thought of within the context of economics as *customers*.

On that note, the use of the dodos as mascots—as tools for branding the museum—was not merely contained within the walls of the institution. As the presence of the gift shop indexed, tourists, myself included, wanted to collect souvenirs. The museum satiated that desire by selling objects featuring the birds, such as magnets, postcards, books, kitchenware, and toys. Much like Dutch sailors before them, tourists further trafficked the birds by purchasing souvenirs boasting their image. Perhaps materially harmless for an extinct species, this practice best symbolized how the dodos continued to exist as things to be bought and sold.

On Tour: Manhattan, New York

As the rain drizzles, I round the corner of West 81st Street and Central Park West and approach the main entrance of the American Museum of Natural History in Manhattan. Guarding the main entrance of the museum is a monument christened the *Equestrian Statue of Teddy Roosevelt*. Sculpted by James Earle Fraser, this bronze statue enshrines former American president and naturalist Theodore Roosevelt as he sits atop a horse. Next to Roosevelt are effigies of an African American and Native American, two figures who appear to guide him through the wilderness. The horse rears as the famed conservationist pulls taught the reins.

James W. Loewen, a historian critical of the rhetoric of display practiced by museums, argued that the *Equestrian Statue* signified the institution's longstanding commitment to white supremacy, an insidious orientation to nonwhite persons that manifested through the museum's many exhibitions. "Inside its doors," criticized Loewen, "the American Museum takes the same stance, putting American Indians and Africans closer to animals, whites furthest removed."[35] Speciesist language aside, Roosevelt does appear as a patriarch, or the master of all beings nonwhite and nonhuman.

At a quarter past one o'clock in the afternoon, I hurry past the troublesome monument and stand at the doors of the Bernard Family Hall of North American Mammals. The entrance to this exhibition doubles as the meeting point for the museum's free public tour, a seventy-five-minute guided excursion through a few of the institution's more noteworthy halls.

"So my name is Ginger," exclaims a voice like that of Fran Drescher. "I'm a tour guide at the museum of natural history!"

"[The museum] has thirty-three million objects in its collection, and less than one percent is on display," Ginger reports. "So every item that is on display is here for a reason. It has significance. There's a story underlying it. So what I'm going to do is share the highlights."

We proceed into the Bernard Family Hall, pausing briefly to talk about the Alaskan Brown Bear before making a beeline to the Hall of Biodiversity.

The Hall of Biodiversity is a multimillion-dollar exhibition that strategically reroutes apprehensions and anxieties about the sixth mass extinction event in natural history *into* displays and *onto* tourists. Publicized as an interactive exhibition detailing "the variety and interdependence of all living things,"[36] the Hall of Biodiversity seeks to extend discussions of extinction from the likes of scientists and policy makers to tourists and their families and friends to raise consciousness about our species' interconnectedness with the more-than-human world.

"We're in the Hall of Biodiversity," announces Ginger. In the dark exhibition, I immediately notice the brightest display in the hall, the "Spectrum of Life Wall." Themed as a cladogram, which is a scientific diagram illustrating relationships between/among various organisms, this massive display is essentially an oversized family tree that exhibits more than 1000 of the museum's collected nonhuman specimens.

Ginger, however, does not park us in front of the bright exhibit. Instead, we pause before a less noticeable display that features a gaggle of seemingly unrelated nonhumans. Billed as "Lives in the Balance: Endangered Species," this floor-to-ceiling display showcases more than a dozen nearly extinct nonhumans alongside a lengthy didactic that reads:

> Today, more species are declining and facing imminent extinction than at any time since the most recent great mass extinction took place 65 million years ago. In an effort to address this crisis, the U.S. Congress in 1973 passed the

> Endangered Species Act (ESA), giving protection to species it classified as either threatened or endangered. An ENDANGERED SPECIES is considered to be at the brink of extinction within the foreseeable future throughout most or all of its habitat range. A THREATENED SPECIES is one that is likely to become endangered in the near future. The ESA's lists of species are in addition to the Red List of Threatened Species, which has been compiled by the International Union for the Conservation of Nature since the 1960s. The animals shown here are a tiny fraction of the plants and animals known to be endangered. If action is not taken to preserve and protect these species and the places where they live, they will eventually become extinct, like the dodo.

Indeed, encased in the display is a complete casting of a dodo (see Figure 10.3).

"So I wanted to stop here, uh, because I mentioned that everything on display is here for a reason," Ginger teases. "Everything in this case is endangered. One exception: The dodo bird is extinct."

Ginger gestures for tourists to have a look for themselves. A nearby couple pauses for a selfie in front of the dodo while others read the grim didactic before following Ginger out of the exhibition.

Figure 10.3 Lives in Balance: Endangered Species.

As we enter the adjacent Paul and Irma Milstein Family Hall of Ocean Life, I ask Ginger, "At other natural history museums, they have the dodo off to the side by [herself]. Do you know why they chose to encase her with other endangered [nonhumans] even though she's the only extinct one?"

Ginger responds, "You know, that's a good question. I don't know if they have a good place for her."

I reveal with a gasp.

"She's also a casting, so . . ." I comment of the dodo's supposed inauthenticity.

Ginger interrupts, "Half is real!"

"Half is real? Ahh! I've been saying she's a casting," I reveal with a gasp. Ginger cuts this conversation short to tend to other tourists and I take solace in my thoughts. With only a few authentic bones of the dodos left in the world, it is always a humbling experience to stand in front of the flightless bird of Mauritius.

At Home: Reflections on the American Museum

Capturing and dispersing the dodos beyond the beaches of Mauritius satiated more than the human desire for pleasure and profit. When the Netherlands claimed Mauritius, extinction was not a readily accepted phenomenon as religious doctrine, namely the Great Chain of Being, dictated that that "every species in existence was made in a single creation event."[37] Consequently, the potential for species eradication evaded the minds of wandering crews as they took copious notes about the feathered figures of their newly discovered island. Little did sailors know, as they jotted down informative descriptions of the birds to be shared with those back at home, they began the longstanding tradition of portraying the dodos as beacons of knowledge. When Dutch sailors failed to recognize and respond to the dodos' pleas for autonomy and survival, the missed communication initiated an ongoing relationship in which humans dominated the birds by way of trafficking them for purposes of knowledge production.

In their own way, all three of the museums represented their trafficked dodos on display as things that produced knowledge. First was the Zoological Museum, an institution that historicized the birds by contextualizing the struggle of their "booty." Next was the Oxford Museum, an institution that discussed the dodos' rise to iconicity through didactics that referenced their contribution to literature and modern science. Lastly was the American Museum, an institution that framed the lost species as advocates against the sixth mass extinction in natural history. Despite each museum's individual stake in the perseveration and public pedagogy of natural history, emergent at each tourist destination was the use of the birds as knowledge-based artifices of history, culture, science, and public policy.

What differentiated the American Museum from the Zoological Museum and the Oxford Museum was that the stateside institution offered a frank description—a main teaching objective—of why curators displayed remnants of the dodos. In 1997, the American Museum hosted an international symposium titled "Humans and Other Catastrophes" where scholars representing an array of disciplinary backgrounds gathered to investigate how humans greatly accelerated the extinction of nonhumans. Together, they affirmed that extinction was occurring "at one hundred times the rate predicted by the fossil record,"[38] a phenomenon that meant humans were responsible for implementing measures to significantly reduce these rates for the sake of biodiversity. The Hall of Biodiversity represented one such measure as the institution sought to condemn human inconsideration of the more-than-human world by ascribing "collective blame" onto tourists through the exhibition.[39]

Cashing in on the idiom "go the way of the dodo," the American Museum deployed the reconstructed skeleton of their dodo as an advocate against the woes of the Anthropocene, namely the Holocene Extinction. As the didactic next to the case of endangered nonhumans read, "If action is not taken to preserve and protect these species and the places where they live [from humans], they will eventually become extinct, like the dodo." In this way, the institution used the birds as advocates—as producers of knowledge about extinction—by deploying their dodo as the figure in a cautionary tale against what ecological philosopher Timothy Morton referred to as *agrilogistics*, or the 10,000-year "time of certain logistics of agriculture that arose in the Fertile Crescent and went viral, eventually requiring steam engines and industry to endure."[40] In doing so, the museum further cemented the ongoing process of humans trafficking the dodos for purposes of knowledge production about complex topics, including the mass extinction of *all* species.

TOWARD AN EXPANDED UNIVERSE OF SPECULATION

Presuming rather than denying that the dodos produced signals aimed at curtailing conquest, predation, and habitat loss, this chapter offered an ethnographic account of three museums of natural history to showcase three rhetorical effects of the sixth persona: the human audience who failed to recognize and respond to nonhuman rhetoric. This chapter argued that when Dutch sailors failed to observe signals produced by the dodos pleading for their autonomy and survival, this missed communication between crews and the birds established the conditions whereby sailors embodied the sixth persona and thus trafficked the dodos for purposes of pleasure, profit, and knowledge. As encounters between tourists and the dodos on display at museums detailed, these rhetorical effects of the sixth persona—of the audience created through missed communication—survived into the twenty-first century, a time marked by a noteworthy loss of the dodos due to extinction.

In the introduction of their overview of rhetorical studies, Sonja K. Foss, Karen A. Foss, and Robert Trapp summarized the prevailing disposition shared by rhetoricians about the study of nonhuman rhetoric: "Although many of the theorists whose ideas we will discuss in this book argue that symbols are uniquely human, we believe that the debate over whether animals use symbols is unresolved and perhaps unresolvable."[41] For rhetoricians invested in perpetuating a belief in human exceptionalism, this claim is far from unreasonable; understanding humans as the unique wielder of rhetoric as symbolic action has long been the dominant paradigm in rhetorical studies, a practice that has successfully guided rhetoricians in their pursuit to better assess communication among humans. Unfortunately, for rhetoricians invested in chipping away at anthropocentric logics, this claim dismisses nonhumans with the simple stroke of a pen, a dogmatic behavior that overtly ignores the many attempts by rhetoricians over the past thirty years to conceptualize an understanding of nonhumans within the study of persuasion. As this chapter suggested, rhetoricians ought not to quibble over *if* nonhumans possess the capacity to persuade, but should instead presume that nonhuman rhetoric abounds around humans who may or may not notice signals. Doing so allows rhetoricians to reopen foreclosed debates about nonhuman rhetoric by assessing the rhetorical effects of internatural communication, a possible avenue for future studies that would likely alter the ways scholars understand the human.

To answer the question that guided this chapter, if rhetoricians were to presume rather than deny that nonhumans possess the capacity to persuade, then humans as a species would cease to be exceptional insofar as rhetoric is concerned. Whereas humans would wield the symbol, nonhumans would wield the signal. Separate and together, humans and nonhumans would produce and disseminate texts to be read by others—within and across species—and rhetoricians could tend to such communication as particular case studies. Perhaps more importantly, if rhetoricians were to reject mechanistic and correlationist views of nonhumans, then humans would no longer be thought of as the primary creators of texts as *all* species would have a stake in the construction of reality. In this way, humans as a species would become a variety of audiences, an endless number of rhetorical personae like that of the sixth persona. Rhetoricians could then tend to specific instances of internatural communication and decipher the rhetorical effects emergent where species meet.

NOTES

1. Anthony Cheke and Julian P. Hume, *Lost Land of the Dodo: The Ecological History of Mauritius, Réunion, and Rodrigues* (New Haven: Yale University Press, 2008), 21–2.

2. Anwar, Janoo, "Discovery of Isolated Dodo Bones [*Raphus cucullatus* (L.), Aves, Columbiformes] from Mauritius Cave Shelters Highlight Human Predation, with a Comment on the Status of the Family Raphidae Wetmore, 1930," *Annales de Paléontologie* 91 (2005): 167–80.

3. Janoo, "Discovery," 169.

4. Cheke and Hume, *Lost Land*, 6–7.

5. Emily Anthes, "The Smart, Agile, and Completely Underrated Dodo," *The Atlantic*, June 8, 2016, accessed January 4, 2017. http://www.theatlantic.com/science/archive/2016/06/the-dodos-redemption/486086/

6. Phaedra C. Pezzullo, "Unearthing the Marvelous: Environmental Imprints on Rhetorical Criticism," *Review of Communication* 16 (2016): 37.

7. Natasha Seegert, "Play of Sniffication: Coyotes Sing in the Margins," *Philosophy & Rhetoric* 47 (2014): 158–78. In this way, Seegert rejected what Stephen J. Lind referred to as the "industry standard" definition of rhetoric as symbolic action. Stephen J. Lind, "Un-Defining Man: The Case for Symbolic Animal Communication," in *Perspectives on Human–Animal Communication: Internatural Communication*, ed. Emily Plec (New York: Routledge, 2013), 163.

8. Seegert, "Play," 163–4.

9. Lynn Worsham, "Toward an Understanding of Human Violence: Cultural Studies, Animal Studies, and the Promise of Posthumanism," *Review of Education, Pedagogy, and Cultural Studies* 35 (2013): 52.

10. Seegert, "Play," 159.

11. This type of rhetorical inquiry ought to be catalogued under the rubric of what Emily Plec called the study of "internatural communication." Emily Plec, "Perspectives on Human–Animal Communication: An Introduction," in *Perspectives on Human–Animal Communication: Internatural Communication*, ed. Emily Plec (New York: Routledge, 2013): 5.

12. See, for example, Diane Davis, "Creaturely Rhetorics," *Philosophy & Rhetoric* 44 (2011): 88–94; Debra Hawhee, "Toward a Beastial Rhetoric," *Philosophy & Rhetoric* 44 (2011): 81–7; George Kennedy, "A Hoot in the Dark: The Evolution of Generic Rhetoric," *Philosophy & Rhetoric* 25 (1992): 1–21.

13. Seegert, "Play," 164.

14. Davis, "Creaturely," 89.

15. Edwin Black, "The Second Persona," *Quarterly Journal of Speech* 56 (1970): 109–19.

16. See, for example, Dana L. Cloud, "The Null Persona: Race and the Rhetoric of Silence in the Uprising of '34," *Rhetoric & Public Affairs* 2 (1999): 177–209; Robin E. Jensen et al., "Theorizing the Transcendent Persona: Amelia Earhart's Vision in *The Fun of It*," *Communication Theory* 20 (2010): 1–20; Charles E. Morris III, "Pink Herring & The Forth Persona: J. Edgar Hoover's Sex Crime Panic," *Quarterly Journal*

of Speech 88 (2002): 228–44; Philip Wander, "The Third Persona: An Ideological Turn in Rhetorical Theory," in *Contemporary Rhetorical Theory: A Reader*, eds. John Lucaites, Celeste Michelle Condit, and Sally Caudill (New York: Guilford Press, 1999), 357–70.

17. Wander, "The Third Persona," 376.

18. Cloud, "The Null Persona," 179.

19. Natasha Seegert, "Rewilding Rhetoric with Animate Others," *Review of Communication* 16 (2016): 77.

20. Seegert, "Play," 163–64.

21. Barbara Kirshenblatt-Gimblett, *Destination Culture: Tourisms, Museums, and Heritage* (Berkeley: University of California Press, 1998), 17–18.

22. See, for example, Eric Aoki, Greg Dickinson, and Brian L. Ott, "The Master Naturalist Imagined: Directed Movement and Simulations at the Draper Museum of Natural History," in *Places of Public Memory: The Rhetoric of Museums and Memorials*, eds. Greg Dickinson, Carole Blair, and Brian L. Ott (Tuscaloosa: University of Alabama Press, 2010), 238–65; Carole Blair, Greg Dickinson, and Brian L. Ott, "Introduction: Rhetoric/Memory/Place," in *Places of Public Memory: The Rhetoric of Museums and Memorials*, eds. Greg Dickinson, Carole Blair, and Brian L. Ott (Tuscaloosa: University of Alabama Press, 2010), 1–54; Aaron T. Phillips, "Eliding Extraction, Embracing Novelty: The Spatio-Temporal Configuration of Natural History," *Environmental Communication* 8 (2014): 452–67.

23. Carole Blair, "Reflections on Criticism and Bodies: Parables from Public Places," *Western Journal of Communication* 65 (2001): 274.

24. Danielle Endres, Aaron Hess, Samantha Senda-Cook, and Michael K. Middleton, "*In Situ* Rhetoric: Intersections Between Qualitative Inquiry, Fieldwork, and Rhetoric," *Cultural Studies <=> Critical Methodologies* 16 (2016): 512. See also Sarah L. McKinnon, Robert Asen, Karma R. Chávez, and Robert Glenn Howard, "Introduction: Articulating Text and Field in the Nodes of Rhetorical Scholarship," in *Text + Field: Innovations in Rhetorical Field Methods* (University Park, PA: The Penn State University Press), 1–21.

25. Moments marked *on tour*, then, function akin to what Michael Middleton, Aaron Hess, Danielle Endres, and Samantha Senda-Cook characterized as "personal vignettes," or "moments of insight from conducting this work including unexpected outcomes, striking interactions with participants, and the lessons learned from placing our bodies in the field." Michael Middleton, Aaron Hess, Danielle Endres, and Samantha-Senda Cook, *Participatory Critical Rhetoric: Theoretical and Methodological Foundations for Studying Rhetoric In Situ* (Lanham: Lexington Books, 2015), xx.

26. Cheke and Hume, *Lost Land*, 10.

27. "Steve & Snot's Test-Tubular Adventure," *American Dad*, Season 9, Episode 1, directed by Joe Daniello (2013; Los Angeles: Twentieth Century Fox Film Corporation, 2015), DVD.

28. Pierre Bourdieu, "Social Space and Symbolic Power," *Sociological Theory* 7 (1989): 17.

29. Perry J. Moore, *A Concise History of Dutch Mauritius, 1598–1710: A Fruitful and Healthy Land* (New York: Kegan Paul International, 1998), 12.

30. Samuel J. M. M. Alberti, "Introduction: The Dead Ark," in *The Afterlives of Animals: A Museum Menagerie*, ed. Samuel J. M. M. Alberti (Charlottesville: University of Virginia Press, 2011), 4.

31. Hugh Edwin Strickland and Alexander Gordon Melville, *The Dodo and Its Kindred* (London: Reeve, Benham, and Reeve, 1848), 7–45.

32. "Learning more…," *Oxford University Museum of Natural History*, accessed January 20, 2016, http://www.oum.ox.ac.uk/learning/pdfs/dodo.pdf.

33. Stickland and Melville, *The Dodo*, 4.

34. "Learning more…"

35. James W. Loewen, *Lies Across America: What Our Historic Sites Get Wrong* (New York: Touchstone, 1999), 31.

36. "Hall of Biodiversity Educator's Guide," *American Museum of Natural History*, accessed July 18, 2015, http://www.amnh.org/our-research/center-for-biodiversity-conservation/resources-and-publications/for-educators/exhibition-guides/hall-of-biodiversity-educator-s-guide/.

37. Mike Viney, "The Nature of Fossils: A Historical Perspective Part 1 of 5," accessed March 18, 2016, http://petrifiedwoodmuseum.org/FossilsPart1.htm.

38. "Overview," *American Museum of Natural History*, accessed June 11, 2015, https://www.amnh.org/science/biodiversity/extinction/Intro/Symposium.html.

39. Torin Monahan, "Please Increase Your Browser's Memory Partition: Individual Action and Collective Blame in the Hall of Biodiversity," *Museum Anthropology* 25 (2001): 42.

40. Timothy Morton, "How I Learned to Stop Worrying and Love the Term *Anthropocene*," *Cambridge Journal of Postcolonial Literary Inquiries* 1 (2014): 259.

41. Sonja K. Foss, Karen A. Foss, and Robert Trapp, *Contemporary Perspectives on Rhetoric: 30th Anniversary Edition* (Long Grove: Waveland Press, Inc., 2014), 2.

BIBLIOGRAPHY

Alberti, Samuel J. M. M. "Introduction: The Dead Ark." In *The Afterlives of Animals: A Museum Menagerie*, edited by Samuel J. M. M. Alberti, 1–16. Charlottesville: University of Virginia Press, 2011.

American Museum of Natural History. "Hall of Biodiversity Educator's Guide." Accessed July 18, 2015. http://www.amnh.org/our-research/center-for-biodiversity-conservation/resources-and-publications/for-educators/exhibition-guides/hall-of-biodiversity-educator-s-guide/.

———. "Overview." Accessed June 11, 2015. https://www.amnh.org/science/biodiversity/extinction/Intro/Symposium.html.

Anthes, Emily. "The Smart, Agile, and Completely Underrated Dodo." *The Atlantic*, June 8, 2016.

Aoki, Eric, Greg Dickinson, and Brian L. Ott. "The Master Naturalist Imagined: Directed Movement and Simulations at the Draper Museum of Natural History." In *Places of Public Memory: The Rhetoric of Museums and Memorials*, edited by

Greg Dickinson, Carole Blair, and Brian L. Ott, 238–265. Tuscaloosa: University of Alabama Press, 2010.

Black, Edwin. "The Second Persona." *Quarterly Journal of Speech* 56 (1970): 109–119.

Blair, Carole. "Reflections on Criticism and Bodies: Parables from Public Places." *Western Journal of Communication* 65 (2001): 271–294.

Blair, Carole, Greg Dickinson, and Brian L. Ott. "Introduction: Rhetoric/Memory/Place." In *Places of Public Memory: The Rhetoric of Museums and Memorials*, edited by Greg Dickinson, Carole Blair, and Brian L. Ott, 1–54. Tuscaloosa: University of Alabama Press, 2010.

Bourdieu, Pierre. "Social Space and Symbolic Power." *Sociological Theory* 7 (1989): 14–25.

Cheke, Anthony, and Julian P. Hume. *Lost Land of the Dodo: The Ecological History of Mauritius, Réunion, and Rodrigues*. New Haven: Yale University Press, 2008.

Cloud, Dana L. "The Null Persona: Race and the Rhetoric of Silence in the Uprising of '34." *Rhetoric & Public Affairs* 2 (1999): 177–209.

Davis, Diane. "Creaturely Rhetorics." *Philosophy & Rhetoric* 44 (2011): 88–94.

Endres, Danielle, Aaron Hess, Samantha Senda-Cook, and Michael K. Middleton. "*In Situ* Rhetoric: Intersections between Qualitative Inquiry, Fieldwork, and Rhetoric." *Cultural Studies <=> Critical Methodologies* 16 (2016): 511–524.

Foss, Sonja K., Karen A. Foss, and Robert Trapp. *Contemporary Perspectives on Rhetoric*: 30th Anniversary Edition. Long Grove: Waveland Press, Inc., 2014.

Hawhee, Debra. "Toward a Bestial Rhetoric." *Philosophy & Rhetoric* 44 (2011): 81–87.

Janoo, Anwar. "Discovery of Isolated Dodo Bones [*Raphus cucullatus* (L.), Aves, Columbiformes] from Mauritius Cave Shelters Highlight Human Predation, with a Comment on the Status of the Family Raphidae Wetmore, 1930." *Annales de Paléontologie* 91 (2005): 167–180.

Jensen, Robin E., Erin F. Doss, Claudia L. Janssen, and Sherrema A. Bower. "Theorizing the Transcendent Persona: Amelia Earhart's Vision in *The Fun of It*." *Communication Theory* 20 (2010): 1–20.

Kirshenblatt-Gimblett, Barbara. *Destination Culture: Tourism, Museums, and Heritage*. Berkeley: University of California Press, 1998.

Kennedy, George. "A Hoot in the Dark: The Evolution of Generic Rhetoric." *Philosophy & Rhetoric* 25 (1992): 1–21.

Lind, Stephen J. "Un-Defining Man: The Case for Symbolic Animal Communication." In *Perspectives on Human-Animal Communication: Internatural Communication*, edited by Emily Plec, 226–244. New York: Routledge, 2013.

Loewen, James W. *Lies Across America: What Our Historic Sites Get Wrong*. New York: Touchstone, 1999.

McKinnon, Sarah L., Robert Asen, Karma R. Chávez, and Robert Glenn Howard. "Introduction: Articulating Text and Field in the Nodes of Rhetorical Scholarship." In *Text + Field: Innovations in Rhetorical Field Methods*, edited by Sarah L. McKinnon, Robert Asen, Karma R. Chávez, and Robert Glenn Howard, 1–21. University Park, PA: The Penn State University Press, 2016.

Middleton, Michael, Aaron Hess, Danielle Endres, and Samantha Senda-Cook. *Participatory Critical Rhetoric: Theoretical and Methodological Foundations for Studying Rhetoric In Situ*. Lanham: Lexington Books, 2015.

Monahan, Torin. "Please Increase Your Browser's Memory Partition: Individual Action and Collective Blame in the Hall of Biodiversity." *Museum Anthropology* 25 (2001): 41–45.

Moore, Perry J. *A Concise History of Dutch Mauritius, 1598–1710: A Fruitful and Healthy Land*. New York: Kegan Paul International, 1998.

Morris III, Charles E. "Pink Herring & The Forth Persona: J. Edgar Hoover's Sex Crime Panic." *Quarterly Journal of Speech* 88 (2002): 228–244.

Morton, Timothy. "How I Learned to Stop Worrying and Love the Term *Anthropocene*." *Cambridge Journal of Postcolonial Literary Inquiries* 1 (2014): 247–264.

Oxford University Museum of Natural History. "Learning More…" Accessed January 20, 2016. http://www.oum.ox.ac.uk/learning/pdfs/dodo.pdf.

Pezzullo, Phaedra C. "Unearthing the Marvelous: Environmental Imprints on Rhetorical Criticism." *Review of Communication* 16 (2016): 25–42.

Phillips, Aaron T. "Eliding Extraction, Embracing Novelty: The Spatio-Temporal Configuration of Natural History." *Environmental Communication* 8 (2014): 452–467.

Plec, Emily. "Perspectives on Human–Animal Communication: An Introduction." In *Perspectives on Human–Animal Communication: Internatural Communication*, edited by Emily Plec, 1–13. New York: Routledge, 2013.

Seegert, Natasha. "Play of Sniffication: Coyotes Sing in the Margins." *Philosophy & Rhetoric* 47 (2014): 158–178.

———. "Rewilding Rhetoric with Animate Others." *Review of Communication* 16 (2016): 77–79.

"Steve & Snot's Test-Tubular Adventure." *American Dad*, Season 9, Episode 1, directed by Joe Daniello. 2013. Los Angeles: Twentieth Century Fox Film Corporation, 2015. DVD.

Strickland, Hugh Edwin, and Alexander Gordon Melville. *The Dodo and Its Kindred*. London: Reeve, Benham, and Reeve, 1848.

Viney, Mike. "The Nature of Fossils: A Historical Perspective Part 1 of 5." Accessed March 18, 2016. http://petrifiedwoodmuseum.org/FossilsPart1.htm.

Wander, Philip. "The Third Persona: An Ideological Turn in Rhetorical Theory." In *Contemporary Rhetorical Theory: A Reader*, edited by John Lucaites, Celeste Michelle Condit, and Sally Caudill, 357–370. New York: The Guilford Press, 1999.

Worsham, Lynn. "Toward an Understanding of Human Violence: Cultural Studies, Animal Studies, and the Promise of Posthumanism." *Review of Education, Pedagogy, and Cultural Studies* 35 (2013): 51–76.

Chapter 11

How Dogs (and Other Nonhuman Animals) Become Interesting

Marilyn M. Cooper

Can we talk with animals? Wittgenstein seems to say no, when he writes, "If a lion could talk, we could not understand him."[1] But I think he is mostly trying to say that a common language is not enough for understanding, that beings talking with one another must more importantly engage with one another. Just before he says this about lions, he comments, "one human being can be a complete enigma to another. We learn this when we come into a strange country with entirely strange traditions; and, what is more, even given a mastery of the country's language. We do not *understand* the people. (And not because of not knowing what they are saying to themselves)."[2] Vicki Hearne also points out that "Wittgenstein does not leap to say that his lion is languageless, only that he is not talking."[3] Hearne goes on to say, "there is a minor mistake of fact here—since lions do talk to some people, and are understood," and she adds, "there are lions and lions."[4] Like Hearne, Derrida draws attention to the plural, or the generic singular, in this question; he says "there is already a heterogeneous multiplicity of the living"; "one will never have the right to take animals to be the species of a kind that would be named The Animal, or animal in general."[5] Vinciane Despret focuses instead on the "we" in "can we talk with animals" and changes the question to "what makes a 'we' for lions?" She argues that "The question is not what *is* a lion, but 'how does one become a lion,' not only in the lion community, not only in the lion species, but also 'how does one become a lion' in the work of scientists constructing what it is to be a lion." As she says of Alex, the grey parrot who talked with Irene Pepperberg, "This is a question of becoming"; it "is not what parrots are but what they might be rendered capable of," especially when they are engaged in working with human animals who are interested in them and willing to try to understand them.[6]

There are lions and lions, and there are humans and humans. If, as Hearne says, most philosophers don't deign to talk with animals, especially not parrots, who, as she says, refuse to let you choose the topic, still many humans who live with animals, animal breeders, ethologists, and animal trainers and people who work with herding dogs, among others, do. How do they escape what Karen Barad calls "the sticky problem of humanity's own captivity within language"?[7] Some animals, such as parrots and great apes, can use human languages; other animals who live with humans, such as dogs, understand human languages in some sense; and some humans who spend a lot of time with other animals come to understand their languages in some sense, but language is not, as Barad argues, the dominant way in which humans communicate with other beings (and things) in the world.

For Barad, the shift from a focus on language as the primary means of communication to a focus on performativity, or intra-action, forestalls the untenable options of traditional realism and social constructionism, both of which estrange humans from the world. Intra-action reconfigures communication as the co-constitution of those in communication, a process Donna Haraway, in her discussion of training her dog Cayenne, describes as "opening up what is not known to be possible, but might be, for all the intra-acting partners."[8] Despret and Bruno Latour; co-constitution explain how some animal ethologists have been acting on the understanding that communication is co-constitution; they set up their experimental apparatuses in a way that gives "all the chances" to the animals to behave differently, that allows them to become interesting.[9] Latour suggests that one way to think of how to do this is to consider the scientists' apparatuses as Whiteheadean propositions, "*offers* ... to relate to another under a certain perspective."[10] If, as Barad says, "knowing is a matter of part of the world making itself intelligible to another part,"[11] rhetors might do well to begin to develop a nonanthropocentric account of communication.

But why, really, does it matter, especially to rhetors, whether or not we can communicate with other animals? On one level, I could say that we are increasingly coming to understand that nonhuman animals are not all that different from human animals[12]—but different enough that learning how to communicate with them may help us see the role of difference in communication positively, as offering new possibilities rather than erecting barriers. But on a deeper level, I would say, with Barad, that doing so enables us to make an "ethico-onto-epistemological" shift in how we conceive of communication, one that will allow us "to meet the universe halfway, to take responsibility for the role that we play in the world's differential becoming."[13] Understanding communication as intra-active co-constitution rather than as a transmission of exact linguistic meanings enables—and requires—us to acknowledge how we affect and are affected by our engagements with others, human and nonhuman, how we participate in the diversifying evolution of the world.

In this chapter, I will take up Barad's "posthumanist performative" understanding of discursive practices, which "insists on understanding thinking, observing, and theorizing as practices of engagement with, and as part of, the world in which we have our being"[14] and Whitehead's notion of propositions, which, as Isabelle Stengers describes them, mark "the opening of a full range of new diverging possibilities for becoming"[15] to argue that theorizing communication as co-constitution makes dogs (and other nonhuman animals) more interesting to rhetors. The crux of this argument is that human animals can and do communicate with other animals, often and more effectively without using language, and that rhetoricians need to pay more attention to the strategies of co-constitution involved in such communication. I focus here mostly on animals that humans other than scientists most often interact with (dogs especially), but I also extend my analysis to sheep, baboons, leopard seals, and grey parrots.

The differences between human and nonhuman animals can seem more obvious than differences among human animals, and certainly they are much at issue when the topic is what makes humans human. In interactions with nonhuman animals, however, humans often seem to forget these differences or seem at a loss about how to cope with them. Haraway comments about how she finally understood a problem she was having in working with her dog Cayenne in agility training: "I set about actually teaching what the release word meant instead of fantasizing that Cayenne was a native English speaker."[16] Humans who live with dogs, including me, habitually think that dogs understand their language. Dogs do learn the meaning of verbal commands, but they respond more naturally to tone of voice and even more to motion, or body language, as I learned in taking herding lessons with my border collie Scout. Early on, Scout and I were in a pen with sheep, and I complained to our trainer Rose, "Scout isn't doing anything I tell her to do!" Rose said, "That's because you're standing still. You're not telling her to do anything. You have to move." I was trying to get Scout to go around the sheep and bring them to me. One important instinct that herding dogs have that aids in training them is how to "balance" the sheep to the herder: how to keep sheep moving toward the herder instead of running off to the side or scattering. To do this, the dog must keep herself on an axis that runs between her, the sheep in the middle, and the herder. This instinct can be honed in training by the herder repeatedly moving off balance in one direction or another. Though no verbal commands are necessary to get a dog with good instincts to balance the sheep, there are ones that are used to tell the dog which way to go—counterclockwise ("away") or clockwise ("come bye"). I was just standing there in the pen telling Scout to come bye, and she quite reasonably just lay there because the sheep were already in balance. When I finally got better at moving in a comprehensible way, I didn't need to say anything to Scout as she repeatedly shifted direction to balance the sheep in relation to my movements. Bruce Fogt comments, "Dogs don't understand our language. They don't communicate that way. No matter how much we talk or shout, they won't understand unless we put motion with it."[17] Fogt's comments emphasize how communication involves more than simply responding to verbal commands. Though Scout learned to recognize "come bye" and "away," they needed to be spoken in the comprehensible context of my moving.[18]

Herding is not just telling dogs what to do. Fogt quotes a trainer on the importance of developing good balance in herding dogs:

> Some people don't care as much about it. They like a dog just to do what they say and they don't let it think on its own. But they run into trouble when they get into a difficult situation because the dog doesn't know how to do anything without being told. I like a dog with good balance that can think on its own.[19]

Quite beyond instincts, dogs learn what to do in different situations and can think through new situations in terms of that learning. Haraway insists that training is not a matter of training to, but of training with, and she emphasizes that in this relationship, humans have an "obligation to respond to the authority of the dog's actual performance."[20] As partners-in-training, humans need to learn to trust that their dogs have learned how to do things, just as the dogs learn to trust their human partners once the humans have become consistently comprehensible (something that took Scout and me a while to get to). Haraway puts it more precisely when she says she and Cayenne achieved "a little coherence"[21] in their intra-action; they were able to coordinate their behaviors.

Agility training and herding are examples of Barad's discursive practices. Discursive practices are not primarily linguistic, nor exclusive to humans, nor "anthropomorphic placeholders for the projected agency of individual subjects, culture, or language."[22] They do not determine intra-actions; instead, they are actual "*ongoing agential intra-actions of the world.*"[23] Like other adherents to what has been called *new materialism*,[24] Barad understands agency to be an action not limited to humans; she says quite directly, "Agency is not aligned with human intentionality and subjectivity."[25] Neither subjects, objects, nor agents precede intra-actions in the world—this is how intra-action differs from interaction which presumes bounded agents who are not affected by their engagement. Intra-actions entangle behavior, physical and social arrangements, and material to make matter matter, to give it meaning. Haraway says that intra-actions mark "the coming into being of something unexpected, something new and free, something outside the rules of function and calculation, something not ruled by the logic of the reproduction of the same."[26] In training, dogs and trainers become intelligible to each other in specific ways; they connect differently, and meanings and possibilities are reconfigured. Authority and trust signal these emerging meanings. As Scout and I worked with sheep, the material and social conditions—the pen, the long line on Scout, the routines of training—caused my movements to take on meanings they didn't have elsewhere, as did her lowered tail, and as our behaviors began to correspond, new possibilities of what we could do with sheep arose.

And then there are the sheep, not incidental figures in the discursive practice of herding. Sheep also have some instincts that aid in herding training; they like to stay together, and sheep who work with trainers quickly begin to follow the trainer. Despret studies the work of animal ethologists—it is in her analysis of Thelma Rowell's research on sheep that she argues that using apparatuses that allow animals to become interesting produces more interesting results. Rowell wanted to know how sheep organize themselves socially. Most research on sheep had studied only the possibility of hierarchical organization and studied it in terms of competition over food. In her earlier research, she had already discovered that mother and daughter ewes maintained long-term relationships, which inspired her to further question what conditions allow them to do so. Rowell set up an apparatus that eliminated the hypothesis that sheep were simply motivated by competition over food by setting out twenty-three bowls for her twenty-two sheep. This allowed more interesting hypotheses to become visible. Instead of seeing sheep as being "strictly determined by its hormones and by hierarchical rules," sheep could be seen as animals that connect with one another "in a spirit of both competition and coordination to invent a solution to several problems."[27] Earlier analyses of fighting among sheep had suggested that that this activity seemed to serve a variety of functions rather than just confirming hierarchies. Similarly, in the condition of having twenty-three bowls, a sheep leaving its bowl and shoving another sheep away from its bowl made visible the concept of supplanting, a "highly complicated conception of hierarchy, in no way comparable to a rigid organization that determines behaviors predictably."[28] The sheep that is approached may or may not accept the supplantation, allowing a number of reasons to be explored through further observation. The twenty-third bowl thus allowed the sheep to become more interesting, to become less like subjects of the research and more like partners in it. As botanist Robin Wall Kimmerer says, "Experiments are not about discovery but about listening and translating the knowledge of other beings."[29]

Derrida asks the question "and say the animal responded?" but he never discusses whether they do respond rather than just react instinctively, as he is more concerned to dispute the distinction between response and reaction in humans as well as in animals. Barad and Whitehead would agree with his argument, but they go on to assume a world in which all animals cannot help but respond/react; all are entangled with others and with entities of all kinds in their ongoing becoming. This approach suggests an analysis of communication different from that which Derrida pursues, a turn away from a concern with the sticky trap of language toward what communicative behaviors animals can become capable of—one could say, away from deconstruction to a new kind of constructivism. It is not necessary to attribute similar capabilities to nonhuman animals in order to suggest that the pragmatics and ethics of communicating with and understanding others is the same regardless of whether the interlocutors are human or nonhuman. The question is not whether nonhuman animals respond,[30] but what is at stake in assuming that humans cannot communicate with them. Haraway argues eloquently for the responsibility humans have to "meet the universe halfway":

> Disarmed of the fantasy of climbing into heads … to get the full story from the inside, we can make some multispecies semiotic progress. To claim not to be able to communicate with and know one another and other critters, however imperfectly, is a denial of moral entanglements (the open) for which we are responsible and in which we respond.[31]

Many individual nonhuman animals, both wild and domesticated, do engage with individual human animals and are understood, though humans are not always responsive to the possibilities that arise in such intra-actions. Primatologist Barbara Smuts was perforce responsive in her field work with baboons in Kenya. She says that her work with them became much more interesting when she abandoned herself to the "far superior knowledge" of the baboons and "moved as a humble disciple, learning from masters about being an African anthropoid," noting that she had to learn the system of baboon etiquette, a task forced on her "by the fact that the baboons stubbornly resisted my feeble but sincere attempts to convince them that I was nothing more than a detached observer, a neutral object they could ignore."[32] She argues that this intra-action was the most valuable part of her research: "the deepest lessons came when I found myself sharing the being of a baboon because other baboons were treating me like one."[33] As with Rowell's sheep, this is an example of how, even inadvertently, research apparatuses can offer nonhuman animals a chance to become interesting partners rather than mute subjects.

Smuts was amused by the baboons' dismissal of her effort at scientific objectivity, but wildlife photographer Paul Nicklen, who was also amused by the behavior of a massive female leopard seal he was photographing in Antarctica, instead did not let the seal become a partner offering him new possibilities for intra-action. His goal was simply to document his belief that "leopard seals are intelligent animals capable of interaction with humans."[34] With the advice of a leopard seal expert, Nicklen tried to remain calm as the seal, as predicted, threatened him by taking his camera and head in her mouth. Nicklen, as instructed, kept photographing. After repeating her threats a few times, the seal swam off and then, as Nicklen says, "the truly unexpected happened. She quickly returned with a live, fresh-caught penguin in her mouth."[35] She released the penguin a few feet from Nicklen, and when he didn't catch it, she repeatedly retrieved it and released it in front of him again. Then she brought him exhausted penguins that she had beaten up a bit. Nicklen says, "Of course, I did not accept these offerings, but kept photographing."[36] Finally, she brought him dead penguins, and when he again refused to accept them, he said, "she appeared frustrated and blew bubbles in my face."[37] Nicklen notes that as a biologist, he is reluctant to anthropomorphize, but he believed that "she was trying to feed me penguins because she realized that I was an absolutely useless predator in her ocean, and my ineptness at securing a meal agitated her."[38] He remarks, "Throughout these encounters, I often flooded my mask because I was laughing so hard that water trickled in the sides."[39] He continued to photograph her for four more days, during which time she rammed his camera once, and once whacked him on the head with a dead penguin. Concerning the two attacks, he "realized she was establishing her boundaries."[40]

Nicklen admits that his understanding of the leopard seal's behavior is incomplete, and he remembers the encounter as an "astonishing gift from the sea."[41] However, despite setting up conditions under which the seal was allowed to become interesting, he shied away from seeing the encounter as opening new possibilities for becoming. In repeatedly bringing him penguins and attacking him when he got too close, the seal offered him propositions, offers to make him attentive to her in a way that would diverge from his usual path. He recognized these as unexpected behaviors, but he was not diverted: he laughed at her, repeatedly got too close, and kept on photographing. He may have succeeded in his goal of making humans think more kindly of leopard seals, and thus more concerned about how climate change will affect them (although the most often shown of his images is a scary shot of her open mouth of sharp teeth as she approached him),[42] but he did not pay attention to her proposition. Unlike Rowell and Kimmerer and Smuts, he learned from the encounter only what he set out to prove: that leopard seals do not necessarily prey on humans and will interact with them.[43]

This approach to intra-action owes a lot to Whitehead's concept of propositions[44] especially as explained by Stengers, and I will draw on her discussion of Whitehead's account of the sixth day in the Christian creation narrative[45] to clarify what's involved in paying attention in situations that offer new possibilities for becoming. Whitehead's account is stated in a single sentence at the end of his discussion of expression: "The account of the sixth day should be written, He gave them speech and they became souls."[46] Expression refers to situations in which "something is infused throughout the environment which will make a difference."[47] Such situations offer propositions, "tales … that may be told."[48] Propositions are not linguistic sentences, verbal statements, or conscious experiences but rather breaks in continuity, and they are not the sole province of human experience. Propositions express

> not only … the experience of words inducing disclosure, a world felt in a different manner, but also … the disruption, or variation of interest, which a rabbit or dog may experience, and finally … the possibility of the kind of disruption of social continuity we may observe when even oysters or trees seem to forget about survival.[49]

The "truly unexpected" action of the seal bringing Nicklen a penguin, a sheep shoving another sheep away from a bowl of food when there were more than enough bowls for all, baboons refusing to ignore Smuts—all are propositions that can be paid attention to—entertained, as Whitehead says. Propositions are not true or false; they are instead lures for thinking. Entertaining them is experienced as

> a tremendous mode of excitement. Like a stone thrown into a pond it disturbs the whole surface of our being. But this image is inadequate. For we should conceive the ripples as effective in the creation of the plunge of the stone into the water. The ripples release the thought, and the thought augments and distorts the ripples.[50]

Entertaining propositions leads to speculative adventures. At least in human animals. And this is where language comes in.

What does Whitehead mean when he says "He gave them speech and they became souls"? Not that the possession of language is what makes humans human. What was given to humans on the sixth day was instead language "as providing a social environment for the speculative reception of propositions."[51] Language presupposes propositions; it does not create them. Propositions need an environment in order to emerge, in the same way that Barad's discursive practices, which are also not essentially linguistic, refer to social and material conditions. Language provides for humans a particular social environment that enables speculative thought:

> We became souls because of the difference that language makes in the rippling consequences of a proposition's impact. Being given language means that when

> a proposition is entertained it is given a social environment such that its impact may be amplified into many divergent, entangled consequences, activating that mode of functioning which is the soul.[52]

Speculative thought is what, for Whitehead, distinguishes human animals, what allows them to become souls. While animals entertain propositions of what may be in feelings "tinged with purposes, hopes, and expression derived from conceptual functioning ... the life of a human being receives its worth, its importance, from the way in which unrealized ideals shape its purposes and tinge its actions."[53] As Stengers says,

> Speculative language should be able to induce not the reaction of the rabbit becoming aware that this grey shade is what we call a wolf, that is a convinced "it matters!", but a speculative adventure entailing questions such as "how does it matter?", "does it really matter?", "what if I accepted that it does not matter?", "how did it come to matter?", unrealized ideas then shaping our experiences.[54]

Humans do also entertain propositions in the same way as other animals do, as disruptions that create possibilities, but humans can further elaborate possibilities through speculative thought.

The presence or absence of speculative thought is what really distinguishes Rowell's entertainment of the proposition offered by her sheep from Nicklen's failure to entertain the possibilities for becoming opened by the leopard seal's proposition. Both Rowell and Nicklen were attentive to their partners in a way, but instead of really listening to the seal, trying to understand the implications of her behavior, Nicklen was simply concerned to prove his point that leopard seals were not vicious animals. Becoming a soul means requiring to understand, and understanding differs from knowing in its orientation to the experience of transformative disclosure that comes from entertaining propositions rather than being oriented to proving statements valid. "We require to understand how mere matter-of-fact refuses to be deprived of its relevance to potentialities beyond its own actuality of realization," Whitehead says.[55] For Rowell, the sheep's behavior set off a line of questioning about how it matters, especially about what conditions allow sheep to cooperate or prevent them from doing so. This disclosure transformed how she thought about sheep and activated other relevant research and observations made available through language, as well as making further potentialities visible.

Becoming a soul doesn't happen all the time. Often we fail to entertain propositions, fascinated by the matter-of-fact or our own preoccupations, as was Nicklen in his encounter with the leopard seal. Stengers remarks, "Sometimes we feel the ripple, sometimes we do not, as when we hear 'the salt, please,' and do not feel it was meant to interrupt us, to express indifference or annoyance against the so very interesting things we were talking about."[56] It is not a moral failing when we do not, but if we lose, or refuse, "this sense of disclosure, we are shedding that mode of functioning which is the soul."[57]

Whitehead offers speculative thought as an addition, not an alternative to the logical thought of science. It is, Stengers says, an attempt to change the problem, just as Barad does in shifting from the concept of interaction to the concept of intra-action and from the metaphor of reflection to the metaphor of diffraction.[58] What is added in both of their theories is the capacity "to pay attention and believe in this world," in the specifics of this world, and "not to judge it in terms of generalities,"[59] a shift that Deleuze and Guattari labeled the empiricist conversion.[60] It is an attempt to remedy, at least in part, the estrangement of humans from the world so that we may, as Barad says, "take responsibility for the role that we play in the world's differential becoming."[61]

What this suggests for rhetors is that we need an understanding of communication that includes but is not specific to linguistic interaction, one that is defined more broadly as performative intra-action that depends for its effectiveness on discursive practices that make entertaining propositions offered by animals, and other beings in the world, possible. We need to believe that we can talk with them, and we need to develop habits that enable us to find them interesting and that lead to speculative thought.

This is what Pepperberg did in her work with Alex. What was striking about her research was that it broke with previous behaviorist studies, offering Alex a chance to behave differently through a methodology that assumed that parrots could learn language like human children do, in social interaction. Pepperberg wanted to be certain that Alex comprehended the meanings of his linguistic behavior instead of just responding to verbal commands. She says, "I was planning to demonstrate in a parrot cognitive processes that only humans and higher primates were considered capable of achieving. You need very special conditions to do that and, equally important, to have people believe what you might be claiming."[62] Alex learned to identify objects; to identify the shapes, colors, size, and number of objects; and to tell what was the same or different about objects. But importantly—and surprisingly, at least to some—in the "back-and-forth, the relationship,"[63] between himself and his trainers, he also learned how to ask for things he wanted, to say "no" when he didn't want to do something, to play jokes on his trainers, to apologize, and to correct, insult,[64] and ask questions of others.[65]

One day, Pepperberg and Alex were working on numbers. She showed him a tray of objects of different materials and colors and asked "How many green wool?" She describes what happened next:

> Alex eyed the tray and looked at me in a way he sometimes did, which I can describe only as wryly. "One," he piped. The answer was "two."
> "No, Alex. How many green wool?"
> That look again. "Four," he said in his charming sing-song double-syllable way: "Foo-wah."
> "C'mon, Alex. How many green wool?"
> It was no use. He kept alternating: "one … four … one … four."
> By now I had realized he was just messing with my head. I knew that he knew the correct answer. "OK, Alex," I said sternly. "You are just going to have to take a time-out." I took him to his room and closed the door.
> "Two … two … two … I'm sorry … come here!" … I immediately heard coming from behind Alex's closed door.[66]

Concerned about whether she could prove to the satisfaction of other scientists that Alex knew what he was doing, Pepperberg was very careful in instances like this not to claim definitively that he did, though it seemed clear to her:

> A simple example: if Alex said "Want grape" and you gave him a banana, he'd spit it right back at you and repeat insistently, "want grape." He wouldn't stop until you gave him a grape. If you were dealing with a child, you would accept without question that he or she really wanted a grape, and that banana simply wouldn't do. But that's not science.[67]

Pepperberg clearly *understood* Alex in these situations, but she didn't *know* that she did. This distinction between speculative and scientific thought may also explain Nicklen's hesitation over his interpretations of the leopard seal's behavior.

In these training sessions, Alex was commonly offered a new proposition, and he sometimes entertained the proposition in surprising ways. Pepperberg wanted to see if grey parrots could solve a problem that Bernd Heinrich's ravens had succeeded at—figuring out how to get an attractive item that was dangling on a string attached to a perch. One of her other parrots succeeded immediately, but when she tried Alex on the problem by suspending an almond on a string for him,

> He looked down at the nut and looked at me. He did nothing. I was wondering what he was thinking. After a few seconds he said, "Pick up nut."
> I was a bit taken aback. I said, "No, Alex, *you* pick up nut."
> He looked right back at me and said, "Pick up nut!" a little more insistently this time.
> I tried to encourage him several more times, but he simply refused.[68]

It took several years and another incident in which a parrot new to training succeeded in the task and another parrot who, like Alex, had been in training a long time, refused for Pepperberg to realize that "once Alex had learned how to label objects and request things, he relished the control it gave him over his environment, the ability to manipulate the people around him."[69]

Because of the repetition needed for statistical validity, scientific experiments can be excruciatingly boring for both human and nonhuman animals engaged in them, and Alex often simply refused to continue responding or played jokes on the trainers, as in the incident of counting green wool objects. Pepperberg relates another counting incident in which Alex's behavior instead seemed to reveal surprising mastery of the concept of zero. Shown a tray of objects of which there were two, three, and six blocks of different colors, Alex was asked, "Which color three?" He responded, "Five." When he continued to answer "five," Pepperberg asked, "OK, smarty pants. What color five?" He answered, "None." To confirm that he meant there was no group of five blocks of the same color, she tested him six more times on trays that lacked a certain number of same-colored blocks, and he answered correctly five times. His surprising answer of "five" when asked for "which color" came after two weeks during which he had refused to answer any questions, and Pepperberg concludes, "It was as if he said to himself, *OK, how can I make this interesting?*"[70] No one knows what an animal might be capable of, which is what makes communicating with them interesting—for both of the animals.

Stengers argues just this, that in intra-action (her example is domestication, but it applies as well to training or communication), an animal's behavior is not determined by functional signs (like language):

> what domestication celebrates is that neither the seduced nor the seducer may be defined in terms of functional signs that make a social, specific world for the [animal]. The expressive sensibilia that are being invented between themselves, laboriously, painfully, indicate that this time the adventure of life designates the individual, or the living person, as an active, vibrant site of the negotiation that takes place within it among fear, curiosity, and attraction. A site that testifies both to the infection and to the canalization through which what is new becomes important.[71]

Alex became a new "we," a parrot in his work with Pepperberg, constructing what a parrot might be capable of if she intra-acted with him. Despret comments on Alex's behavior:

> Alex talks because Pepperberg desires it and demands it of him, ... She was able to negotiate with Alex over what in speech could interest him. Alex talks because for diverse reasons his desire overlaps with that of Pepperberg.[72]

Communication with animals, human or nonhuman, involves "the coming into existence of new propositions … [that mark] the opening of a full range of new diverging possibilities for becoming."[73] It requires entertaining propositions by negotiating overlapping desires, being open to the possibilities suggested by behaviors, and being willing to work to understand the meanings enacted by those behaviors. In her response to J. M. Coetzee's lectures, Wendy Doniger says that animals do speak. The problem is that we all too often "refuse to grant them the dignity of *listening* to them"[74]— of entertaining the possibilities their behaviors suggest—whether they are human or nonhuman. When, like Derrida and Nicklen, we refuse to entertain the possibilities that arise in intra-action, we refuse to acknowledge our role in the co-constitution of the cosmos.

NOTES

1. Ludwig Wittgenstein, *Philosophical Investigations*, 3rd ed., trans. G. E. M. Anscombe (New York: Macmillan, 1958), 223.
2. Wittgenstein, *Philosophical Investigations*, 223.
3. Vicki Hearne, *Animal Happiness* (New York: Skyhorse Press, 1994), 169.
4. Hearne, *Animal Happiness*, 167, 168.
5. Jacques Derrida, *The Animal that Therefore I Am*, ed. Marie-Louise Mallet, tr. David Wills (New York: Fordham University Press, 2008), 31.
6. Vinciane Despret, "The Becomings of Subjectivity in Animal Worlds," *Subjectivity* 23 (2008), 127.
7. Karen Barad, *Meeting the Universe Halfway: Quantum Physics and the Entanglement of Matter and Meaning* (Durham, NC: Duke University Press, 2007), 137.
8. Donna Haraway, *When Species Meet* (Minneapolis: University Minnesota Press, 2008), 223.
9. Vinciane Despret, "Sheep Do Have Opinions," in *Making Things Public: Atmospheres of Democracy*, ed. Bruno Latour and Peter Weibel (Cambridge: MIT Press, 2005), 360.
10. Bruno Latour, "A Well-Articulated Primatology: Reflections of a Fellow Traveler," in *Primate Encounters: Models of Science, Gender, and Society*, ed. Shirley C. Strum and Linda Marie Fedigan (Chicago: University of Chicago Press, 2000), 372.
11. Barad, *Meeting the Universe Halfway*, 185.
12. Studies in cognitive ethology, a field initiated by Donald R. Griffin *(Animal Minds.* Chicago: University of Chicago Press, 1992), have demonstrated that nonhuman animals exhibit self-awareness, form conscious plans, and communicate semantic information to each other (vervets, chickens, and honeybees, for example) and to humans (honeyguides and dogs).
13. Barad, *Meeting the Universe Halfway*, 381, 396.
14. Ibid., 133.

15. Isabelle Stengers, "Whitehead's Account of the Sixth Day" (paper presented at the Stanford University Whitehead Symposium, April 21, 2006), 24. www.geco.site.ulb.ac.be/Telecharger-document/16-Whiteheads-Account-of-the-Sixth-Day.html Accessed December 7, 2011.

16. Haraway, *When Species Meet*, 227.

17. Bruce Fogt, *Lessons from a Stock Dog* (Sidney, OH: Working Border Collie, 2003), 51.

18. Similarly, the border collie Chaser famously learned to identify 1022 different toys and perform various actions (bring, nose, paw) regarding them, an ability ethologist Ádám Miklósi attributed to three years of intensive training (Jessica Griggs, "Border Collie Takes Record for Biggest Vocabulary," *New Scientist*, December 20, 2010. www.newscientist.com/article/mg20827921.900-border-collie-takes-record-for-biggest-vocabulary/). But as I am arguing, to focus only on language in this interaction is to ignore the co-constitution of Chaser and trainer John Pilley in the game they create in their intra-action.

19. Fogt, *Lessons from a Stock Dog*, 32.

20. Haraway, *When Species Meet*, 224.

21. Ibid., 228.

22. Barad, *Meeting the Universe Halfway*, 149.

23. Ibid., 148–49; emphasis in the original.

24. Cf. Scot Barnett and Casey Boyle, eds., *Rhetoric Through Everyday Things* (Tuscaloosa: University of Alabama Press, 2016); Diana Coole and Samantha Frost, eds., *New Materialisms: Ontology, Agency, and Politics* (Durham, NC: Duke University Press, 2010); Paul Lynch and Nathaniel Rivers, eds., *Thinking with Bruno Latour in Rhetoric and Composition* (Carbondale: Southern Illinois University Press, 2015); and particularly Jane Bennett, *Vibrant Matter: A Political Ecology of Things* (Durham, NC: Duke University Press, 2010), 9–11; Tim Ingold, *Making: Anthropology, Archaeology, Art and Architecture* (London: Routledge, 2013), 95–97; Bruno Latour, *Politics and Nature: How To Bring the Sciences into Democracy,* tr. Catherine Porter (Cambridge: Harvard University Press, 2004), 73–75; and Thomas Rickert, *Ambient Rhetoric: The Attunements of Rhetorical Being* (Pittsburg: University of Pittsburgh Press, 2013), 202–13.

25. Barad, *Meeting the Universe Halfway*, 177.

26. Haraway, *When Species Meet*, 223.

27. Despret, "Sheep Do Have Opinions," 367.

28. Ibid., 368.

29. Robin Wall Kimmerer, *Braiding Sweetgrass: Indigenous Wisdom, Scientific Knowledge, and the Teachings of Plants* (Minneapolis, MN: Milkweed, 2013), 158.

30. Or as Tim Ingold says, correspond: "To correspond with the world, in short, is not to describe it, or to represent it, but to *answer to it* … to mix the movements of one's own sentient awareness with the flows and currents of animate life" (*Making*, 108).

31. Haraway, *When Species Meet*, 226.

32. Smuts, "Reflection," 109–10.

33. Ibid., 110.

34. Paul Nicklen, *Polar Obsession* (Washington, DC: National Geographic Society, 2009), 155.

35. Nicklen, *Polar Obsession*, 157.
36. Ibid.
37. Ibid.
38. Ibid.
39. Ibid.
40. Ibid.
41. Ibid., 158.
42. See Nicklen's Ted Talk of March 2011.
43. Though this was his most memorable encounter, Nicklen also got in the water and photographed thirty other leopard seals without untoward incident.
44. Latour ("A Well-Articulated Primatology," 372–76) and Haraway *(When Species Meet*, 243–45) refer to it explicitly, and Despret refers to Latour.
45. The sixth day is when God created humans, and also, as Haraway reminded Stengers, beasts, cattle, and "everything that creeps on the ground." Thus, though Whitehead allows that both human and nonhuman animals entertain propositions, Stengers admits that his account is "quite a partial one," and, like Barad, she argues in a footnote to her talk for reconsideration of "the long and many-faceted history of our cohabitation with the other creatures" and invention of "new modes of thinking and feeling the togetherness of our lives ("Whitehead's Account of the Sixth Day," 29–30).
46. Alfred North Whitehead, *Modes of Thought* (New York: Free Press, 1938), 41.
47. Whitehead, *Modes of Thought*, 20.
48. Stengers, "Whitehead's Account of the Sixth Day," 24.
49. Ibid., 24.
50. Whitehead, *Modes of Thought*, 36.
51. Ibid., 28.
52. Ibid., 26.
53. Ibid., 27.
54. Stengers, "Whitehead's Account of the Sixth Day," 28–29.
55. Whitehead, *Modes of Thought*, 83.
56. Stengers, "Whitehead's Account of the Sixth Day," 27.
57. Whitehead, *Modes of Thought*, 62.
58. Barad, *Meeting the Universe Halfway*, 89.
59. Stengers, "Whitehead's Account of the Sixth Day," 28.
60. Giles Deleuze and Félix Guattari, *What Is Philosophy?* tr. Hugh Tomlinson and Graham Burchell (New York: Columbia University Press, 1994), 75.
61. Barad, *Meeting the Universe Halfway*, 396.
62. Irene M. Pepperberg, *Alex and Me: How a Scientist and a Parrot Discovered a Hidden World of Animal Intelligence—and Formed a Deep Bond in the Process* (New York: Harper, 2008), 71.
63. Pepperberg, *Alex and Me*, 72.
64. Having had graduate students say "You turkey" to him when he seemed to behave stupidly, Alex said "You turkey" to a toy parrot when it didn't respond to his request for it to tickle his neck.
65. This is an important way in which Alex's communicative behavior differs from that of the border collie Chaser.

66. Pepperberg, *Alex and Me*, 134–35.
67. Ibid., 107.
68. Ibid., 151.
69. Ibid., 151.
70. Ibid., 191; emphasis in the original.
71. Isabelle Stengers, *Thinking with Whitehead: A Free and Wild Creation of Concepts*, tr. Michael Chase (Cambridge: Harvard University Press, 2011), 438.
72. Despret, "The Becomings of Subjectivity in Animal Worlds," 127.
73. Stengers, "Whitehead's Account of the Sixth Day," 24.
74. Wendy Doniger, "Reflection," in J. M. Coetzee, *The Lives of Animals*, ed. Amy Gutmann (Princeton: Princeton University Press, 1999), 105; emphasis in the original.

BIBLIOGRAPHY

Barad, Karen. *Meeting the Universe Halfway: Quantum Physics and the Entanglement of Matter and Meaning*. Durham, NC: Duke University Press, 2007.

Barnett, Scot, and Casey Boyle, eds. *Rhetoric Through Everyday Things*. Tuscaloosa: University of Alabama Press, 2016.

Bennett, Jane. *Vibrant Matter: A Political Ecology of Things*. Durham, NC: Duke University Press, 2010.

Coole, Diana, and Samantha Frost, eds. *New Materialisms: Ontology, Agency, and Politics*. Durham, NC: Duke University Press, 2010.

Deleuze, Gilles, and Félix Guattari. *What Is Philosophy?* Tr. Hugh Tomlinson and Graham Burchell. New York: Columbia University Press, 1994.

Derrida, Jacques. *The Animal that Therefore I Am*. Ed. Marie-Louise Mallet. Tr. David Wills. New York: Fordham University Press, 2008.

Despret, Vinciane. "The Becomings of Subjectivity in Animal Worlds." *Subjectivity* 23 (2008): 123–39.

———. "Sheep Do Have Opinions." *Making Things Public: Atmospheres of Democracy*. Ed. Bruno Latour and Peter Weibel. Cambridge: MIT Press, 2005. 360–68.

Doniger, Wendy. "Reflection." J. M. Coetzee, *The Lives of Animals*. Ed. Amy Gutmann. Princeton: Princeton University Press, 1999. 93–106.

Fogt, Bruce. *Lessons from a Stock Dog*. Sidney, OH: Working Border Collie, 2003.

Griffin, Donald R. *Animal Minds*. Chicago: University of Chicago Press, 1992.

Griggs, Jessica. "Border Collie Takes Record for Biggest Vocabulary." *New Scientist*, December 20, 2010. www.newscientist.com/article/mg20827921.900-border-collie-takes-record-for-biggest-vocabulary/

Haraway, Donna J. *Modest_Witness@Second_Millennium.FemaleMan©_Meets_OncoMouse™*. New York: Routledge, 1997.

———. *When Species Meet*. Minneapolis: University Minnesota Press, 2008.

Hearne, Vicki. *Animal Happiness*. New York: Skyhorse Press, 1994.

Ingold, Tim. *Making: Anthropology, Archaeology, Art and Architecture.* London: Routledge, 2013.

Kimmerer, Robin Wall. *Braiding Sweetgrass: Indigenous Wisdom, Scientific Knowledge, and the Teachings of Plants.* Minneapolis, MN: Milkweed, 2013.

Latour, Bruno. "A Well-Articulated Primatology: Reflections of a Fellow Traveler." In *Primate Encounters: Models of Science, Gender, and Society.* Ed. Shirley C. Strum and Linda Marie Fedigan. Chicago: University Chicago Press, 2000. 358–81.

———. *Politics and Nature: How To Bring the Sciences into Democracy.* Tr. Catherine Porter. Harvard University Press, 2004.

Lynch, Paul, and Nathaniel Rivers, eds. *Thinking with Bruno Latour in Rhetoric and Composition.* Carbondale: SIU Press, 2015.

Nicklen, Paul. *Polar Obsession.* Washington, DC: National Geographic Society, 2009.

Pepperberg, Irene M. *Alex and Me: How a Scientist and a Parrot Discovered a Hidden World of Animal Intelligence—And Formed a Deep Bond in the Process.* New York: Harper, 2008.

Rickert, Thomas. *Ambient Rhetoric: The Attunements of Rhetorical Being.* Pittsburgh: University of Pittsburgh Press, 2013.

Stengers, Isabelle. *Thinking with Whitehead: A Free and Wild Creation of Concepts.* Tr. Michael Chase. Cambridge: Harvard University press, 2011.

———. "Whitehead's Account of the Sixth Day." Paper delivered at the Stanford University Whitehead Symposium, April 21, 2006. www.geco.site.ulb.ac.be/Telecharger-document/16-Whiteheads-Account-of-the-Sixth-Day.html. Accessed December 7, 2011.

Smuts, Barbara. "Reflection." J. M. Coetzee, *The Lives of Animals.* Ed. Amy Gutmann. Princeton: Princeton University Press, 1999. 107–20.

Whitehead, Alfred North. *Modes of Thought.* New York: Free Press, 1938.

Wittgenstein, Ludwig. *Philosophical Investigations*, 3rd ed. Tr. G. E. M. Anscombe. New York: Macmillan, 1958.

Chapter 12

How to Understand a Parrot's Words and What You Can Learn from Him

Early Indian Writers on Animal Speech

Andrea Gutiérrez

When taking up the question of the animal,[1] most thinkers from ancient times until the modern day have had the speculative preoccupation of establishing human/animal difference, typically carried out using anthropocentric criteria.[2] Only in recent years have a handful of continental philosophers problematized earlier attempts at asserting difference, since, in fact, humans *are* animals.[3] South Asian writers throughout history have also attempted to sort out animal/human distinctions, often using the lens of animal speech. Asserting difference was not a striking priority for pre-modern South Asian writers, who largely recognized some amount of permeability between ideas of animals and of humans.[4] However, some thinkers did assert varying degrees of difference between animal and human language. Some of these cases that did not reduce difference expressed other priorities and established alternative paradigms for assessing language. These other early Indian paradigms prioritized transcendent speech, mimetic capacities, and the ability to transmit knowledge, as well as recognizing the crucial role of the listener(s) in interpreting nonhuman language. I argue that these alternative, non-anthropocentric paradigms—of nonvocalized speech, animals' sounds conveying meaning, parroted speech as a teaching tool, and agentive listeners—can offer us novel approaches to exploring speech and language. It will come as no surprise to those familiar with South Asia that these historical Indian paradigms were very early forerunners of some recent trends in rhetorical studies.

Before delving into my analysis, I must express a few caveats. I write on a purported human/animal divide with some hesitation, as I risk granting validity to the idea, but I do so pragmatically, for convenient categorization using patterns of thought familiar to readers. Secondly, I treat mainly bird voice, and often parrots.[5] These animals received heightened attention as the lens through which ancient and medieval Indian thinkers frequently explored this topic, due to their extraordinary capacities in language and speech production. Certainly, different animals enjoyed varying levels of consideration among thinkers. Birds, noticeably articulate by anthropocentric standards and often able to reproduce human language, most frequently appear in such writing. My selection of texts certainly reflects the parrot's exceptional linguistic position in the animal kingdom (including varying linguistic capabilities among humans). My discussion often applies to bird or even parrot speech in particular, but certain ideas apply more broadly to animal language. Finally, my discussion sometimes focuses on voice, speech, and utterances, terms taken from the Sanskrit words *vāc*, *vākya*, and *uccāraṇa* as appearing in these texts. These words can indicate "language" but are not synonymous with it. The texts refer to voice or utterance because the authors, who conceived of language primarily in oral contexts, privileged speech or voice in their discussions of language.[6]

The variety of my evidence suggests that questions on the extent of difference (or nondifference) between animal and human voice were pertinent in wide-ranging disciplines. I have selected material from many genres in order to demonstrate the varying attitudes asserting human/animal distinctions and to show that animal language was a concern for numerous writers of pre-modern India. My priority is to highlight alternative Indian paradigms for language and speech that expand beyond the pedestrian notion of spoken human language, including soundless or unuttered speech, voiced sound units of animal speech, and the communicative power of parroted speech. Early Indian examples among nonhuman and human animals (and others!) are akin to modes of communication described by Alex C. Parrish in his *Adaptive Rhetoric*,[7] namely, forms of animal signaling and communication that, once identified, do not easily allow ideas of human exceptionalism to remain.

I begin with some extraordinary examples of nonhuman speech from religious and yogic texts. The yogic discussion delineates how to understand animal sounds and language in a process that models how these writers conceived that language was generally understood. The religious and yogic examples illustrate paradigms that acknowledge typical human limitations to understanding nonhuman language. Then, a discussion from a text on Sanskrit grammar attempts to position "human" in relation to "animal" voice and defines the qualities of each in regard to articulation. Later, this same discourse of articulate animal language appears within the context of fiction, a genre that allows for human/animal permeability. Next, an early linguist's theory subverts traditional schematized hierarchies of language ability. Finally, a portion of an epistemological debate on sources of valid knowledge expands upon the possibility of animals using language to teach. These examples all bolster my argument that alternative paradigms for speech and new assessments of agency can help us appreciate animal language better.

These early Indian texts communicated alternative paradigms that often involved a nonhuman speaker *and* a (usually) human listener, perceiver, or audience. While most pre-modern Indian writers still privileged human language, the thinkers that I present here did not discount possibilities and processes of nonhuman voice and language.

SCREAMING [SILENTLY], BEYOND THE REACH OF ORDINARY HUMANS

South Asian paradigms concerning speakers and voice challenge Western paradigms, as for example in the *Jaiminīya Brāhmaṇa*, a pre-Common Era religious text:

> "When people in this world offer no oblation and lack true knowledge, but cook for themselves *animals that cry out*, those animals take the form of men in the other world and eat in return." "How does one avoid that?" "When you offer the first oblation with the voice, that is how you avoid it and are free of it." … "When people in this world offer no oblation and lack true knowledge, but cook for themselves *rice and barley, which scream soundlessly*, that rice and barley take the form of men in the other world and eat in return." "How can one avoid that?" "When you offer the last oblation with the mind, that is how you avoid it and are free of it."[8]

All living beings cry, even rice and barley. Rice and barley are living beings because they retain the potential to germinate at any point, using the life force that remains inside them long after their harvesting. Rice and barley cry soundlessly because of the limitations of our human sense organs (*indriya*s) in perceiving their voices. According to this Hindu religious text, all living beings make sounds, cry, or communicate, whether or not humans can understand it.

An additional recognition of animal sounds that also reduces the human/animal divide appears in *Yogasūtra* 3.17 (ca. 400 CE) on the special yogic power of understanding the meaning of animal noises. It is implicit in this *sūtra* (verse) that, if animal noises have meaning that *some* people can understand, these noises are language, albeit language that humans do not commonly understand. The esoteric yogic method in this section of the text is to attain broad mastery of an active practice of concentration.[9] Then, the *yogin* can apply this focused mental mastery to any number of specific areas, including the interpretation of animal sounds that do not follow the conventions of human language.[10] The commentators explaining this verse discuss language, how sounds have meaning due to conventions, and how the human mind is able to perceive sound units as language, but one must not forget that the extensive discussion's purpose is the deciphering of animal noises.[11]

Both the original commentary on the *Yogasūtras* and Vācaspati Miśra's later commentary (ca. tenth century CE) discuss language, sound, and conventional meaning as mentioned in this *sūtra*. Both commentators' discourses bring up the notion of speech or voice as based on sound units of syllables or phonemes. The original commentary on this verse begins, "In the context of this verse, it is the sound units of speech that hold the meaning."[12] In other words, the semantic value of speech is contained in the sound units. This idea will contrast with the Sanskrit grammarians' discussion of animal speech presented in my next section (Debates on Articulation), but syllabic units of sound are prominent there as well. Regardless, one key aspect of this commentary is that "the division (of concept, object or meaning, and sound) is known through convention."[13] This indicates that the meaning or understanding of a verbal expression depends on the convention that a certain social structure has dictated, often involuntarily. A word or noise means what the social group in question wants and conventionally understands it to mean.

According to these commentators, an animal language has conventions of meanings that are used for communicating via sound units (or noise). The social structure of that animal species (or genus, perhaps) has determined those conventions of meaning. If a *yogin* can make the proper divisions to deconstruct a speech act and can then reassemble those parts using yogic joint restraint, he can understand that animal language. Vācaspati Miśra was a polymath who wrote on a number of Hindu schools of thought, and therefore his comments sometimes reflect a broader perspective commonly held among the various orthodox schools. He followed the original commentary's idea that sound units contain the meaning in speech and added, "this speech is meaningful only in the syllables or sounds as established in common belief *and not in the verbal expression*."[14] For him, it was not the expression of speech that imparts meaning but the sound itself. The act of speaking was less important.

These commentators took the opportunity of exploring speech to better comprehend how we humans understand language. Their emphasis on treating any language, human or not, as conventionally understood sound codes that communicate meaning is reminiscent of recent scholars' view of animal language as semiotic codes. Jakob von Uexküll, Jesper Hoffmeyer, Thomas Sebeok, and Giorgio Agamben have all described animal language and animal sign codes as semiotic systems rather than semantic incidences of discourse, as we sometimes consider human language to be.[15] Interestingly, the *Yogasūtra* commentators, both ancient and medieval, largely concord with our contemporary scholarly discussions of biosemiotics and philosophy of language more broadly speaking.

DEBATES ON ARTICULATION: WHEN SANSKRIT GRAMMARIANS TALK ABOUT SPEAKING

The *Yogasūtra* commentaries on language discussed above promoted permeability between concepts of human and nonhuman, both in our shared linguistic features and in one common method to understand the languages of both. Other Sanskrit commentaries on language asserted greater separation between human and animal, particularly in different beings' capacities for articulate speech. One section of the grammatical commentaries on Patañjali's *Mahābhāṣya*, a second-century BCE text on Pāṇini's *Aṣṭādhyāyī* (outlining Sanskrit grammar in concise rules and dating to the fourth century BCE), explored the question of what designated "human" as opposed to "animal" voice. Later commentaries (spanning over a millennium) on the *Mahābhāṣya* end up affirming difference and suggest that animals articulate differently from humans but are not speechless. These writers attempt to delineate animal speech capacities in their explanations of one of Pāṇini's grammar rules, which specified that a particular verbal form should be used "when [describing the] '*vyakta*' (i.e., manifest or articulate) voice[s] [of humans] speaking together."[16] This rule indicated that one should, thus, use a *different* verbal form when describing the sounds of *animals* "speaking" or calling out together.

Following the establishment of this grammar rule, the commentators debated what defines human or animal voice according to what "*vyakta*" meant in the original rule, where the term was used to distinguish particularly *human* voices speaking together. "*Vyakta*" ("manifest") can mean "audible," "apparent," or "articulate," so the writers discussed whether Pāṇini meant "articulate" when he used this word. Further, they asked precisely *what* is articulate if we say human voice is articulate. Patañjali argued that some bird speech seems to be articulate (for example, a rooster who says, "cock-a-doodle-doo!)"[17] but requires the use of a different verbal form. He sought to explain why animal voice, seemingly articulate, cannot be described using the same verbal structure as that used for human articulate voice. Similarly, the parrot or mynah's voice appears to be intelligible; we can understand the utterances of such birds. Hence the commentators negotiated the precise meaning of "articulate speech" and who has it.

Patañjali determined that "*vyakta*" must mean not only "manifested and intelligible" but "*especially* articulate in voice," for him, meaning that the voice must be able to articulate all of the sounds (phonemes or syllabic units) in the Sanskrit alphabet.[18] Of course, here he gestured to a perceived "lack"[19] in animals' speech, observed according to purely anthropocentric criteria concerning one human alphabet. Finally, he concluded that this grammar rule must refer to "those beings whose sounds are articulate in their speech," meaning that not only does the voice have to be articulate in sound production but also articulate in expression (or idea).[20] In other words, an articulate voice is one that can articulate articulately.

When referring to the rooster's articulation of "cock-a-doodle-doo," Patañjali declared, "this is an imitation of them"[21] (i.e., a human imitation of what roosters say). He stated that "cock-a-doodle-doo!" is only an approximation of what roosters actually vocalize, but he inadvertently revealed that humans are *incapable* of producing the exact vocalizations of roosters. We can only produce an imperfect imitation, a resemblance. He did not appreciate the lack in *humans* of fully articulating rooster vocalizations, although this is not surprising for someone working under the assumption that the Sanskrit language is the perfect and only paradigm. Nonetheless, he did not seem to appreciate that the articulations of this particular bird are beyond the scope of the human organs of speech.

One commentator on Patañjali, Kaiyaṭa (ca. eleventh century CE), wrote that parrots, mynahs, and others do not *innately* have an articulate voice.[22] This may surprise us, as such birds obviously can produce articulate voicing innately if they are able to do so at all. Kaiyaṭa continued: "the range of some of the sounds (in parrots, mynahs, and so on) is because of the will and effort of humans," meaning that the intention and dominion of humankind must be present for parrots to articulate articulately. On one hand, Kaiyaṭa credited animals' speech abilities to humans, removing a great deal of birds' own agency. On the other hand, with this statement he observed that parrots speak human languages only in captivity, which Western biologists only noted in the twentieth century.

The many commentators wrote at various historical periods, and a comparison of two later commentators' language (Kaiyaṭa's and Nāgeśa's) reveals an ideological shift possibly due to their chronological distance. Kaiyaṭa (eleventh century CE) had asserted that birds do not innately have these capacities of articulation, while Nāgeśa, seven centuries later, writes that "animals lack an articulate voice like that of humans, etc."[23] Without reading too much into brief statements, Nāgeśa (seventeenth–eighteenth century CE) projects some of the same dichotomous discourse of man versus beast that European rationalists including René Descartes and Gottfried Wilhelm Leibnitz express at the same historical moment.[24] In Europe, this discourse hypervalued the rational mind, man, and the agency of humans, which resonates with Nāgeśa's concurrent discourse.

Overall, this linguistic debate concluded that animal voice is relatively inarticulate. However, this discourse would please cognitive ethologists today in that Patañjali questioned pre-established distinctions of human and animal and did not unthinkingly commit what ethologist Frans de Waal calls "anthropodenial."[25] De Waal uses this term to indicate that humans underestimate commonalities between human animals and other animals, as in humans denying that other beings can think or behave in ways that humans can. Patañjali himself problematized human notions about language and what we think about animal voice. So, while he ultimately privileged human speech, he defined such speech in the context of other animal speech and he examined other people's definitions, as did the later commentators.

OUT OF THE CAGE AND INTO THE—

This grammatical-linguistic discourse of articulate beings must have had some currency in its day, as it appears recognizably intact in a satirical passage from a seventh-century CE novel by Bāṇa, *Kādambarī*. This counterexample reproduces the animal linguistic discourse for its own comic purposes but confirms that this discussion informed the ideas of the literate Brahmanical society of northern India. The author was clearly conversant with this discourse, as evinced in his description of parrot Vaiśampāyana's speech, which the king has just observed when he comments:

> You've heard this *bird's clearness in enunciating syllables* and the sweetness in his tone … It is indeed a great marvel that this bird gives forth *a voice in which the division of letters is not muddled up*, and *syllable lengths, nasals, and grammatical purity are clearly differentiated*, a *voice in which every syllable is perfectly clear, a voice of distinction.* Moreover, in his case, there is something more. Though *he's an animal, his behavior is intelligent,*[26] *like that of a man* of refinement … It's generally the case that birds, and animals in general, have their knowledge limited to fear, food, copulation, sleep and wakefulness. This, then, is a great wonder.
>
> (Minister Kumārapālita replies:) *What is surprising in this? For certain kinds of birds—parrots, mynahs and others*—as Your Majesty well knows, *can repeat what they've heard spoken.* In that respect it's hardly very remarkable if either in consequence of training received in *a previous birth* or *by the efforts of some man a special refinement is produced.* Another thing is that formerly these birds used to have voices which enunciated perfectly clearly, just like men's. But through Agni's curse, parrots' talking became indistinct, and elephants' tongues were turned back to front.[27]

In Bāṇa's *Kādambarī*, this parrot tells the substories within the main narrative as his own subnarrative. In the above passage from the frame story's main narrative, the author artfully summarizes many ideas that arose in the grammarians' debate on the capacities of bird voice and articulate speech, albeit using humor to illustrate and lighten the complex relations between bird and human. He admits that birds speak through the efforts of man (Kaiyaṭa's "effort of humans" in the Patañjali commentary), indicating that language acquisition is learnt and not entirely innate. He mentions the karmic ideology of animals having perfected themselves in former births with the soul passing from body to body, often from human to animal and vice versa. Finally, Bāṇa mocks the presumably rationalist dichotomy of human refinement and intelligence versus "birds, and animals … limited to … food, copulation, sleep and wakefulness," testifying to this dichotomy's existence in South Asian discourses as well as Western ones.

In contrast with the grammar commentators' assertions—which maintained difference and alterity—of the limitations of animal voice, *Kādambarī*, a work of fiction, rejects human–animal difference. It promotes the permeability of ideas surrounding human and nonhuman animals since parrot Vaiśampāyana had actually been human in his prior life; toward the novel's conclusion, he is again reborn as human.[28] Furthermore, this parrot is fully articulate, as much as any human voice is, according to the requisites of the previous linguistic debate (i.e., manifesting all phonemes of the Sanskrit alphabet, intelligible, and articulate in expression beyond sound production). Being a work of fiction, Bāṇa's novel can create an environment that eliminates human-animal alterity regarding the parrot's use of language, literary creation, speech capacities, and faculties of knowledge.

MORE PRE-MODERN PARADIGMS FOR SPEECH

Such situations may only be fully realizable in fiction, but there were premodern writers who theorized other realities about speech, as did one of the earliest linguists, whose writing later became influential in movements of Śaivite *tantra* and *vedānta*: Bhartṛhari (ca. 450–500 CE, falling roughly between linguists Patañjali and Kaiyaṭa). While Bhartṛhari was not interested in animals *per se*, his well-known work on linguistic theory concerned, in part, forms of speech. Bhartṛhari overturned what we might consider the hierarchy of articulate beings and expressed a different paradigm of what was valued in speech, reflecting classifications that were supposedly extant since the late Vedic period, according to Nāgeśa's later writing.[29] The four classifications of speech or voice range from *least* to *most* manifest: 1) supreme voice (also called "brilliance"); 2) seeing voice ("seeing" in the sense of "understanding"); 3) middle voice; and 4) uttered voice.[30] Early philosophers of language typically considered the least manifest forms of voice to be immanent formations of speech or language, and also the most representative of *Brahman,* the universal spirit. In his *Vākyapadīya*, Bhartṛhari wrote that supreme voice is *śabdabrahman*,[31] *Brahman* as word or sound, associated with divine, revealed sound. This divine speech is, incidentally, the *least* manifest of all forms of speech, while uttered voice, at the opposite end of the spectrum, represents the most manifest, mundane, and least divine speech.[32]

Nāgeśa, who was a follower of Bhartṛhari's linguistic theory, related Bhartṛhari's classification of "uttered voice" to his own discussion of articulate voice in his grammar commentary on Patañjali (described in my section on Debates on Articulation).[33] In another work, Nāgeśa located the four aforementioned classifications of voice in four centers or *cakra*s within the body: "Supreme voice is located at the root *cakra* (at the base of the spine), seeing (= "understanding") voice is at the navel, the middle voice is known to be placed in the heart, and uttered voice is at the position of the throat."[34] According to this theory, uttered voice, apart from being uttered with the breath and vocal cords, is not particularly remarkable.

In contrast, "'brilliance' (i.e., supreme voice) was common to all beings, birds, and beasts; it was regarded as a flash of intelligence and also as constituting the meaning of a sentence,"[35] according to Bhartṛhari. This level of speech, the least manifest, conveys the most fundamental layer of semantic value. With an understanding of this aspect of Nāgeśa's commentary, we see that the entire earlier grammar discussion of "articulate" versus "inarticulate" is not the hierarchy of voice that one might imagine. If we expand our concept of language to include the idea that articulate voice might not indicate superior voice, the previous analysis of animal speech by Patañjali's commentators needs reconsideration. We need not attribute any particularly exemplary qualities to articulate speech except that it is articulate within the context of one language's phonetic system.

A LITTLE BIRD TOLD THEM

Thus far, some ideologies have valued divine aspects or transcendental meaning in unvoiced speech, while others have prioritized the refined technical execution and sophisticated thought of articulate beings. Still other schools of thought—in this case, medieval logicians—emphasized other evaluations of speech, at least within their epistemological discussion of means of valid knowledge. As always, there is a spectrum of opinions and arguments among and within the different schools of thought. However, in this discussion, the medieval school of New Logic (*Navya Nyāya*) headed by Gaṅgeśa (fourteenth century CE) largely concurred with the Mīmāmsakas, the Vedic exegetes—interpreters of language and religious law, meaning that at least two schools of Indian philosophy agreed on points in this discussion. The case concerns sources of testimony from speakers such as parrots as means of valid knowledge, in other words, whether or not a parrot can be a teacher or a testifier. Interestingly, here the parrot himself can be viewed as speaking utterances that might persuade (human) listeners of his statements' relation to factual reality. This contrasts with the grammatical examples in which *humans* attempted to convince others of their arguments concerning bird speech.

Gaṅgeśa's New Logic and some preceding schools of philosophy (at least the Vaiśeṣikas, according to the argument under discussion here) agree that parrots have knowledge of words and of the sequence of words and are able to construct sentences that can lead a listener to valid cognitions.[36] Higher knowledge of the relation of certain words to fact and prior knowledge of certain facts (which might be confirmed by perception) are not required when uttering statements that can impart knowledge. An understanding of the relation of certain spoken words to fact(s) comes about on the part of the listener in part because of correct semantic coherence within the sentence(s) uttered by a parrot or other being.[37] The listener also comes to understand the matter at hand through inferring what is perceptually observable (in some cases) or through inferring the relation of facts to linguistically coherent spoken statements (with predictable word collocation and so on).[38] This is comparable to a child reading aloud from a book on quantum physics. The child may not understand the intricacies of particle movement or the relation of the words to confirmable facts (or theoretical ideas) about our physical world, but scientists listening to the child reading could still learn from this recitation, as long as there are semantically coherent sentences in the text.

In this situation, it is irrelevant if one knows whether or not the speaker is a good testifier or a trustworthy informer. The *person listening* uses his knowledge of the meaning of words to determine the validity of a statement, and then decides whether or not he can infer something or learn from such a statement. A parrot is able to construct meaningful sentences from word knowledge alone, but the listener must have knowledge of the meaning of words (which the bird may or may not have) in order either to have knowledge acquisition from a statement or to determine that the statement is fallacious.[39] Any statements "parroted" by a parrot which are authoritative "due to their conforming with the original (past) utterance" can result in knowledge.[40]

Implicit here is the parrot's impeccable ability to mimic another person's voice, remember the utterance, and reproduce it with phonetically recognizable repetitions. Interestingly, the logician in this argument compares a parrot's utterances to someone reciting the Vedas without knowing the meaning of the passages being recited.[41] This sort of recitation was a common practice in the oral preservation of Vedic texts. Some individuals were entrusted with the duties of flawless memorization of portions of a text so that they would be able to reproduce it on demand, while other individuals would study the language, ideas, philosophy, and so on to be expert in these areas. The burden of memorizing texts of very great length required the division of tasks in the interest of full textual preservation. This comparison of parrots to Vedic reciters parallels the frequent appearance in early works from South Asia of parrots as teachers of the Vedas and other texts, and of parrots being taught the Vedas.[42]

This portion of the debate on valid testimony reproduces the argument of the Prabhākaras (followers of one school of *Mīmāṃsā* [exegesis] led by Kumārila Bhaṭṭa), but Gaṅgeśa's logicians concurred with this. One does not need to be an expert or even know the truth to teach something and communicate knowledge. Further, as seen in this case, not only fiction promotes the value of birds' speech acts. Here, philosophers of logic promoted bird teachers. In my earlier examples, philosophers of language either equated bird speech with the divine or took the question of animal voice seriously enough to write at length on interpreting animal language (the *Yogasūtras*) and on animals' articulateness and articulation (*Mahābhāṣya*). So, if early Indian writers might value birds as potential teachers due to their abilities to memorize and repeat tracts of text, are there other nonanthropocentric ways to appreciate nonhuman speech acts?

ARTICULATING AGENCY

I referred to agency earlier and I focus on it here as it has been a key concern in rhetorical theory, although traditionally a human-centered concern. Per Karen A. Foss, "Rhetorical theory cannot be divorced from questions about human agency, the role of symbols in the creation of the human world, and the power of audiences to coconstruct that world."[43] Rhetoricians such as Carolyn R. Miller have described agency as the intersection between speaker (or language communicator) and audience and as socially and ideologically constructed, meaning that a given cultural context determines the attribution of agency to certain subjects and might deny it to others.[44] Alex Parrish has emphasized the necessity of recognizing the vital role of audience in communicative acts.[45] For the premodern Indian context, audience certainly was a factor in constructing situations of speech agency. It fell to yogic *listeners* of animal speech to interpret correctly the speech acts of animals whose sounds conveyed meaning. In the epistemological discussion of good testimony and apt teachers, the *listener* needed knowledge of words' meanings (and sometimes contextual data grasped from the surroundings) to determine the validity of a parrot's statement and if one could acquire knowledge and learn from the bird. The parrot or other animal's agency depended in part on the human listener's interpretive skills and in part on the listener's attribution of validity to the bird's speech.

Rhetoricians have also signaled the importance of body, space, and material concerns for agency in speech acts, while still generally privileging human bodies and spaces.[46] While analyzing these texts, I have found that recent scholarly contributions from the area of new materialism offer perhaps the most useful model for approaching agency and subjectivity in speech acts.[47] Nonanthropocentric new materialists such as Karen Barad indicate that "Agency is an enactment … So agency does not involve choice in any liberal humanist sense; rather, it is about the possibilities and accountability entailed in reconfiguring material-discursive apparatuses of bodily production, including the boundary articulations and exclusions that are marked by those practices."[48] In *Kādambarī*, parrot Vaiśampāyana's "material-discursive apparatuses" include his organs of articulation and his story-telling abilities, key devices in the narrative of reuniting lost lovers. The parrot has agency in his enactment (or performance) of story-telling due to the material-discursive apparatuses he has available, produced by and in his body. Following the new materialist view, parrots and other birds can show agency simply through their verbalized repetitions.

Also in *Kādambarī*, in a passage describing a religious hermitage of sages, Bāṇa wrote that the parrots recited a prayer that "they'd picked up from incessantly hearing it."[49] Ultimately, it is irrelevant if these birds *chose* to recite or if they simply imitated Brahmin priests. The performative value of this sort of prayer or recitation might be equally valuable for ritual efficacy as long as it is uttered. In other words, the result is effected because of the sound of the mantra recitation itself. Moreover, these birds are agents of their recitations, according to new materialism. Bird utterances in ritual performance do differ from the educational or epistemological value of bird utterances discussed by Gaṅgeśa, but Barad would treat both as assertions of agency, regardless of any intentionality in speaking. Further, parrot Vaiśampāyana's story-telling is not solely renarration; his iterations are new creations, as his tellings add new layers of contextual meaning and variation. Birds in fiction might be unique in vocalizing utterances that express *creative* agency. However, as I have developed throughout this chapter, such a situation is not necessary for avian speech acts to have value or agency.

Even when speech acts are verbatim repetitions of earlier sources, "parroted" speech has value for early South Asian literary cultures in its mimesis as such. The value of mimicry in premodern traditions lies in the accurate memorization of a text for the sake of its preservation and in this being a tool for transmitting verbal testimony to others as a means of acquiring valid knowledge (i.e., teaching). In fact, for a re-evaluation of speech acts, it is important to remember that within this tradition, intact recitation was *more* significant than new creation. In such a context, new materialism's model for understanding agency is very apt, since it deprioritizes intention and acknowledges the complex relations between matter and discourse that create meaning. This model can account for considering bodily processes—flawlessly repeated utterances—as voiced agency in both *Kādambarī* and Gaṅgeśa's debate on means of knowledge. Equally well, this model can account for agency in the bodily production of articulately sounded phonemes, as in the *Mahābhāṣya* grammar discussion.

The new materialist approach to engaging agency through speech iterations has applications on various levels. On the micro level, a speaker has agency effecting ritual results through the recitation of prayer or mantra. On a broader level, a speaker has agency in the memorization, preservation, and reproduction of works with great academic, religious, and literary value. Finally, in certain circumstances, the narrator is not simply a voice "parroting" information to be passed on but also has the potential to be a creative agent in and of himself. Such narrative iterations reformulate, recreate, and make newly relevant the stories embedded within larger texts, a phenomenon found in many literary traditions.

CONCLUSION: ARTICULATING DIFFERENCE

Nonhuman animals articulate differently from humans but are not speechless or without language; passages from the *Yogasūtra* and *Mahābhāṣya* commentaries assert that nonhuman animals have language and speech. Oftentimes, nonhuman articulations of voice are beyond ordinary human perception or understanding. In these cases, early religious texts can illuminate nonanthropocentric ways of considering inaudible voice, either as grains crying out beyond our sensory perception or as unuttered speech, the supreme level of voice that communicates all knowledge and is representative of *Brahman*, the universal divine. In determining what is real and what has value, these religious approaches to voice thwart Western anthropocentrism. Yogic texts can help us comprehend animal language, if we can reach such an advanced level of yogic concentration. While highlighting distinct priorities, such as the accurate reproduction of texts for posterity and for teaching, these works suggest that *animals* demonstrate ideal models for *human* behavior and qualities. The writers never label these qualities as "human" or "animal," but they present their ideals in *animal* form.

In some of these premodern texts, the writers negotiate interspecies relations and human-animal dynamics that transcend impermeable statuses. Texts such as the *Yogasūtra* commentaries, *Kādambarī*, and Gaṅgeśa's *Tattvacintāmaṇi* promote permeability rather than rigid boundaries dividing ideas of "human" from "animal." While later commentators of the *Mahābhāṣya* attempt to position humans differently from animals, most of the texts in this chapter altogether ignore a hierarchical positioning of humans over animals. Ultimately, all of these writers use animal speech to articulate their own paradigms for understanding themselves and their environments that included animals. Most works suggest that certain birds and humans alike can repeat texts accurately for preservation, articulate intelligibly and correctly, communicate meaningful ideas via semiotic codes or statements, and even teach, if listeners can appreciate the relations between facts and the words representing those facts. Interestingly, both the *Yogasūtra* commentaries and the *Tattvacintāmaṇi* place the responsibility on the *listener* to interpret and infer correctly based on sets of knowledge. The speaker has value nonetheless, but it falls to the listener to assess and interpret the situation. I reiterate Parrish in stating that a shift in responsibility to the listener or audience is necessary for rhetorical studies today. If we are to revise our anthropocentric modes of understanding and evaluating animal language, it falls upon us listeners, with our human limitations, to negotiate the gap in our understanding of nonhuman language and voice in order to come closer to what rhetorical animals mean.

NOTES

1. Jacques Derrida used the phrase "the question of the animal" in his analysis of Western philosophy's approach to understanding "human" and "animal." See Jacques Derrida, *The Animal that Therefore I Am*, ed. Marie-Louise Mallet, tran. David Wills (New York: Fordham University Press, 2008). Cary Wolfe and others in animal studies have used the phrase more recently. See Cary Wolfe, "Human, All Too Human: 'Animal Studies' and the Humanities," *PMLA* 124.3 (2009): 564–75. Derrida and Wolfe depart significantly from earlier Western thinkers who grappled with the problem of what is "human" and what is "animal" in the human animal.

2. Acknowledgements: I would like to thank Joel Brereton for his wisdom and teachings, Stephen Phillips for assisting me in interpreting the *Tattvacintāmaṇi* parrot passage, Katarzyna Beilin for awakening me to humans' encounter with our animal selves many years ago, and the editors of this volume for their assistance.

3. See, in particular, Jacques Derrida, *The Animal that Therefore I Am*, and "'Eating Well,' or the Calculation of the Subject: An Interview with Jacques Derrida," in *Who Comes after the Subject?* ed. Eduardo Cadava, Peter Connor, and Jean-Luc Nancy (New York: Routledge, 1991), and Giorgio Agamben, *The Open: Man and Animal*, tran. Kevin Attell (Stanford, CA: Stanford University Press, 2004).

4. This permeability between human animal and nonhuman animal is supported by the theory of karmic rebirth within Hindu ideology, in which humans and animals take rebirth as either humans or animals depending upon the acts one has performed. Incidentally, one basic difference in conceiving "human" and "animal" in Sanskritic cultures was that animals' bodies are horizontal when moving, indicated by the common word for an animal (including a bird) "*tiryañc,*" different from vertically erect humans.

5. What I and other scholars have translated or described as a parrot is technically a parakeet. While it is a misnomer, *parrot* is the conventional term appearing across the literature; thus I have maintained this nomenclature.

6. I indicate in my translations and text if an argument pertains specifically to voice or speech or addresses language more generally.

7. Alex C. Parrish, *Adaptive Rhetoric: Evolution, Culture, and the Art of Persuasion* (New York: Routledge, 2014), especially chapters 3 and 4.

8. *Jaiminīya Brāhmaṇa* 1.43, tran. Wendy Doniger O'Flaherty in *Tales of Sex and Violence: Folklore, Sacrifice, and Danger in the Jaiminīya Brāhmaṇa* (Chicago: University of Chicago Press, 1985), 33–34. Probable composition of this text was at least by mid-first millennium BCE. Note that the verb used here for crying out (*ākrand*) was typically used for human shouts or cries of pain. It was not primarily the sound that an animal makes. Emphasis added.

9. The original commentary (*Vyāsabhāṣya*) on verse 3.4 defines this active practice of three-part concentration and joint restraint thus: "joint restraint practiced on one thing consists of joining three parts: concentration, reflection or absorption, and abstract meditation." *Pātañjalayogadarśanam: Tattvavaiśāradī-Yogavārttiketiṭīkādvayopetaṃ Vyāsabhāṣyam*, vol. 3 (Vārāṇasī: Kāśī Hindū Viśvavidyālaya, 1992), 1083.

10. *Yogasūtra* 3.17. My translation: "From the superposition of concept, object (or meaning), and sound onto one another, there arises a co-mingling of the three. From yogic joint focus restrained on the separate parts of that co-mingling, there arises knowledge of the noises of all beings." Ibid., 1221.

11. Ibid., 1221–38 and on.

12. Ibid., 1222.

13. Ibid.

14. Ibid., 1226.

15. Carlo Brentari, *Jakob von Uexküll: The Discovery of the Umwelt between Biosemiotics and Theoretical Biology* (Dordrecht/New York/London: Springer Publishing, 2015); Jesper Hoffmeyer, "The Semiome: From Genetic to Semiotic Scaffolding," *Semiotica* 198 (2014): 11–31; Thomas Sebeok, *The Sign and its Masters* (Austin: University of Texas Press, 1979) and Matthew Calarco, *Zoographies: The Question of the Animal from Heidegger to Derrida* (New York: Columbia University Press, 2008), 84.

16. Pāṇini *sūtra* 1.3.48.

17. The rooster really says "coo-coo" in Sanskrit ("*kukku*"), but I have translated the typical human English rendering of the rooster's speech. *Vidhiśeṣarūpaṃ*, vol. 2 of *Vyākaraṇamahābhāṣya of Patañjali with the Commentary (Bhāṣyapradīpa) of Kaiyaṭa Upādhyāya and the Supercommentary (Bhāṣyapradīpoddyota) of Nāgeśa Bhaṭṭa*, ed. with notes and variants M. M. Pandit Shivdatta Sharma (Delhi: Chaukhamba Sanskrit Pratishthan, 1988), 165.

18. Ibid. Interestingly, in his effort to make articulate speech the domain of only humans, Patañjali unintentionally *excludes* some humans from this group: those who, because of irregularities in their organs of vocal production or because of different capacities, are not able to produce all of the sounds represented in the Sanskrit alphabet.

19. This is the lack (privation) that Martin Heidegger theorized in *Sein und Zeit*, although this lacking did not indicate a hierarchical or evaluative ranking of beings. Martin Heidegger, *Sein und Zeit*, German ed. Thomas Rentsch (Berlin: Akademie Verlag, 2001), 194 and surrounding pages, according to Jacques Derrida's later analysis in *The Animal that Therefore I am*, 155.

20. *Vidhiśeṣarūpaṃ*, vol. 2 of *Vyākaraṇamahābhāṣya of Patañjali*, 165. "A voice must not only be articulate in sound production, but also articulate in expression."

21. Ibid.

22. Ibid.

23. Ibid.

24. Descartes' language-test argument and action-test argument (testing response in applying knowledge to new situations) discredited animals' reasoning abilities and linguistic capacities. In fact, Descartes similarly used birds such as magpies and parrots as examples in which adequate speech organs are not sufficient to produce declarative speech, "that is, so as to show that they understand what they say." René Descartes, *A Discourse on Method*, tran. John Veitch (London: Dent, 1912), 45. In "The Principles of Nature and of Grace," Leibnitz prioritized intellect (i.e., reason) and humankind's capacity for inference, neither of which he thought was observable in "brute" animals, and both of which he thought were necessary for executing one's will (i.e., intentional agency) based upon reasoning intellect. Gottfried Wilhelm Leibnitz, *The Philosophical Works of Leibnitz*, ed. George Martin Duncan (New Haven: The Tuttle, Morehouse & Taylor Company, 1908), 300–302. Later rationalist and enlightenment thinkers followed this trend of thought.

25. Elliot Sober, "Comparative Psychology Meets Evolutionary Biology: Morgan's Canon and Cladistic Parsimony," in *Thinking with Animals: New Perspectives on Anthropomorphism*, ed. Lorraine Daston and Gregg Mitman (New York: Columbia University Press, 2005), 85. The discussion of anthropodenial first appeared in Frans De Waal, "Anthropomorphism and Anthropodenial: Consistency in our Thinking about Humans and Other Animals," *Philosophical Topics* 27 (1999): 255–80.

26. Note the inflected gendered pronoun for animals in Sanskrit (here, "he"). Sanskrit requires inflecting the gender of animals, with neuter reserved for neuter beings (e.g., some mythological creatures and exceptions). This contrasts with the English convention of using neuter gender "it" even in cases when the gender of the animal is known, a decidedly anthropocentric languaging of animals that removes animals' personhood. See Nina Varsava, "The Problem of Anthropomorphous Animals: Toward a Posthumanist Ethics," *Society & Animals* 1 (2013): 1–17, for anthropocentric nongendered languaging of animals.

27. Bāṇa, *Princess Kadámbari,* tran. David Smith (New York: Clay Sanskrit Library, New York University Press, 2009), 53–55. Emphasis added.

28. Bāṇa no doubt gave parrot Vaiśampāyana this name as a humorous reference to the legendary human Vaiśampāyana, narrator of the *Mahābhārata.* Parrot Vaiśampāyana's human reincarnations in *Kādambarī* are not literally synonymous with the Vaiśampāyana of the *Mahābhārata*, but in both texts they are Brahmin sages.

29. R. C. Dwivedi, "Bhartṛhari and Kashmir Śaivism," *Annals of the Bhandarkar Oriental Research Institute* 72–73 *Amṛtamahotsava* 1.4 (1991–1992): 105.

30. Ibid. In this case, speech and voice are essentially synonymous terms.

31. Ibid., 101.

32. Ibid., 96.

33. *Vidhiśeṣarūpaṃ*, vol. 2 of *Vyākaraṇamahābhāṣya of Patañjali*, 165.

34. *Paramalaghumañjūṣā*, 11. My translation of the Sanskrit passage cited by Dwivedi, "Bhartṛhari and Kashmir Śaivism," 105.

35. Ibid., 104. This discourse is not restricted to the distant past but is apparent in much literature from the subcontinent, including the *Kuyiṟ Pāṭṭu* (*Songs of the Cuckoo*), a twentieth-century Tamil poetry cycle of Subramania Bharati. This work reveals the poet's obviously *advaita vedānta* leanings. The cuckoo speaks in human language in poem three (lines 11–12), Bharati wonders if he himself might take bird form in his next birth (poem one, line 27), there is a lightning *flash* of sweet taste in the bird's presence (poem one, line 18), and the cuckoo's song of "coo-coo-coo" brings the whole world's meaning to Bharati's mind, akin with Kashmiri Śaivism's understanding of "brilliance" (poem one, lines 33–34). Cuppiramaṇiya C. Paratiyār, *Kuyiṟ Pāṭṭu* (Madras: South India Saiva Siddhanta Works Publishing Company, 1965).

36. Argument IV.i.16. Gaṅgeśa, *Word: The Śabdakhaṇḍa (Vol. IV) of the Tattvacintāmaṇi, with Introduction, Sanskrit Text, Translation, & Explanation*, V. P. Bhatta, vol. 1 (Delhi: Eastern Book Linkers, 2005), 123 (Sanskrit) and 286 (English). In this same point, the Vaiśeṣikas express that they have argued this elsewhere in their treatises. I have paraphrased my own translation, since the argument refers to some technical linguistic aspects outside the scope of this chapter. For example, the probans and probandum refer to aspects of language and semantics from previous arguments and Gaṅgeśa ellipses portions discussed earlier in this work. This complicates a more literal translation of this section.

37. Argument IV.i.17. Gaṅgeśa, *Word: The Śabdakhaṇḍa (Vol. IV) of the Tattvacintāmaṇi*, 123 (Sanskrit) and 286 (English).

38. Argument IV.i.18. Ibid. In other words, we as listeners can infer and come to understand something from having heard a linguistically correct statement and by establishing the relation of such a statement to observable fact(s).

39. Argument IV.i.42. Ibid., 126 (Sanskrit) and 311 (English).

40. Ibid.

41. Ibid.

42. To name a few, see the Pāli *Tittirajātaka* 438, in *The Jātaka, or Stories of the Buddha's Former Births,* English edition, vol. 3, tran. H. T. Francis and R. A. Neil (London: The Pali Text Society, 1973), 319–23; and Tamil *caṅkam* poem *Perumpāṇā-ṟṟuppaṭai*, lines 300–310 (ca. second century CE) in *Pattuppāṭṭu: Ten Tamil Idylls*, 2nd ed., tran. J. V. Chelliah (Madras: The South India Saiva Siddhanta Works Publishing Society, 1962), 123 (Engl. tran.), 122 (Tamil original).

43. Karen A. Foss, "Rhetorical Theory," 853–57 in *Encyclopedia of Communication Theory*, ed. Stephen W. Littlejohn and Karen A. Foss, vol. 2. (Thousand Oaks, CA: SAGE Reference, 2009), 856. Further, the components of a rhetorical situation per Lloyd Bitzer include the rhetor and the degree of agency available to him; the audience and the constraints available to them; the discourse, message, or symbols; the demand calling for communication; and the larger contexts. Ibid., 853.

44. Carolyn R. Miller, "What Can Automation Tell Us about Agency?" *Rhetoric Society Quarterly* 37.2 (2007): 150–53.

45. Parrish, *Adaptive Rhetoric*, 118–23.

46. Cheryl Geisler, "How Ought We to Understand the Concept of Rhetorical Agency? Report from the ARS," *Rhetoric Society Quarterly* 34.3 (Summer 2004): 14–15.

47. New materialism is a recent interdisciplinary turn that works from the first principle that one cannot isolate matter from other components of the universe such as mind, and that elements such as mind or emotion have material components. In rejecting the dialectics of mind versus matter, reason versus emotion, and rationality versus animality, the turn offers much to animal studies, in my opinion, as well as informing topics as varied as sexual differing and subjectivity.

48. Karen Barad, "Matter Feels, Converses, Suffers, Desires, Yearns and Remembers," Interview with Karen Barad, Ch. 1.3., *New Materialism: Interviews & Cartographies,* Rick Dolphijn and Iris van der Tuin, editors and introduction, Series: New Metaphysics (Open Humanities Press, Michigan Publishing, University of Michigan Library, 2012). Accessed December 30, 2015. http://quod.lib.umich.edu/o/ohp/11515701.0001.001/1:4.3/--new-materialism-interviews-cartographies?rgn=div2;view=fulltext.

49. Bāṇa, *Princess Kadámbari*, 177.

BIBLIOGRAPHY

Agamben, Giorgio. *The Open: Man and Animal.* Translated by Kevin Attell. Stanford, CA: Stanford University Press, 2004.

Barad, Karen. "Matter Feels, Converses, Suffers, Desires, Yearns and Remembers." Interview with Karen Barad. Ch. 1.3. *New Materialism: Interviews & Cartographies*. Edited and introduction by Rick Dolphijn and Iris van der Tuin. Series: New Metaphysics. Open Humanities Press, Michigan Publishing, University of Michigan Library, 2012. Accessed December 30, 2015. http://quod.lib.umich.edu/o/ohp/11515701.0001.001/1:4.3/--new-materialism-interviews-cartographies?rgn=div2;view=fulltext.

Bāṇa. *Princess Kadámbari.* Translated by David Smith. New York: Clay Sanskrit Library, New York University Press, 2009.

Brentari, Carlo. *Jakob von Uexküll: The Discovery of the Umwelt Between Biosemiotics and Theoretical Biology*. Dordrecht/New York/London: Springer Publishing, 2015.

Calarco, Matthew. *Zoographies: The Question of the Animal from Heidegger to Derrida*. New York: Columbia University Press, 2008.

Derrida, Jacques. *The Animal that Therefore I Am*. Edited by Marie-Louise Mallet. Translated by David Wills. New York: Fordham University Press, 2008.

———. "'Eating Well,' or the Calculation of the Subject: An Interview with Jacques Derrida." In *Who Comes after the Subject?* Edited by Eduardo Cadava, Peter Connor, and Jean-Luc Nancy. New York: Routledge, 1991.

Descartes, René. *A Discourse on Method.* Translated by John Veitch. London: Dent, 1912.

De Waal, Frans. "Anthropomorphism and Anthropodenial: Consistency in our Thinking about Humans and Other Animals." *Philosophical Topics* 27 (1999): 255–80.

Dwivedi, R. C. "Bhartṛhari and Kashmir Śaivism." *Annals of the Bhandarkar Oriental Research Institute* 72–73 *Amṛtamahotsava* 1.4 (1991–92): 105.

Foss, Karen A. "Rhetorical Theory." Vol. 2 of *Encyclopedia of Communication Theory*, edited by Stephen W. Littlejohn and Karen A. Foss, 853–57. Thousand Oaks, CA: SAGE Reference, 2009.

Gaṅgeśa. *Word: The Śabdakhaṇḍa (Vol. IV) of the Tattvacintāmaṇi, with Introduction, Sanskrit Text, Translation, & Explanation.* Vol. 1, introduction, translated, and edited by V. P. Bhatta. Vol. 1. Delhi: Eastern Book Linkers, 2005.

Geisler, Cheryl. "How Ought We to Understand the Concept of Rhetorical Agency? Report from the ARS." *Rhetoric Society Quarterly* 34.3 (Summer 2004): 9–17.

Heidegger, Martin. *Sein und Zeit.* German edition by Thomas Rentsch. Berlin: Akademie Verlag, 2001.

Hoffmeyer, Jesper. "The Semiome: From Genetic to Semiotic Scaffolding." *Semiotica* 198 (2014): 11–31.

Jaiminīya Brāhmaṇa. Translated by Wendy Doniger O'Flaherty. In *Tales of Sex and Violence: Folklore, Sacrifice, and Danger in the Jaiminīya Brāhmaṇa*, edited by O'Flaherty, Wendy Doniger. Chicago: University of Chicago Press, 1985.

Leibnitz, Gottfried Wilhelm. *The Philosophical Works of Leibnitz.* Edited by George Martin Duncan. New Haven: The Tuttle, Morehouse & Taylor Company, 1908.

Miller, Carolyn R. "What Can Automation Tell Us about Agency?" *Rhetoric Society Quarterly* 37.2 (2007): 137–57.

Paratiyār, Cuppiramaṇiya C. *Kuyiṟ Pāṭṭu.* Madras: South India Saiva Siddhanta Works Publishing Company, 1965.

Parrish, Alex C. *Adaptive Rhetoric: Evolution, Culture, and the Art of Persuasion.* New York: Routledge, 2014.

Pattuppāṭṭu: Ten Tamil Idylls. 2nd Edition. Translated by J. V. Chelliah. Madras: The South India Saiva Siddhanta Works Publishing Society, 1962.

Pātañjalayogadarśanam: Tattvavaiśāradī-Yogavārttiketiṭīkādvayopetaṃ Vyāsabhāṣyam. Vol. 3. Vārāṇasī: Kāśī Hindū Viśvavidyālaya, 1992.

Sebeok, Thomas. *The Sign and its Masters.* Austin: University of Texas Press, 1979.

Sober, Elliot. "Comparative Psychology Meets Evolutionary Biology: Morgan's Canon and Cladistic Parsimony." In *Thinking with Animals: New Perspectives on Anthropomorphism*, edited by Lorraine Daston and Gregg Mitman. New York: Columbia University Press, 2005.

The Jātaka, or Stories of the Buddha's Former Births. English edition. Vol. 3, translated by H. T. Francis and R. A. Neil. London: The Pali Text Society, 1973.

Varsava, Nina. "The Problem of Anthropomorphous Animals: Toward a Posthumanist Ethics." *Society & Animals* 1 (2013): 1–17.

Vyākaraṇamahābhāṣya of Patañjali with the Commentary (Bhāṣyapradīpa) of Kaiyaṭa Upādhyāya and the Supercommentary (Bhāṣyapradīpoddyota) of Nāgeśa Bhaṭṭa. Vol. 2, *Vidhiśeṣarūpaṃ.* Edited with notes and variants by M. M. Pandit Shivdatta Sharma. Delhi: Chaukhamba Sanskrit Pratishthan, 1988.

Wolfe, Cary. "Human, All Too Human: 'Animal Studies' and the Humanities." *PMLA* 124.3 (2009): 564–75.

Chapter 13

The Rhetoric of Nonanthropocentric Rhetoric

Kristian Bjørkdahl

In 1974, the *Journal of Social History* published an article by one Charles Phineas, a veterinary historian at Boxer College, entitled "Household Pets and Urban Alienation." Inspired by the emerging discipline of social history, the author issued a programmatic call for a history of marginalized groups. Phineas believed that such an endeavor would open up "promising possibilities,"[1] but at the same time, he worried it might also be fraught with difficulty. There was a mean *catch-22* at the heart of it all: Marginalized groups were not in a position to write their own histories, precisely because they were marginalized.

> Homosexuals deserve a history, but a movement in this direction has not materialized, perhaps because homosexuals lack political muscle, perhaps because of more personal tensions among historians. Left-handers, another large group long subject to intense social discrimination, merit attention, but again their collective consciousness has lagged.[2]

Phineas nevertheless felt that a history of marginalized groups was both attainable and desirable, and boldly suggested that we should take on "the ultimate in the history of the inarticulate": The history of pets. Seeing as pets had become "a vital focus of modern society," they were now of "undeniable" academic importance. Phineas had already begun to consider the methodological minutiae of writing such a history: "A study of manuals on animal care, municipal ordinances, city registries and veterinarians' literature could produce more than one interesting essay."[3] But here, too, Phineas saw trouble:

> Without political power or claims, pets will hardly attract the interest of radical social historians. And at the other pole of academe, university administrators will be under no pressure to add courses on the history of pets, until such a time at least as obedience schools are merged with standard undergraduate fare.[4]

Despite these difficulties, Phineas "hope[d] for some serious work" in this field and concluded with a pathetic appeal to the journal's readers: "The history of pets remains too much the history of their masters, revealing more about the owning society than the owned. We must and can go further [...] Only then will the full historical power of domestic animals be unleashed."[5]

On the face of it, Phineas' article was an example of profound disciplinary innovation and daring. Historians could indeed move with the times! The article was, however, a spoof, and the author's name a pseudonym. Rather than say anything serious about the history of animals, this author wanted to ridicule the prospect of "social history" altogether, not least what he considered its propensity to treat quite marginal topics as central. To Phineas and his conservative colleagues, the idea of a history of animals was as ridiculous as a history of homosexuals or of left-handers! Phineas was not at all a proponent of disciplinary innovation and daring; he was a satirist upon that very idea.

What makes this spoof article interesting more than 40 years later is that the mock visions of this fake author have largely materialized—with scholars in some respects even venturing beyond his *faux* prophesies.[6] What is more, this development has not been unique to the discipline of history. With the coming of the "animal turn" in the humanities and social sciences,[7] there has been a surge of interest in animals and their experiences, in human-animal relationships, and in the symbolic and practical uses of animals by humans. There are now journals devoted to human-animal relations and courses in animal studies, as well as introductory textbooks, newsletters, discussion groups, and all the other markers of an academic field on its way to adulthood. And while obedience schools have still to merge with undergraduate fare, Donna Haraway's books are by now familiar items on undergraduate reading lists, which is pretty much the same thing.[8]

A retrospective vindication over Phineas' ridicule came in an aptly titled text from 2002, "A Left-Handed Blow: Writing the History of Animals," by historian and animal studies scholar Erica Fudge.[9] Here, Fudge reiterated Phineas' prospect, but now in earnest—thus embodying the thrust of the animal turn toward "taking animals seriously."[10] Fudge was concerned about whether and how humanity's predisposition for anthropocentrism can be challenged by a history of animals. She argued that this would require "holistic" history writing, which she defined as the attempt to "write a history which refuses the absolute separation of the species; refuses that which is the silent assumption of humanist history." The idea of a history of animals was not, Fudge argued, "merely a 'fad' in the ever widening reach of historical scholarship," but a development of "existing debates in the discipline as well as in the wider world of human relationships with nature." The history of animals is, she concluded, "a necessary part of our reconceptualization of ourselves as human."[11]

Fudge illustrates well for history what I, here, want to take up in the case of rhetoric, namely a nonanthropocentric reconfiguration of the discipline. If the human-animal distinction is part of the self-image of rhetoric—and I believe it is—what would it mean to dismantle that distinction? Would we have to deconstruct "rhetoric" as we know it? What could and should be the way to go about reconfiguring rhetoric beyond the exclusively human?

RECONFIGURING RHETORIC

Efforts to enact nonanthropocentric reconfigurations of rhetoric have a peculiar recent history. In one sense, rhetoric joined the animal turn puzzlingly late. Only in the last few years have "animal rhetorics" begun to crystallize as a specialization within the discipline—with a 2011 special section of *Philosophy & Rhetoric* providing one important impetus. By this time, scholars in history, sociology, philosophy, law, literature, and so on had been doing "animal studies" for two decades or more.[12] From another perspective, however, rhetoric had too. One could even argue that rhetoric was among the first academic disciplines to "reconceptualize" itself within debates about "human relationships with nature." This particular reconceptualization can be pinpointed with rare accuracy, specifically to the date of publication for George A. Kennedy's article, "A Hoot in the Dark: The Evolution of General Rhetoric."[13]

My approach in what follows will be to look at responses to Kennedy's famous article as a proxy of the animal turn in rhetoric. Although most recent scholars of animal rhetorics are certainly not true disciples of Kennedy, the signal importance of his article is nevertheless undeniable. When the text was published, in 1992, Kennedy was a respected veteran of rhetorical studies, with several notable histories of rhetoric to his name, not to mention a recent translation of Aristotle's *Rhetoric*. He was, in other words, not some random professor with an agenda, but rather a representative—or so everyone thought *pre*-1992—of rhetoric's mainstream. The academic moves of someone like him would presumably have repercussions in the field. But Kennedy's "hoot" fell largely on *confused* ears. The response—or rather, lack thereof—recalled first and foremost a certain nervousness, as Debra Hawhee tellingly narrates: "There were whispers, sidelong glances, and muttering, all of which bespoke a slight panic about his—and the field's—direction. What was Kennedy doing? Was our distinguished leader, translator of the Sage himself, going off some deep end and taking the discipline with him?"[14] Despite what, in retrospect, seems like a momentous, even revolutionary, proposal, Kennedy's article did not become the object of much debate. It was well known, no doubt, but did not cause any sustained interest in "animal rhetorics."

When such a field finally materialized, a couple of decades later, Kennedy's article was seen, as Diane Davis writes, to be "a seminal essay that should also have been pathbreaking."[15] This paradox is what interests me here. Why has Kennedy's article come to be seen as seminal, but marginal? Why was it *not* pathbreaking?

In this chapter, I will follow the traces of Kennedy's article through the webs of academic culture, as a reception historian interested in the rhetorical dynamics of cultural change. As a backdrop for this study, I posit Richard Rorty's idea that conceptual and intellectual change can be mapped onto Kuhn's famous theory of scientific revolutions, in other words, that such change tends to take the form of a battle between the different vocabularies of an old and a new generation—vocabularies that speak to different language games, and which hence do not "connect."[16] My approach will be to bring this theory down on the ground, as it were, and confront it with empirical material.[17] This orientation is inspired by the "late" Quentin Skinner's rhetorical history of concepts,[18] where the central idea is to study the rhetorical moves that actors make in order to enact—or in principle restrain—change. My goal with this is to see how Kennedy's call for a reconfiguration of rhetoric was taken up by other rhetorical scholars, and to see how they in turn have rejected and refused, recast and refined, his proposal. Possibly, this might also tell us something about the future prospects of a nonanthropocentric rhetoric.

A satisfactory story of Kennedy's article must account for the fact that it, at first, was largely sidelined, while it over time acquired a status as a paradoxical classic—that it became, as Davis claims, a seminal essay that failed to break any paths. My argument will be that Kennedy's text went against the grain of two opposite incentives of the academic field to which it was addressed.[19]

First was the incentive of *conservation*, which is attached to the project of maintaining an academic legacy. On the way to a career in academia, any academic makes investments—of time, effort, and identity—in particular problems, theories, themes, schools, or the like. Once that investment is made, that person will be reluctant to give it up, because that would mean to forfeit both professional identity and self-image as well as a hard-won status in the field.[20] For those who had their investments tied up with maintaining rhetorical tradition somehow, Kennedy's proposal presumably represented too much of a move away from the problems, theories, themes, and schools in which they had invested.

Second is the incentive of *innovation*. If the academic field prizes tradition, it also rewards those who proclaim to be taking the discipline in new directions, driving the forefront of research, forming the field's *avant-garde*. Scholars who had invested in this progressive project, thought that Kennedy's proposal offered a promising stimulus, but at the same time, they found reactionary elements lodged at the heart of his seemingly revolutionary call. Consequently, their response has typically been to first acknowledge his contribution and then to distance themselves from it.

If these two factors in combination explain both the relative neglect that Kennedy's article was met with—especially in the first period—and the attitude that many of his more recent commentators take toward him, we must also note that this has left the legacy of "A Hoot…" in a peculiar situation. Kennedy offered too much of a novelty for the conservatives, but was considered old hat by the avant-garde—so his call fell between two stools. As I will argue in conclusion, that is no place to leave Kennedy's contribution.

HALF-HEARTED ECHOES OF A HOOT IN THE DARK

The argument of Kennedy's piece should be familiar enough to readers of this book, but to refresh memory, Kennedy sketches a theory of a "general rhetoric," which in his very broad view sees rhetoric as prior to speech and, thus, prior to human beings. Kennedy's basic goal is to account for the fact that what we know as "rhetoric" has emerged from within an evolutionary process, where all sorts of signaling behaviors fulfill pragmatic, evolutionary functions. When understood broadly, Kennedy argues, "communication" is such a basic aspect of many—if not most—species' strategies of coping in their environments that a type of "proto-rhetoric" can be said to be performed not only by other animals but perhaps even by plants. He offers an ample number of examples of nonhuman "rhetoric," drawing comparisons along the way with familiar human forms. With this basic understanding of rhetoric, Kennedy arrives at his oft-cited definition: "Rhetoric in the most general sense may perhaps be identified with the energy inherent in communication: the emotional energy that impels the speaker to speak, the physical energy expended in the utterance, the energy level coded in the message, and the energy experienced by the recipient in decoding the message."[21] Maybe one can even identify a basic rhetoric unit, Kennedy speculates, which he creatively calls a "rheme." Kennedy combines this evolutionary strand, however, with various elements of classical rhetoric. For instance, he construes confrontations between stags as analogous to deliberative rhetoric, and gives corresponding examples for the judicial and epideictic genres. He also connects animal rhetoric to the five canons, with his argument of the primacy of *actio*. The ultimate objective of his piece was to find out what we talk about when we talk about rhetoric. The great surprise, to his fellow rhetoricians, was that Kennedy went not just to his own specialist area of knowledge—classical rhetoric—but also to evolutionary sources for his answer.[22]

How was this call received by the community of rhetorical scholars? The first stop in the reception history of "A Hoot in the Dark" is in fact the only text that qualifies as a direct *response* to Kennedy's article, namely Jo Liska's "The Role of Rhetoric in Semiogenesis."[23] For a long time, this response was the only real, prolonged engagement with Kennedy's proposal, although even this is an overstatement, considering that hers was a six-and-a-half-page "Discussion Note." Liska was "entirely sympathetic to [Kennedy's] suggestion that understanding human codes will be facilitated by examination of nonhuman codes," but she thought his definition "far too broad to be conceptually useful." Her aims were consequently to "examine Kennedy's conceptualization, identify its excesses, and provide an alternative perspective that imposes some constraints" on what "may be considered rhetorical and what should not."[24] More concretely, Liska argued that Kennedy's definition had nothing to distinguish "rhetoric" from "all behavior on the part of all living organisms."[25] In contrast, she argued that rhetoric emerges at a particular point on a "continuum of sign behavior," and that it should be understood, more specifically, as "the manipulation of signs in the service of social influence."[26]

Liska's article is the great exception to the rule of the early reception, not only because she investigates Kennedy's claims in some depth, but also because she actually takes in interest in animals. However, we may note that even Liska had a conservative objective. She sought to impose "constraints" by way of identifying Kennedy's "excesses" and thought his approach far "too broad." Despite her fundamentally sympathetic attitude, her article can be read as an attempt to police the boundaries of rhetoric by insisting on what "may be considered rhetorical and what should not."

After Liska's head-on treatment, scholars tended to make much briefer mentions of Kennedy and also to sideline the animal issue almost entirely. George Khushf is one example. He mentioned "A Hoot…" briefly in a piece from 1995, in which his main objective was to provide "a phenomenological account of the transformation of consciousness brought about by writing," a shift that, as he saw it, involved replacing "the self-other relation of rhetoric […] by a self-enclosed world of text leading to the reinterpretation of rhetoric as ornamentation."[27] Khushf's argument was that "the increasing grammacentrism of the West […] leads to the metaphysical doctrines that Derrida criticizes, and that Derrida's attempt to formulate a nonmetaphysical structure of writing must involve a recovery of the rhetorical dimension of language."[28] As Khushf saw it, the importance of Kennedy's contribution, vis-à-vis Derrida's view, was that Kennedy understood rhetoric to be a phenomenon prior not just to speech, but also to writing. Khushf also underlined the importance of Kennedy's diachronic perspective as opposed to Derrida's synchronic view. However, Khushf never moved beyond a rudimentary engagement with Kennedy's article and—as importantly, in the present context—he did not really consider the implications for our view of animals, or for the academic practice of rhetorical studies.

Khushf's use of Kennedy might seem marginal, but for my narrative, his is precisely the type of response that should concern us. Khushf reads "A Hoot…" in the way that Michel de Certeau once argued that readers in general read, that is, as *poachers*;[29] Khushf takes what he needs for his own purposes—"beating the text into shape," as Richard Rorty once wrote[30]—and leaves the rest. What he leaves, in short, are the animals. They have no central part to play in Khushf's drama. To be clear, I am not suggesting that there is anything wrong with this reading strategy. "Poaching" is an apt term for how we read a lot of the time, and also quite legitimate; if we could not adapt what we read to our purposes, we would never progress. The point here is rather that programmatic calls like Kennedy's are not automatically taken as such. Even when they are not directly opposed, revolutionary proposals are often watered down and appropriated for other, more moderate, purposes.

Similar marginalizations have been common throughout the reception. In many cases, Kennedy's article has been used to provide incidental support for the point that rhetoric is more than language, more than symbol, which the author in question has then proceeded to develop within some other frame than Kennedy's. For instance, in a rare explication of Charles Sanders Peirce's notion of rhetoric, Roberta Kevelson wrote that "Peirce's Rhetoric does not restrict its scope and domain to words; it is no less concerned with spatial relations, configurations which have persuasive force,"[31] and also that Peirce's notion of rhetoric is a "radical, instinctual thrust- and-grasp force: It is an energy with basic, biological goals."[32] In this connection, she mentions Kennedy's idea that rhetoric is a form of energy, but only very briefly, for support and, perhaps, to give her own contribution an air of updated relevance. The same tendency is on display in Ulrike Zinn Jaeckel's article "Hecklers and the Communication Triangle," where the author—but again, only very briefly—appropriates Kennedy's ideas about the relation between senders and receivers to her own agenda: "[W]e can … explore the article's implications for the speaking and the listening roles,"[33] she says in a brief reference to "A Hoot…," but she does not go on to explore those implications. The fact that one such implication could be a study of animal rhetorics is not mentioned by any of these authors.

Another way to take proposals like Kennedy's, obviously, is to challenge them. This is indirectly what Mari Lee Mifsud does when she juxtaposes Kennedy's view in "A Hoot…" with that of another veteran rhetorician, Henry Johnstone Jr. Mifsud argued that, while Johnstone would agree with Kennedy that animals—and also machines—do "communicate" in one sense, the question of *in what sense* was precisely the issue: "Johnstone […] stops short of identifying the nonhuman and human senses of communication in the way that Kennedy does. He contends that when one collapses the distinction between these two senses of communication, one fails to question the assumption that communication is primarily a kind of competence." As she argues: "For Johnstone, though, human communication is unique and exists on a level different from that of bees or chess-playing machines. For Johnstone, communication is a *consciousness*, not a competence."[34] As she explains, consciousness in this context denotes freedom. Unlike animals and machines, say Big Blue, for which the energies of input and output are equivalent, a human actor—like, say, Kasparov—is conscious of his own deliberations, of the calculations he does and the motivations he has, including the various temptations he might be under. And to be conscious is to be free to qualitatively weigh and prioritize among these various concerns.

Here is implied a direct challenge of the relevance of Kennedy's idea that rhetorical energy is a basic evolutionary phenomenon that predates speech, writing, and the emergence of human beings. "Rhetoric" is uniquely human or, at any rate, exclusive to conscious beings, Mifsud's version of Johnstone insists. Mifsud does not construe Johnstone's resistance to Kennedy's call as absolute; he acknowledges, after all, that bees and machines also communicate. He is just saying that these are distinct from human rhetoric, and therefore, that they are phenomena that fall—and *should* fall—outside the scope of rhetorical studies. From this perspective, then, Kennedy was not necessarily wrong in any fundamental way, but at the same time, there were no significant implications of his intervention. Any prospect of "animal rhetorics" would require those inverted commas.

In his 1998 article "Rhetoric as a Course of Study," David Fleming introduces Kennedy in a reflection on the present use of the term "rhetoric" and its implications for rhetorical education.[35] Kennedy's intervention is here rendered as an extremity toward *naturalizing* rhetoric, where, as Fleming notes, "the word's range of application becomes virtually unlimited. Kennedy has gone so far as to claim that rhetoric is 'manifest in all animal life and existed long before the evolution of human beings.'"[36] This point of Kennedy's is not taken up by Fleming, however, who immediately adds: "But universalization occurs even when the word is confined to our species."[37] Fleming latches parts of his discussion onto the Big Rhetoric/Little Rhetoric debate,[38] where he saw Kennedy's contribution as a radical expression of the former—which Fleming construes as a view that considers rhetoric a "transdisciplinary hermeneutic method whose object of analysis, persuasion, was located everywhere and in everything."[39] The subtitle of Kennedy's article, with its reference to a *general rhetoric*, gave it away; Kennedy apparently envisioned a bigger, more thinly distributed, "rhetoric" than anyone had been able to conjure up so far. The reflex of scholars like Fleming was to resist this movement. True, Fleming did not opt for the reverse extreme, but wanted rather to revive a notion of rhetoric "that is neither so *big* that it loses cultural specificity, nor so *little* that it fails to attract adherents."[40] He did complain, however, that the universalist features of this "new 'rhetoric' [...] have served to separate it from its own history," and that "it is perhaps unsurprising [that] there would come attempts to restrain its globalization and dispersal."[41]

Fleming's piece illustrates quite explicitly the argument I am making, that certain scholars wanted to resist the move toward a general rhetoric. Fleming has been a proponent of rhetoric as an integrated part of (higher) education, he is the author of a book about the recent history of composition,[42] and he has on at least one other occasion taken on the role of restrainer of new rhetoric's universalist impulse. In "Can Pictures Be Arguments?" his contention, against the avant-garde tendency to accept the notion that "images argue," was that pictures should *not* be thought of as arguments.[43] The point here is not to single out Fleming as a particularly reactionary figure. I highlight him here because he constitutes such an explicit example of the impulse to *restrain* Kennedy's proposal. Fleming was not alone in this regard. Even Liska wanted to moderate and restrain Kennedy's proposal, as did Mifsud/Johnstone. For their part, Khushf, Kavelson, and Jaeckel did not engage in head-on challenge or counterargument, but they managed to restrain Kennedy by other routes—notably through appropriation and marginalization.

The most striking evidence for the effects of the conservative incentive in this case is, however, more speculative and also negative in form. To be more specific, it consists in the overwhelming lack of response, that is, in the conspicuous *neglect* that was offered in response to Kennedy's article. Neglect is certainly not an uncommon response to an academic article; in fact, most academic output is hardly read by anyone at all. But it is a much stranger response in a case like this, where the author is a respected Grand Old Man of rhetoric, who issues a programmatic call, which presumably will have revolutionary implications for the discipline. When an almost united academic community largely failed to enter into dialectic with such a call, that is testimony, I would argue, that something more than coincidence is at play.

The assumption I want to apply in making sense of this is that scholars are prone to react to calls like Kennedy's with an eye to their own academic "investments."[44] An academic invested in an established scholarly tradition would naturally want to resist innovations that challenge or, actually, *devalue* these investments. There are several reasons why we should think that such motives are operative in this case. We are talking here about a proposal to rethink the meaning of several key words of our culture, including "rhetoric," "animals," and "humanity," and we are asked, more specifically, to rethink the idea that humans are the only "rhetorical animal." Words and concepts like these are far more than signifiers pointing to signifieds. We can think of them, rather, as semantic banks. A word like, say, "Human" is a container into which we have deposited various valuable meanings over a long period of time. Together with other key words, it contains the life savings of our cultural identities. In the same way that we only grudgingly part ways with the contents of our bank accounts, we are not keen on throwing away what is in these semantic banks. Without them, we have—and are—nothing.

If we can imagine such an effect to be operative at this level of generality, the threat to a person's self-image is presumably all the more acute when that person's professional identity is also deeply intertwined with these entities. This is precisely what we have in the case of rhetorical scholars, many of whom have seen themselves as guardians of a peculiarly *human* capacity. According to the founding fathers of this discipline, rhetoric is what makes us human.[45] The conservative reflex comes not just from meanings that we attach to concepts, however, but also from the more practical aspects of work. A rhetorical scholar, for instance, does typically not think of herself as someone who reads—or has any need to read—evolutionary theory. She is certainly not someone who does experiments. For the average rhetorician, these are distractions. They constitute a slippery slope toward Big Rhetoric, where the already challenged disciplinary core of rhetoric can no longer hold.

To be clear, I am not arguing that scholars necessarily occupy positions in an academic exchange consciously and strategically *in order to* achieve a certain social status. The line between internal and external motivations is, I believe, not so clear-cut. Rather, the social positions of scholars are bound up with what types of scenarios they are willing to take seriously, what arguments they are likely to find reasonable, what they will count as evidence, and so on.

If this speculation offers one way to make sense of conservative responses to Kennedy's article, which I am arguing was predominant in the first part of the reception, a later article by James Arnt Aune signals the transition to the next part, which has been marked more by an incentive of innovation. Aune notes that Kennedy compares animal signaling with epideictic rhetoric, and then states: "A renewed interest in the biology of human nature, communication, and politics would be a useful antidote to the postmodernist overemphasis on the malleability of the human subject, assuming that evolutionary psychology can escape its historic association with scientific racism."[46] What is interesting here is that Aune allows, and even encourages, Kennedy's interest in "the importance of evolutionary biology for the understanding of human behavior." With that same formulation, however, he restricts the scope of Kennedy's program exclusively to *human* behavior, and thus performs a subtle redefinition of the importance of Kennedy's article. Aune is drawn in two directions at once: He sees Kennedy as providing an impetus for disciplinary innovation—the study of the biology of communication—but at the same time, he wants to use this form of innovation to safeguard rhetoric from unwanted "slippages" into other, more threatening innovations. Aune is interesting, however, not just for being in a bind between innovation and conservation, but for how he construes the nature of the threat and Kennedy's program as a fortification. For when he envisages that Kennedy's contribution could be an "antidote to the postmodernist overemphasis on the malleability of the human subject," the later reception of Kennedy's article has muted Aune's hopes. In the recently emerging field of animal rhetorics, "postmodernist" antiessentialism and malleability of the subject has in fact been one very notable tendency, if not even *the* defining feature.

SEMINAL, AND *NOT* PATHBREAKING

For an interesting example of how the avant-garde enters the reception history of Kennedy's article, we can backtrack from Aune to an article by Cornelia Wells from 2003, in which she cites Kennedy in the context of what she calls "fragmatics"—an intentionally improvisionary and *ad hoc* approach to rhetoric and its history. Wells certainly demonstrates what most would associate with the term "postmodernist," and the phenomenon of *malleability* is in fact what motivates her article. Wells cites Kennedy's point that "the meaning is the interpretation given to the communication by another animal," and asks: "If then the energy required to 'code' and 'decode' messages [. . .] is always open to mischarge or mis(and back)fire or otherwise miss the mark, why are we talking or rhetoricizing [. . .] in the first place?"[47] For Wells, the implication of the "naturalism" of the likes of Kennedy and Richard Rorty seems to be that we have no incentive to try to hold things in place, to strive for structure and stability. This means that rhetoric comes to be seen as the "eternally *ad hoc*," always partial and in progress. Wells's aim is to work out the practical, academic implications of this ontological insight in the form of an improvisionary historiographic practice, and in this, she takes her cue from among others Victor Vitanza and his "antimethodology [. . .] which endlessly disrupts and desynthesizes all vestiges of totalitarian (totalizing or systematic) knowledge."[48] The idea that instability runs through all our exchanges inspires also her style, which is experimental, to say the least—and piecemeal, staccato, stop-start, to be more specific.

Wells never really uses Kennedy to develop a concept of *animal* rhetorics, and we should keep in mind that her article was published some time before the more recent wave of reception to "A Hoot. . ." Nevertheless, her approach foreshadows later developments by placing Kennedy's article in a setting dominated not by a conservative incentive but by the opposite, *innovation*. To apply the terminology of one of Wells's leading lights, Richard Rorty,

Wells's desire to transgress "normal discourse" and instead experiment with "abnormal" or "revolutionary discourse" is apparent in practically every sentence of her article.[49] In what appears to be a full reversal of Aune's scenario, Wells sees Kennedy not as offering an antidote to malleability, but rather as providing resources with which to develop rhetoric along precisely those lines.

A willingness to take on academic experimentation and innovation is the foremost characteristic, I would argue, of the more recent part of the reception. If the term *postmodernist* remains too troubled to use, recent references to Kennedy's article are typically made within the frameworks of scholars who—in comparison with Kennedy—are certainly more *avant-garde*. To mention but a few cases, Diane Davis cites Kennedy's ideas from within what is essentially a Derridaean gaze, as does Seegert, though somewhat less faithfully.[50] Pflugfelder uses Kennedy as a sort of bridge to Latour,[51] while Thomas Rickert, in developing his version of "ambient rhetoric," engages with such thinkers as Kristeva, Derrida, Deleuze, Heidegger, and Latour but adds that his "line of thought also bears an affinity to George Kennedy's important essay 'A Hoot in the Dark,' where Kennedy claims that rhetoric is energy."[52] Victor Vitanza seeks to integrate ideas from Kennedy with others from Deleuze.[53] And Emily Plec writes, in her intro to her edited volume, that "Deleuze and Guattari's essay 'Becoming-Animal' [...] provides a way of thinking about communication that, in some ways, echoes [...] Kennedy's definition of rhetoric as essentially 'a form of mental and emotional energy.'"[54]

The recent part of the reception is too voluminous to allow a detailed examination here, but for the sake of example, I will look more closely at a few. The first stop is Diane Davis's 2011 article "Creaturely Rhetorics," which is not a coincidence, considering that Davis has been cited quite widely and is now seen as a leading figure in this field. Davis argues that Kennedy's merit was to show "that rhetoric has a 'natural' dimension to which rhetorical studies has not attended."[55] She agrees "wholeheartedly [...] that rhetoric, at its most elemental, takes place at the level of the creature," but argues that Kennedy's contribution involves "a recuperative, deflective move that protects a fairly standard anthropocentric approach to rhetorical theory." Then she adds what one might consider a dialectic advance from Kennedy:

> I would like to argue instead that the "mental and emotional energy" that Kennedy calls "rhetoric" is itself dependent on an always prior rhetoricity, an affectab*ility* or persuad*ability* that is due not to any creature's specific genetic makeup but to corporality more generally, to the *exposedness* of corporeal existence. To be affectable, persuadable, is to be always already affected, persuaded, which means: always already *responsive*. Rhetoric is not first of all an essence

> or property "*in* the speaker" (a natural function of biology) but an underivable obligation to respond that issues from an irreducible relationality.[56]

Although she embraces Kennedy's push, Davis's commitments are closer to those of Derrida in "The Animal that Therefore I Am," one important implication of which is that:

> [I]t's not just a matter of acknowledging that certain nonhuman creatures display capacities that rhetoricians, along with philosophers and scientists, tend to insist are specifically human. It's at least as important for rhetorical studies to ask whether it's certain that *humans* display these capacities—or at least if we display them in all the purity that contemporary theories of rhetoric presume.[57]

With a rhetorical term, Davis here applies a strategy of *paramologia*, which is incidentally quite characteristic of a strand in the recent reception of Kennedy. She admits that Kennedy is on to something when he construes rhetoric as taking "place at the level of the creature," and more generally, with his quest for rhetoric "at its most elemental," but insists that his gaze does not allow us to overcome tradition. Because of his "recuperative, deflective" moves, Davis says, Kennedy's contribution represents a "fairly standard anthropocentri[sm]." With Derrida, Davis's concern is to overcome essentialism, so that "rhetoric" is no longer understood as "an essence or property '*in* the speaker,'" but rather as "an affectab*ility* or persuadab*ility* that is due … to corporality more generally."

One could object that this countermove is somewhat strange, first because Kennedy so clearly stresses reception and the relation between senders and receivers, and second because the capacity to be *affected by* rhetoric appears no less an essential characteristic—nor, for that matter, any less of a relational obligation—than the capacity to *utter* it in the first place. After all, hoots in the dark are bound to be mute if there is no one there to respond. Uttering/signaling and being affected by uttering/signaling are two sides of the same rhetorical coin—and neither is possible without the other.[58] Davis is set, however, on transgressing the frame laid down by Kennedy, and this move seems motivated by what is in fact a somewhat different point, namely the Derridaean one that not even human subjects display the capacities so often assumed to be characteristic of our species—presumably, the capacity for intentional and effective use of discourse. For Davis, it seems, Kennedy fails to question the idea of rhetoric itself. When we get to the Derridaean *avant-garde*, the effort to extend rhetoric to animals is really of no effect, since here, rhetoric as we know it—the intentional use of language and other symbols for particular purposes—is no longer seen to exist. What remains, one could say, is only a fragment of the idea of "communication," specifically the general conditions of *rhetoricity*, *affectability*, *exposedness*, *responsiveness*.

Someone willing to place her bets on this kind of argument clearly does not work under an incentive of conservation. Contrary to the response in the first period, Davis is not invested in an established scholarly tradition; to the contrary, she wants to disassemble it. ("Deconstruction!" is, after all, the imperative of Derrida and his followers). Conversely, Davis has no reason to resist disciplinary innovations, but is driven rather by an aim of furthering them. This incentive of innovation arises from her particular academic investments, just as the opposite incentive arose for the conservatives. As a scholar, Davis has invested in developing the intersection between rhetoric and continental philosophy, which presents not just a new angle on, but also a radical departure from and challenge to, rhetorical tradition.[59] An irony of history is that work performed under the incentive of innovation has become so ubiquitous that one could argue that it is in fact what now constitutes "the establishment."[60]

Another, telling, tendency in the recent reception is an explicit treatment of the status and reception of Kennedy's article itself—that is, the theme that occupies me here. Elsewhere, Davis notes that "George Kennedy's 'A Hoot in the Dark' is the grand exception" to the rule that rhetoric has tended to posit an abyss between humans and animals. She then adds that, "His sociobiological challenge to humanist conceptions of rhetoric was never seriously taken up by the field."[61] Natasha Seegert refers to "A Hoot…" as a "lamentably undervalued article" that "made the rather controversial claim that nonhuman animals, indeed even nonsentient plants, produce rhetoric." She laments that Kennedy's article has typically only been cited for its definition of rhetoric, and then proceeds to position her own contribution as a sort of response to Kennedy's call:

> The time has come to listen to, not just talk about, nonhuman animals and thereby overturn the anthropocentric narrative that limits rhetoric to humans. Kennedy's hoots started the conversation. Rather than relegate those hoots to the dark, let's invite the voices of the more-than-human-world into the liminal space of twilight where the coyote sings and from there welcome them to bound into broad daylight.

Despite Seegert's pathos, she does not enter into any sustained engagement with Kennedy's work but, like Davis, goes to Derrida for both her vocabulary and her problematic orientation. Finally, Debra Hawhee, in "Toward a Bestial Rhetoric," writes that "'A Hoot in the Dark' was indeed untimely," and that its "untimeliness helps to account for the responses it drew, its illegibility to scholars in rhetoric, and the fact that it has not been cited all that frequently."[62] She elaborates the reception of his article as one of "befuddlement."[63] Hawhee nevertheless argues that, "In its own way […] Kennedy's fleeting dalliance with animality leaves us with a few [...] remarkable theoretical possibilities and paths for future research, tantalizing suggestions of what animality can do to—and for—rhetorical theory."[64] Hawhee's combination of a somewhat depreciative tone—"Kennedy's fleeting dalliance with animality"—with the idea that the piece is nevertheless a key contribution to the field, is indicative of the later reception of "A Hoot..." In fact, these examples all display the very same argumentative pattern that I identified in Davis above: Kennedy's article was a important piece, but it must nevertheless be adjusted, upended, improved, and transcended. This argument never develops into an academic dialectic, however, since most of Kennedy's commentators do not engage with his piece beyond a perfunctory reference and acknowledgement. After the mandatory acknowledgement, these commentators start again, now within another—typically more *avant-garde*—frame.

If we can gauge the recent state of animal rhetorics from the second special issue of *Philosophy & Rhetoric* on "Extrahuman Rhetorical Relations," it might seem that Kennedy has already been transcended by this *avant-garde*. While the one-page "Editor's note" states that "Kennedy's article was a harbinger of sorts, heralding the uptake of new materialists, object-oriented ontologists, posthumanists, and others who are seeking to theorize rhetoric in terms that exceed a humancentric account and to ground it in claims that go beyond and decenter the word or symbol,"[65] the article is thereafter mentioned only once in all the twelve essays of that special issue, and in that instance, in a short footnote.

HEROES, KOOKS, AND THE GRAND OLD MEN IN BETWEEN

My argument so far has been that the paradoxical fate of Kennedy's article—seminal but marginal—can be explained by two opposed incentives of this academic field, *conservation* and *innovation*. In short, while Kennedy's proposal was too radical for the conservatives, it was not radical enough for the avant-garde. In some ways, this reading of the reception of "A Hoot..." echoes Richard Rorty's idea of "normal" and "abnormal" discourse—which he in turn takes from Thomas Kuhn's idea of normal and revolutionary science. Rorty's basic contention is that intellectual change does not take place by way of rhetoric, strictly speaking. Rather, it simply emerges as a result of prolonged periods of "muddling through," that is, a largely aimless tossing around of new and revolutionary concepts.[66] Just as I have argued, some actors, depending on their position in social space, stand to benefit from furthering revolution. To engage in revolutionary discourse is always a gamble, however. Those Rorty calls "strong poets," that is, those who invent new language games for us to play, can never be sure of success—they can never be sure that their new vocabulary catches on. One reason for this, obviously, is the resistance of conservatives who are invested in maintaining the present vocabulary. This "older generation" will typically reject innovations that conflict with their intellectual worldview, just as Kuhn proposed of scientists invested in the old paradigm. All the avant-garde can do, then, is gamble. They have to work like poets, and experiment with language that makes no literal sense for those invested in the established language game. The course of history, rather than rhetoric or any other science, will tell whether such innovators will end up as heroic "revolutionaries" or as "kooks."[67]

This view is apposite to the case at hand, since it seems to put some flesh on the bones of the two incentives I have described. The "investments" I have talked about should, on Rorty's view, be seen relative to language games. By learning to speak a particular language, we make an investment, and because truths are made, as Rorty argues, with reference to such games, to speak another language game is not to speak falsehoods, but to address another truth-making "game."[68] Rorty's view, however, is historically quite thin, and he is not even interested in considering the historical *minutiae* of intellectual change.[69] When we consider closely the case of Kennedy, it becomes quite apparent that he falls neither into the category of the "hero" nor that of the "kook." To the contrary, Kennedy's fate in this context is testimony to how one can, in a sense, be both at the same time. The neatness of Rorty's distinction between normal and abnormal discourse, and hence also of my own distinction between the incentives of conservation and innovation, betrays the nuances of how the new—actually and/or potentially—connect to the old. As I have showed in Kennedy's case, there were various strategies for acknowledging but distancing oneself from his contribution, leaving him in the odd position of being *seminal, but marginal*.

Kennedy's "A Hoot in the Dark" has fallen between two stools. Neither kook nor hero, Kennedy has become, in this context, a Grand Old man locked in a *limbo* of which nothing comes. Despite the unique daring and promise of his article, "A Hoot…" appears as infertile as a mule.

When Kennedy's essay first appeared, some of his fellow rhetoricians speculated whether it was, like Charles Phineas' spoof piece, only a mockery—a "beak in cheek" practical joke on his colleagues.[70] But "A Hoot in the Dark" differed from the fake veterinary historian's article on two accounts. Unlike Phineas, Kennedy was serious. But unlike Phineas, Kennedy's program was largely not followed up by the discipline's later development. Whereas Phineas' program was fiction made fact, Kennedy's remained a largely unrequited hoot in the dark.

THE QUALITIES OF KENNEDY

Having studied the responses of others, I can now perhaps voice my own, and it is, quite simply, that we should not allow the story told so far to become the fate of "A Hoot in the Dark." The way I see it, Kennedy's article has unique qualities that should be rescued from the twilight zone between conservative resistance on the one hand and the avant-garde's innovative specializations on the other. The value of "A Hoot in the Dark" comes, first, from the fact that Kennedy opens the rhetoric to a wider range of scientific input than usual. As he writes in his essay, it is the product of his reading, for the "last year or two [...] research on animal communication, a subdivision of social biology, or the more exotic field of zoo-semiotics, a subdivision of linguistics."[71] Second, in incorporating these perspectives into his own, Kennedy in turn communicates widely, that is to a large, nonspecialized audience of rhetorical scholars. To address such an audience is, in a sense, to speak to the core of the discipline—and that is precisely what Kennedy did, latching animal signaling onto classical rhetorical ideas like the genres and the canons. In this way, he managed to paint a picture of a "Big Rhetoric" which avoided the negative connotations, as Fleming said, of a "transdisciplinary hermeneutic method whose object of analysis, persuasion, was located everywhere and in everything." Kennedy's rhetoric was not, in fact, "everywhere and in everything." True, it was a phenomenon distributed widely across different creatures, but so distributed, it had specific, observable functions.

Just as the moves of his commentators can be explained in part by looking at their positions in social space, so Kennedy's move was made possible by his particular position. Already in the essay's opening sentence, he establishes his own status: "After spending much of my professional life teaching rhetoric, I began to wonder what I was talking about."[72] Not only this, but Kennedy's project in the essay as such, reveals the freedom of a respected veteran professor, who no longer works under incentives to either innovate or conserve, but who can leisurely think aloud about what rhetoric is and could be.

There are, however, examples that one does not need to be in this position to incorporate animal signaling and human rhetoric into a single scheme. As an exception to the rule that very few scholars have actually engaged in any depth with Kennedy's article, Alex Parrish's book *Adaptive Rhetoric: Evolution, Culture, and the Art of Persuasion* attempts to develop a "biocultural paradigm" which starts from the idea that "the art of persuasion is a series of behaviors that are the products of both biology and culture, and that we must study them with both of these forces in mind."[73] With reference to the reception of Kennedy, Parrish notes that many thought "one of rhetoric's most influential voices had gone a bit batty," but then adds, "Indeed, he had not.

I would argue that opposite, in fact—that 'Hoot' and the first part of *Comparative Rhetoric* may one day rank as Kennedy's greatest accomplishment, even beyond his field-shaping work in classical rhetoric."[74]

In his book, Parrish furthers the impetus found in Kennedy toward a program of study he calls "adaptive rhetoric." In brief, this is an attempt to develop a naturalist view, wherein rhetoric is defined as "the intentional communicative act of an animal whose purpose is to inform, or to manipulate the behavior of, one or more members of a real or imagined category of hearers called 'audience,'" and seen more broadly as a set of adaptive, evolutionary behaviors that "allows animals to avoid costly conflicts, to organize themselves, to coordinate [...], and to communicate desirable traits," and so on.[75] Parrish proceeds to elaborate this program with the help of many resources besides Kennedy, ranging from the usual suspects in both classical and modern rhetoric to ethology to studies of animal signaling. The point here is not so much the particulars of that elaboration as the fact that Parrish, like Kennedy, joins together into a single view observations, reflections, sources of information, theories—in general, *input*—from both "biological" and "cultural" fields of knowledge. Instead of doing what he argues rhetoricians have always been doing, which is to "split the causes of persuasive behavior in two and [label] one of these aspects taboo,"[76] Parrish seeks seamless integration of these various forms of knowledge.

The effect of this move is, I would argue, a nonanthropocentric rhetoric more worthy of that name. Instead of assuming that rhetoric and adjacent disciplines already contain the resources appropriate for a nonanthropocentric reconfiguration of rhetoric, Kennedy and Parrish alike open up to forms of knowledge that are already oriented to the more-than-human. Their way of getting at the behaviors of various sorts of rhetorical animals is neither obstructed by the tradition's assumption of a human-animal distinction, nor led astray by the avant-garde's allergy of that same tradition. If "animal rhetorics" is to become something more than "a 'fad' in the ever widening reach of [...] scholarship," this track should not be neglected. To put it differently: Among all the innovations in this field, the most attractive of them might still be the old novelty of the seminal essay that should—indeed—have been pathbreaking.

NOTES

1. Charles Phineas, "Household Pets and Urban Alienation," *Journal of Social History* 7 (1974): 339.
2. Ibid.
3. Ibid., pp. 338–9.

4. Ibid., p. 349.

5. Ibid., p. 343.

6. See Joshua Specht, "Animal History after Its Triumph: Unexpected Animals, Evolutionary Approaches, and the Animal Lens," *History Compass* 14 (2016): 326–36.

7. Harriet Ritvo, "On the Animal Turn," *Daedalus* 136 (2007): 118–22.

8. See Donna Haraway, *A Companion Species Manifesto* (Chicago: The University of Chicago Press/Prickly Paradigm, 2003), for how to blend cultural studies with *agility* training.

9. The reader will now be on guard, but "Fudge" is in fact *not* a pseudonym. See Erica Fudge, "A Left-Handed Blow: Writing the History of Animals," in Nigel Rothfels, ed., *Representing Animals* (Bloomington: Indiana University Press, 2002), pp. 3–18.

10. David DeGrazia, *Taking Animals Seriously: Mental Life and Moral Status* (Oxford: Oxford University Press, 1996).

11. Fudge, "A Left-Handed Blow," p. 5.

12. The journal *Society & Animals*, for instance, had its first year of publication in 1993. The journal's tenth anniversary special issue provided a state of the field by way of statements from psychology and clinical psychology, Darwinism and evolutionary science, sociology and criminology, anthropology, political science, history, geography, economics, postcolonial studies, feminism, veterinary science, ethics, and religion.

13. George A. Kennedy, "A Hoot in the Dark: The Evolution of General Rhetoric," *Philosophy & Rhetoric* 25 (1992): 1–21.

14. Debra Hawhee, "Toward a Bestial Rhetoric," *Philosophy & Rhetoric* 44 (2011): 81.

15. Diane Davis, *Inessential Solidarity: Rhetoric and Foreigner Relations* (Pittsburgh: University of Pittsburgh Press, 2010), p. 155.

16. See, for instance, Richard Rorty, *Contingency, Irony, and Solidarity* (Cambridge: Cambridge University Press, 1989).

17. My procedure has been systematic, but simple: I have used the "cited by" function in Google Scholar and Web of Science as a proxy for the reception history of Kennedy's article. This generates 105 and 23 articles, respectively—most of the latter sample being included in the former. A number of these articles are theses or dissertations, which I have not described further, since they are typically not an integrated part of the scholarly exchange. I have also passed over articles whose references to "A Hoot…" are so incidental as to tell almost nothing—cases, for instance, where the article is cited in what I have considered to be an inessential footnote or the like.

18. Quentin Skinner, *Visions of Politics. Volume I: Regarding Method* (Cambridge: Cambridge University Press, 2002).

19. The idea that social incentives structure academic moves is widespread in the sociology of knowledge. For one version of that program, which helpfully incorporates the idea that identity and self-image are notable aspects of the moves and motivations of academics (as they presumably are also of everyone else), see Neil Gross, *Richard Rorty: The Making of an American Philosopher* (Chicago: The University of Chicago Press, 2008).

20. To be clear, I am not suggesting there is a one-to-one relation between such investments and the actual moves that actors make in social space. Rather, scholars are more like players on a football field, where each scholar strategizes to take advantage of the room available to her, given her concrete background, strengths, and current placement (relative to everyone else). This way of seeing takes a lot from Pierre Bourdieu, who uses the metaphor of footballers on a field repeatedly. See, for instance, Pierre Bourdieu and Loïc J. D. Wacquant, *An Invitation to Reflexive Sociology* (Chicago: The University of Chicago Press, 1992), pp. 21–22, where he borrows it from Merleau-Ponty.

21. Kennedy, "A Hoot in the Dark," p. 2.

22. Recently, scholars have turned their attention to how Kennedy's translation of Aristotle's *Rhetoric* foreshadowed his interest in the rhetoric of animals, not least by attributing to Aristotle a more generous view of animals' communicative capacities than previously assumed, given his famous definition in *The Politics*. See Greg Goodale, *The Rhetorical Invention of Man: A History of Distinguishing Humans from Other Animals* (Lanham: Lexington Books, 2015), especially chapter 2; and Debra Hawhee, *Rhetoric in Tooth and Claw* (Chicago: The University of Chicago Press, 2017), especially chapter 1. As far as I can understand, however, this "discovery" is more recent, and did not play any notable part in the recorded reception of Kennedy's article, which is what concerns me here.

23. Jo Liska, "The Role of Rhetoric in Semiogenesis: A Response to Professor Kennedy," *Philosophy & Rhetoric* 26 (1993): 31–8.

24. Liska, "The Role of Rhetoric in Semiogenesis," p. 31.

25. Ibid., p. 32.

26. Ibid., p. 34.

27. George Khushf, "Grammacentrism and the Transformation of Rhetoric," *Philosophy & Rhetoric* 28 (1995): 30–44.

28. Ibid., 30.

29. Michel de Certeau, *The Practice of Everyday Life* (Berkeley; University of California Press, 1984).

30. Richard Rorty, *The Consequences of Pragmatism* (Minneapolis: The University of Minnesota Press, 1982).

31. Roberta Kevelson, "Codes, Crypts and Incantations: Charles Peirce's Rhetorical Turn," *symplokē* 4 (1996): 175

32. Kevelson, "Codes, Crypts and Incantations," p. 177.

33. Ulrike Zinn Jaeckel, "Hecklers and the Communication Triangle," in Theresa Enos and Richard McNabb, eds, *Making and Unmaking the Prospects for Rhetoric*, p. 169. Mahwah: Lawrence Erlbaum, 1996.

34. Mari Lee Mifsud, "On the Idea of Reflexive Rhetoric in Homer," *Philosophy & Rhetoric* 31 (1998): 45–6.

35. David Fleming, "Rhetoric as a Course of Study," *College English* 61 (1998): 169–91.

36. Ibid. p. 170.

37. Ibid.

38. See Alan Gross and William Keith, eds, *Rhetorical Hermeneutics: Invention and Interpretation in the Age of Science* (Albany: State University of New York Press, 1996).

39. David Fleming, "Becoming Rhetorical: An Education in the Topics," in Joseph Petraglia and Deepika Bahri, eds, *The Realms of Rhetoric: Inquiries into the Prospects for Rhetoric Education*, pp. 93–94. Albany: State University of New York Press, 2003.

40. Ibid., p. 173.

41. Ibid., p. 170.

42. David Fleming, *From Form to Meaning: Freshman Composition and the Long Sixties, 1957–1974* (Pittsburgh: University of Pittsburgh Press, 2011).

43. David Fleming, "Can Pictures Be Arguments?" *Argumentation and Advocacy* 33 (1996): 11–22.

44. See Gross, *Richard Rorty* and Pierre Bourdieu, *Homo Academicus* (Stanford: Stanford University Press, 1988).

45. See the introduction to this book.

46. James Arnt Aune, "Modernity as a Rhetorical Problem: Phronēsis, Forms, and Forums in Norms of Rhetorical Culture," *Philosophy & Rhetoric* 41 (2008): 416–17.

47. Cornelia Wells, "Toward a Fragmatics, or Improvisionary Histories of Rhetoric, the Eternally Ad Hoc," *Philosophy & Rhetoric* 36 (2003): 293. The first quote cites Kennedy.

48. Wells, "Toward a Fragmatics," p. 282.

49. See Richard Rorty, *Philosophy and the Mirror of Nature* (Princeton: Princeton University Press, 1979).

50. Diane Davis, "Creaturely Rhetorics," *Philosophy and Rhetoric* 44 (2011): 88–94.

51. Ehren Helmut Pflugfelder, "Rhetoric's New Materialism: From Micro-Rhetoric to Microbrew," *Rhetoric Society Quarterly* 45 (2015): 441–61.

52. Thomas Rickert, *Ambient Rhetoric: The Attunements of Rhetorical Being* (Pittsburgh: Pittsburgh University Press, 2013), p. 162.

53. See http://alex-reid.net/2013/03/vitanzas-big-rhetoric-and-some-more.html. Accessed April 18, 2017.

54. Emily Plec, "Perspectives of Human-Animal Communication: An Introduction," in Emily Plec, ed., *Perspectives of Human-Animal Communication: Internatural Communication*, p. 4. New York: Routledge, 2013.

55. Davis, "Creaturely Rhetorics," p. 89.

56. Ibid.

57. Ibid., p. 92.

58. It is not so clear to this reader that Kennedy actually does what Davis and others accuse him of doing, which is to offer an *essentialist* understanding of animals, humans, or rhetoric. Kennedy notes, in fact, that his ideas should offer "comfort to the pragmatist school of philosophers," Kennedy, "A Hoot in the Dark," p. 7, which is arguably a gesture away from essentialist positions. More generally, one could argue that an evolutionary approach like Kennedy's is incompatible at least with any strong essentialism, since evolution is, after all, *change*.

59. See, for instance, Diane Davis, *Breaking Up (at) Totality: A Rhetoric of Laughter* (Carbondale: Southern Illinois University Press, 2000).

60. I am indebted to Alex Parrish for this point.

61. Diane Davis, "Autozoography: Notes Toward a Rhetoricity of the Living," *Philosophy & Rhetoric* 47 (2014): 549, n4.

62. Debra Hawhee, "Toward a Bestial Rhetoric," *Philosophy & Rhetoric* 44 (2011): 82.

63. Ibid., p. 81.

64. In suggesting what those paths are, however, Hawhee spends most of her time suggesting that Kennedy's uses of Aristotle were "rather predictable and schematic" (p. 83), an argument she has since developed in more detail in Debra Hawhee, *Rhetoric in Tooth and Claw* (Chicago: The University of Chicago Press, 2017).

65. N. N. "Editor's Note," *Philosophy & Rhetoric* 47 (2014): vii.

66. I paraphrase from Rorty, *Contingency, Irony, and Solidarity*, pp. 6–7.

67. I paraphrase from Rorty, *Philosophy and the Mirror of Nature*, p. 339.

68. See Rorty, *Contingency, Irony, and Solidarity*, p. 18.

69. For an elaboration of this argument, see Kristian Bjørkdahl, *Enlarging the* Ethnos*: Rorty, Redescription, and the Rhetorical Labor of Moral Progress*. PhD dissertation. Oslo: University of Oslo, 2016.

70. Hawhee, "Toward a Bestial Rhetoric," p. 1, citing Victor Vitanza.

71. Kennedy, "A Hoot in the Dark," pp. 5–6.

72. Ibid., p. 1.

73. Alex Parrish, *Adaptive Rhetoric: Evolution, Culture, and the Art of Persuasion* (New York: Routledge, 2014), p. 2.

74. Ibid., p. 25.

75. Ibid., p. 3.

76. Ibid., p. 2.

BIBLIOGRAPHY

Aune, James Arnt. "Modernity as a Rhetorical Problem: Phronēsis, Forms, and Forums in Norms of Rhetorical Culture," *Philosophy & Rhetoric* 41 (2008): 402–20.

Bjørkdahl, Kristian. *Enlarging the* Ethnos*: Rorty, Redescription, and the Rhetorical Labor of Moral Progress*. Ph.D. dissertation. Oslo: University of Oslo, 2016.

Bourdieu, Pierre and Loïc J. D. Wacquant. *An Invitation to Reflexive Sociology*. Chicago: The University of Chicago Press, 1992.

Bourdieu, Pierre. *Homo Academicus*. Stanford: Stanford University Press, 1988.

Davis, Diane. "Autozoography: Notes Toward a Rhetoricity of the Living," *Philosophy & Rhetoric* 47 (2014): 533–53.

———. *Breaking Up (at) Totality: A Rhetoric of Laughter*. Carbondale: Southern Illinois University Press, 2000.

———. "Creaturely Rhetorics," *Philosophy and Rhetoric* 44 (2011): 88–94.

———. *Inessential Solidarity: Rhetoric and Foreigner Relations*. Pittsburgh: University of Pittsburgh Press, 2010.

de Certeau, Michel. *The Practice of Everyday Life*. Berkeley: University of California Press, 1984.

DeGrazia, David. *Taking Animals Seriously: Mental Life and Moral Status*. Oxford: Oxford University Press, 1996.

Fleming, David. "Becoming Rhetorical: An Education in the Topics." In Joseph Petraglia and Deepika Bahri, eds, *The Realms of Rhetoric: Inquiries into the Prospects for Rhetoric Education*, pp. 93–94. Albany: State University of New York Press, 2003.

———. "Can Pictures Be Arguments?" *Argumentation and Advocacy* 33 (1996): 11–22.

———. *From Form to Meaning: Freshman Composition and the Long Sixties, 1957–1974*. Pittsburgh: University of Pittsburgh Press, 2011.

———. "Rhetoric as a Course of Study," *College English* 61 (1998): 169–91.

Fudge, Erica. "A Left-Handed Blow: Writing the History of Animals." In Nigel Rothfels, ed., *Representing Animals*, pp. 3–18 Bloomington: Indiana University Press, 2002.

Goodale, Greg. *The Rhetorical Invention of Man: A History of Distinguishing Humans from Other Animals*. Lanham: Lexington Books, 2015.

Gross, Alan and William Keith, eds. *Rhetorical Hermeneutics: Invention and Interpretation in the Age of Science*. Albany: State University of New York Press, 1996.

Gross, Neil. *Richard Rorty: The Making of an American Philosopher*. Chicago: The University of Chicago Press, 2008.

Haraway, Donna. *A Companion Species Manifesto*. Chicago: The University of Chicago Press/Prickly Paradigm, 2003.

Hawhee, Debra. *Rhetoric in Tooth and Claw*. Chicago: The University of Chicago Press, 2017.

———. "Toward a Bestial Rhetoric," *Philosophy & Rhetoric* 44 (2011): 81–7.

Jaeckel, Ulrike Zinn. "Hecklers and the Communication Triangle." In Theresa Enos and Richard McNabb, eds, *Making and Unmaking the Prospects for Rhetoric*. Mahwah: Lawrence Erlbaum, 1996.

Kennedy, George. "A Hoot in the Dark: The Evolution of General Rhetoric," *Philosophy & Rhetoric* 25 (1992): 1–21.

Kevelson, Roberta. "Codes, Crypts and Incantations: Charles Peirce's Rhetorical Turn," *symplokē* 4 (1996): 175–88.

Khushf, George. "Grammacentrism and the Transformation of Rhetoric," *Philosophy & Rhetoric* 28 (1995): 30–44.

Liska, Jo. "The Role of Rhetoric in Semiogenesis: A Response to Professor Kennedy," *Philosophy & Rhetoric* 26 (1993): 31–8.

Mifsud, Mari Lee. "On the Idea of Reflexive Rhetoric in Homer," *Philosophy & Rhetoric* 31 (1998): 41–54.

N.N. "Editor's Note," *Philosophy & Rhetoric* 47 (2014): vii.

Parrish, Alex. *Adaptive Rhetoric: Evolution, Culture, and the Art of Persuasion*. New York: Routledge, 2014.

Pflugfelder, Ehren Helmut. "Rhetoric's New Materialism: From Micro-Rhetoric to Microbrew," *Rhetoric Society Quarterly* 45 (2015): 441–61.

Phineas, Charles. "Household Pets and Urban Alienation," *Journal of Social History* 7 (1974): 338–43.

Plec, Emily. "Perspectives of Human–Animal Communication: An Introduction." In Emily Plec, ed., *Perspectives of Human–Animal Communication: Internatural Communication*. New York: Routledge, 2013.

Rickert, Thomas. *Ambient Rhetoric: The Attunements of Rhetorical Being*. Pittsburgh: Pittsburgh University Press, 2013.
Ritvo, Harriet. "On the Animal Turn," *Daedalus* 136 (2007): 118–22.
Rorty, Richard. *Contingency, Irony, and Solidarity*. Cambridge: Cambridge University Press, 1989.
———. *Philosophy and the Mirror of Nature*. Princeton: Princeton University Press, 1979.
———. *The Consequences of Pragmatism*. Minneapolis: The University of Minnesota Press, 1982.
Skinner, Quentin. *Visions of Politics. Volume I: Regarding Method*. Cambridge: Cambridge University Press, 2002.
Specht, Joshua. "Animal History after Its Triumph: Unexpected Animals, Evolutionary Approaches, and the Animal Lens," *History Compass* 14 (2016): 326–36.
Wells, Cornelia. "Toward a Fragmatics, or Improvisionary Histories of Rhetoric, the Eternally Ad Hoc," *Philosophy & Rhetoric* 36 (2003): 277–300.

Index

About the Contributors

Kristian Bjørkdahl has a PhD in rhetorical studies from the University of Oslo, Norway, and is currently a postdoc fellow at the Centre for Development and the Environment at the University of Oslo. Bjørkdahl works within rhetoric and reception studies, and his previous research includes studies of American neopragmatism, the animal movement, the discourse of development aid, science advice, and public challenges to science, as well as the practice of nonreading—especially of "paperwork."

Marilyn M. Cooper is emerita professor of humanities at Michigan Technological University and companion to Scout, a border collie in training. She has published articles in *CCC*, *College English*, and *JAC* and is currently completing a book manuscript titled *The Animal Who Writes.*

T. Jake Dionne (MA, Syracuse University) is a doctoral candidate in the Department of Communication and an instructor for Program for Writing & Rhetoric at the University of Colorado Boulder. His research and pedagogy comments on the importance of considering (and showing consideration toward) the more-than-human world.

Ellen W. Gorsevski (PhD, Pennsylvania State University) is associate professor in the Department of Communication and School of Media and Communication at Bowling Green State University (BGSU). She is also affiliated faculty with BGSU's Critical and Cultural Studies Program and the undergraduate minor in Peace and Conflict Studies. Dr. Gorsevski researches and teaches about contemporary peacebuilding rhetoric featuring political, social, and environmental advocacy for peace and justice. Her research emphasizes undiscovered, less well known, or often forgotten rhetorical records of diverse activists in sociopolitical and environmental justice movements. Her recent articles analyzing the discourse and leadership communication of nonviolent activists for social, political, and environmental justice have appeared in *Quarterly Journal of Speech*, *Western Journal of Communication*, *Journal of Communication and Religion*, and *Environmental Communication*. Her books are *Peaceful Persuasion: The Geopolitics of Nonviolent Rhetoric* (SUNY Press, 2004) and *Dangerous Women: The Rhetoric of the Women Nobel Peace Laureates* (Troubador Publishing, Ltd., 2014).

Iklim Goksel received her PhD in language, literacy, and rhetoric from the University of Illinois at Chicago, She earned her bachelor's degree in English from Bilkent University in Turkey and a master's degree in English and written communication from Eastern Michigan University. Her scholarly work centers in classical and modern theories of rhetoric, contemporary writing pedagogies, animal studies, and feminist theory.

Dustin A. Greenwalt is a lecturer in the Penn State Department of Communication Arts & Sciences. He received his doctorate from the Department of Communication Studies at the University of Georgia in 2015.

David R. Gruber is a senior lecturer in the School of English and Media Studies at Massey University, Auckland. His research spans rhetoric, science communication, and technical communication. He has a particular interest in the neurohumanities and representations of neuroscience. He has published in *Public Understanding of Science*, *Rhetoric Society Quarterly*, and *Technical Communication Quarterly*, among other journals.

Andrea Gutiérrez is completing her PhD in Asian cultures and languages at the University of Texas at Austin. Her work engages various topics of animal voice and language about animals in early Indian texts as well as the foodways and food history of South Asia. She draws on her study of Sanskrit, Tamil, and Prakrit languages to enrich her research on primarily Hindu and Buddhist traditions.

Susan Hafen (PhD, Ohio University, 1995) is a professor in the Communication Department at Weber State University, where she has taught communication theory, organizational communication, interpersonal and small group communication, interviewing, training and development, research methods, and the senior research seminar since 2003. Previously she taught at the University of Wisconsin–Eau Claire (1995–2003), where she also taught human resources and training and development, diversity training, intercultural communication, and public speaking. Prior to her academic career, Dr. Hafen worked in the corporate business world for a dozen years as a trainer and human resource manager. She was on the start-up team for Kimberly Clark's diaper plant in Utah and Mobil Oil's first coal mine in Wyoming, and she worked at Potomac Electric Power Co. in Washington, DC. Dr. Hafen has published articles and book chapters on guilt and shame in organizations, diversity training, cultural identities, workplace gossip, service learning, and classroom pedagogy.

Kelin Loe is a PhD candidate in rhetoric and composition at the University of Massachusetts Amherst. Her work traces the powers shifting through sensation and affective economies that constitute communicative capitalism, hoping to build space and agency for the posthuman subject, particularly for bodies marked as nonhuman animal, disabled, or otherwise othered. Her research is informed by her own experiences being a neurodiverse scholar. She received her MFA in poetry (2012) and her MA in rhetoric and composition (2014) from the University of Massachusetts Amherst. She is the author of the collection of poems *These Are the Gloria Stories* (Factory Hollow Press, 2014).

Alex C. Parrish (PhD, Washington State University, 2012) is assistant professor of writing, rhetoric, and technical communication at James Madison University in Harrisonburg, Virginia. His first book is *Adaptive Rhetoric: Evolution, Culture, and the Art of Persuasion* (Routledge, 2013). In it, he argues for a more scientific approach to rhetorical theory, one that takes human persuasion as one form, among many, of animal rhetoric. His next book project, under contract with the University of Chicago Press, looks deeper into the relationship between animal rhetorics and the many ways different species are able to sense the world and thus communicate.

Emily Plec (PhD, University of Utah, 2002) is professor of communication studies at Western Oregon University, where she teaches courses in rhetoric, media, intercultural and environmental communication. Her scholarship focuses on communication and social justice, with emphases on the rhetoric of racism in sports, the discourse of women leaders, prison communication, and environmental communication pedagogy and practice. She edited the book, Perspectives on Human Communication: Internatural Communication (Routledge, 2013) and her recent scholarship has appeared in Rhetoric Society Quarterly, Environmental Communication, and Communication and Critical/Cultural Studies as well as edited collections such as this one.

Jennifer Saltmarsh is a PhD student in the University of Pittsburgh English Department. She received an MA in English composition at San Francisco State University and a BA in English literature from UC Berkeley.

Hayley Zertuche earned her PhD in rhetorics, communication, and information design at Clemson University. Her research interests include visual rhetorics, critical animal studies, new materialisms, biosemiotics, bioart, and electrate and material pedagogies.

www.ingramcontent.com/pod-product-compliance
Ingram Content Group UK Ltd.
Pitfield, Milton Keynes, MK11 3LW, UK
UKHW041423091026
15587UKWH00023B/267

9 781498 558457